Z6.20

INTRODUCTION TO SOLAR TECHNOLOGY

▲▼ Addison-Wesley Publishing Company • Reading, Massachusetts • Menlo Park, California
London • Amsterdam • Don Mills, Ontario • Sydney

INTRODUCTION TO SOLAR TECHNOLOGY

Marian Jacobs Fisk
Solar Research & Engineering Corp.

H. C. William Anderson
Solar Research & Engineering Corp.

Cover photo by Marshall Henrichs
Courtesy of Acorn Structures, Inc., Concord, MA
Solar Series 2000 Model
Arlene and Francis Fedele, owners and builders

Library of Congress Cataloging in Publication Data

Fisk, Marian Jacobs
  Introduction to solar technology.

  Includes index.
  1. Solar energy. I. Anderson, H. C. William, joint
author. II. Title.
TJ810. J3   621.47   80-29599
ISBN 0-201-04713-6   AACR2

Copyright © 1982 by Addison-Wesley Publishing Company, Inc.

All rights reserved. No part of this publication may be reproduced, stored in a retrieval system, or transmitted, in any form or by any means, electronic, mechanical, photocopying, recording, or otherwise, without the prior written permission of the publisher. Printed in the United States of America. Published simultaneously in Canada.
ISBN 0-201-04713-6
ABCDEFGHIJ-DO-898765432

Dedicated to **Harvey M. Anderson**
Walter C. Jacobs
L. Marian Jacobs

PREFACE

As traditional energy sources become more scarce, it is only natural for us to look upward, to the sun, for an alternative source of energy. The sun is largely responsible for almost all of our conventional energy sources. Photosynthesis of plants and algae supported ancient life that later became fossil fuels. Even hydroelectric power would not be possible without the evaporation of water. Yet solar energy itself has not been fully accepted as a safe, efficient, and understandable energy source. Until recently the unrestricted availability and low cost of traditional energy sources have precluded much practical application of solar energy. The time has come, however, for all of us to become more aware of the direct contributions solar energy can make to our lives and our economy.

This comprehensive book on solar technology explores many applications of solar energy and will be of interest to students, homeowners, builders, architects, and engineers. While it is suitable for a college textbook, it is also a valuable and readable reference text for anyone who wishes to learn about how solar energy systems work.

CONTENTS Chapter 1 gives the reader a practical technological background. Solar-radiation measurement and heat transfer are introduced and covered to the extent necessary for understanding the rest of the book.

Chapter 2 briefly explains the overall performance of solar heating systems and how the various components are connected and interrelated. The significance of collector efficiency tests is also discussed.

Chapters 3, 4, and 5 delve into the specifics of the solar heating system, the collector, storage unit, and design and controls. Chapter 3 details the collector components and the various choices in collector design. Chapter 4 explains how solar heat storage works and discusses some details of storage-unit construction. Chapter 5 explores the system-design choices and details the control elements necessary for a successful system.

A method for optimum system sizing is presented in Chapter 6. A simplified yet accurate heat-load calculation method is included also. The sizing technique indicates which collector area and system size will prove most economical over the years.

The principles of passive solar heating are covered in Chapter 7. Chapter 8 examines the actual location and design of a solar building and the solar heating system.

Chapter 9 is devoted to other applications of solar energy, including domestic water heating, swimming pool heating, and cooking. Also discussed are concentrating collectors and some of their uses, especially that of providing heat for heat engines and air conditioning units. Many solar-electricity generation methods are also covered.

Chapter 10 looks into the future of solar energy in our country with respect to government and utility intervention. Where to go to find more solar information is included in the appendix (Appendix H), which serves as a fitting close to our introductory text.

ORIENTATION Our goal was to write a readable yet comprehensive introductory text about solar energy and its uses. Most other solar texts are either so simplified that vital explanations are omitted, or assume extensive reader knowledge. We expect that many of our readers will not have such extensive knowledge or that they will have forgotten much that was learned during their formal education.

Our book is primarily for people who are dissatisfied by overly simple explanations but are not ready for a purely scientific treatment of the subject. Because a knowledge of advanced mathematics is not necessary for an understanding of how solar energy systems work and because including such mathematics almost automatically discourages lay readers, no calculus is used in this text. We believe this treatment is appropriate for all readers except students in advanced college courses. Some problems using calculus can easily be included by the instructors of such courses if desired.

The solar energy field is not advanced; it is subject to rapid progress and large shifts in direction. Because of these continual changes our aim is to give the reader an intuitive understanding of and an overall perspective on how solar energy systems work. Quantitative information is often not available but is included where possible. Qualitative aspects of solar energy systems are emphasized to improve readers' basic and overall understanding.

The text is virtually self-sufficient. Everything that should be known or understood is included, for example, our introductory coverage of heat and heat transfer, which will be a review for many but may be new material to others. The topic is covered only to the depth necessary for understanding later chapters.

The book combines theoretical considerations about system design with practical rules of thumb for actual systems. The reader is taken through the entire operation and sizing of the solar heating system, from the sun to the collector, the storage medium, and the house. This is not a do-it-yourself guide to solar heating, but rather a sensible explanation of why and how solar energy systems work.

While our emphasis is on solar heating systems, other solar applications are also covered. Many of the basics of solar heating systems apply to every other type of solar energy application. This book provides a good background for further study.

FEATURES

We are both practicing solar engineers. This text, therefore, includes practical features in a usable format often not included in other, solely academic, texts.

A comprehensive discussion of both active and passive solar devices provides the reader with the basic knowledge needed to understand the workings of most solar heating systems.

Examples are worked out in many chapters. These problems illustrate necessary computation skills and calculation techniques. Also, problems at the end of each chapter review important topics and introduce new applications for material covered in the chapter. A problem-solution guide is available.

Many important equations and sizing guidelines are summarized and illustrated. This information is easy to locate for reference and problem solving.

Complete information for the step-by-step heat-load calculation and optimum system-sizing methods is included on the calculation sheets. It should be necessary to read the instructions in the chapter and work through the example only once. Thereafter, all necessary information is available from the calculation sheets. Sheets for calculations in both metric and English units are included.

Complete references and a bibliography are included with each chapter to point the truly curious reader in the right direction. Information on some very difficult topics not covered fully in this text can, therefore, be easily found.

Solar energy is here today, ready to be used. We hope *Introduction to Solar Technology* brings readers closer to making solar energy work for them.

Yakima, Washington   Marian Jacobs Fisk
January 1982   H. C. William Anderson

CONTENTS

ENERGY 1 — 1

1.1 Solar Energy and the Earth — 3
    1.1.1 The Sun and the Earth — 3
    1.1.2 Measuring Solar Radiation — 8
    1.1.3 Solar Radiation at the Earth's Surface — 9

1.2 Heat and Heat Transfer — 14
    1.2.1 Heat and Temperature — 14
    1.2.2 Introduction to Heat Transfer — 15
    1.2.3 Conduction — 16
    1.2.4 Convection — 19
    1.2.5 Radiation — 22
    1.2.6 The Overall Heat-Transfer Coefficient, U — 28
    1.2.7 Heat-Transfer Numbers — 30

1.3 Energy Conversion — 31

References — 35
Bibliography — 35
Problems — 35

xi

SOLAR HEATING SYSTEMS   2   37

    2.1 How Solar Heating Works   38

    2.2 Energy Flow and Efficiency in Flat-Plate Collectors   46

        2.2.1 What is Collector Efficiency?   46
        2.2.2 Determining Collector Efficiency   50

    2.3. Factors Affecting the Performance of Solar Heating Systems   55

        References   56
        Bibliography   56
        Problems   56

DESIGN FACTORS FOR FLAT-PLATE COLLECTORS   3   59

    3.1 Frames, Boxes, and Insulation   61

    3.2 Glazing   68

    3.3 Absorber Plates and Heat-Transfer Fluids   75

        3.3.1 Absorber Plates for Air Collectors   75
        3.3.2 Absorber Plates for Liquid Collectors   83

    3.4 Surface Coatings   90

    3.5 Fluid-Flow Paths in Collectors and Arrays   95

    3.6 Breaking the Rules   102

        References   104
        Bibliography   105
        Problems   105

STORING SOLAR HEAT   4   107

    4.1 How Sensible Heat Storage Works   109

    4.2 Rock Storage   111

        4.2.1 How Rock Storage Works   111
        4.2.2 Sizing Rock Storage   117
        4.2.3 Construction of Rock Storage Units   118
        4.2.4 Airflow Direction in the Pebble Bed   123
        4.2.5 Selection of Rocks   123
        4.2.6 Cooling with Rock Storage   124

    4.3 Water Storage   124

        4.3.1 How Water Storage Works   124
        4.3.2 Sizing Water-Storage Tanks   128
        4.3.3 Construction of Water-Storage Units   129

    4.4 Phase-Change Storage   132

        4.4.1 How Phase-Change Storage Works   132
        4.4.2 Substances Used for Phase-Change Storage   135
        4.4.3 Problems of Phase-Change Salts   137
        4.4.4 Phase-Change Storage Versus Rock and Water Storage   139
        4.4.5 Sizing Phase-Change Storage   140
        4.4.6 Construction of Phase-Change Storage Units   142
        4.4.7 Cooling with Phase-Change Storage   142

    4.5. Other Types of Energy Storage   144

        4.5.1 Salt Ponds   144
        4.5.2 Annual Heat Storage   146
        4.5.3 Chemical Storage   146

        References   147
        Bibliography   148
        Problems   149

DESIGN OF ACTIVE SYSTEMS   5   150

    5.1 Basic Concepts   151

        5.1.1 Conventional Heating Systems   151
        5.1.2 Air-Type Solar Heating Systems   151
        5.1.3 Liquid-Type Solar Heating Systems   155

    5.2 Elements of Solar Heating and Cooling Systems   156

        5.2.1 Basic System Elements   157
        5.2.2 Fluid-Control Elements   162
        5.2.3 System-Control Elements   166
        5.2.4 Protective Components   170

CONTENTS

5.2.5 Other Machinery Available	172
5.3 System Design	173
5.3.1 Specification	174
5.3.2 Equipment Layout	177
5.3.3 Controls Design	191
5.4 A Word About Obsolescence	200
References	201
Bibliography	201
Problems	202

SIZING ACTIVE SYSTEMS 6 204

6.1 Why is Sizing Difficult?	205
6.2 Heat Load	208
6.2.1 What is Heat Load?	208
6.2.2 Calculated Heat Load	213
6.2.3 Simple Heat-Load Calculation	224
6.3 Economic-Heating-Performance Sizing	226
6.3.1 Introduction	226
6.3.2 Calculation of Economic Factors	231
6.3.3 Determination of Actual Yearly Solar Fraction and Total-System Cost	234
6.3.4 Sizing Example	238
6.4 Summary	242
References	243
Bibliography	243
Problems	244

PASSIVE SOLAR HEATING 7 246

7.1 What is Passive Solar Heating?	246
7.2 Thermal Mass for Simple Passive Systems	250
7.2.1 Massive Walls and Floors	250
7.2.2 Trombe Walls	251
7.2.3 Water Walls	254
7.2.4 Solid Walls	260
7.2.5 Combination of Materials	260
7.2.6 Phase-Change Materials as Thermal Mass	260
7.3 Sizing Simple Passive Systems	262
7.4 Other Passive Solar Heating Systems	264
7.4.1 Roof Ponds	264
7.4.2 Thermosiphoning Air-Type Collectors	266
7.4.3 U-Tube Thermosiphon	268
7.4.4 Greenhouses	268
7.5 Living in a Passively Heated Building	272
References	274
Bibliography	274
Problems	274

SOLAR-BUILDING DESIGN 8 275

8.1 Solar Building Types, and Their Special Considerations	276
8.2 Making Best Use of the Site	281
8.3 Heat Loss and Solar Gain	286
8.4 Placement of Solar Collection Areas and Solar Components	292
8.4.1 Residential Buildings	296
8.4.2 Commercial/Industrial Buildings	298
8.4.3 Mechanical Components	300
8.5 Legal Factors	302
8.6 The Design of Solar Buildings	304
References	306
Bibliography	306
Problems	306

OTHER APPLICATIONS OF SOLAR ENERGY 9 308

9.1 Traditional Applications	309
9.1.1 Solar Water Heating	309

- 9.1.2 Swimming Pool Heating 313
- 9.1.3 Water Purification 315
- 9.1.4 Solar Cooking 316
- 9.1.5 Agricultural Drying 318
- 9.2 Newer Tools and Applications 318
  - 9.2.1 Concentrating Collectors 319
  - 9.2.2 Solar-Powered Heat Engines 325
  - 9.2.3 Solar Absorption and Adsorption Air Conditioners 329
  - 9.2.4 Solar Cells 331
  - 9.2.5 Wind Power 335
  - 9.2.6 Biomass 336
- 9.3 Future Tools and Applications 336
  - 9.3.1 Solar Power Towers 337
  - 9.3.2 Ocean Thermal Energy Conversion 338
  - 9.3.3 Solar Satellites 339
  - 9.3.4 Other Solar Energy Converters 339
- References 341
- Bibliography 342
- Problems 345

10 FACTORS AFFECTING THE FUTURE OF SOLAR ENERGY USE 347
- 10.1 Government 347
- 10.2 Peak Load 350
- 10.3 Solar Energy and Utilities 353
- References 354
- Bibliography 355

A Conversion Factors 359

B Derivation of the Economics for Collector Sizing 361

C Table of Design Conditions 367

D Inflating-Discounting Factors 376

E Solar-System Standards Organizations 383

F Published Standards, Codes, and Performance Criteria 386

G Summary of State Solar Legislation 388

H How To Find More Information 407

I Who's Who in Solar Energy 414

Index 419

INTRODUCTION TO SOLAR TECHNOLOGY

1 ENERGY

There can never be a true "energy shortage," for, as Einstein demonstrated, everything in the universe is energy. Light, heat, matter —everything except space itself—is just energy in one form or another.[1] What is more, of all the known forms of energy, matter is the most concentrated (a 1-kg [2.2-lb] rock represents the same amount of energy as the burning of 2.5 million cubic meters or 650 million gallons of gasoline.) Energy lies all around us in inconceivably vast quantities. But, while the earth itself is composed of so much energy that we can never complain about an energy shortage in the literal sense, we still may be justified in our concern about useful energy supplies. Political and monopolistic games are being played with many of the "nonrenewable" fuels such as oil and natural gas. And whether or not these energy resources are truly being exhausted now, as many contend, it is evident that at some point in the future they will become short enough in supply to become uneconomical. They will be replaced.

---

1. Einstein's famous formula, $E = mc^2$, states that energy, E, is equivalent to mass, m. The c^2 is the square of the speed of light, which may be thought of as a proportionality constant.

The energy in matter itself is difficult to use. Ordinary mass is extremely stable and the unstable materials—the nuclear materials—are hard to handle and pose safety problems. Nuclear fusion is a possibility, but still requires several miraculous "breakthroughs" that have been slow in coming. More promising is the energy that is all around us in the dynamic forces of nature: the wind, tides, waves, rivers, geothermal hot spots, and the sun. The problem is that we have not been forced to find the technological means to convert these natural energies into usable forms because it has been too easy simply to dig or pump our energy out of the ground. The problem is not a shortage of energy itself, but a shortage of technology for converting the energy that lies all around us into usable forms. Energy-conversion technology is the real issue, and solar energy is one of the brightest and most promising frontiers in energy conversion. In this chapter we describe sunlight, heat, and some of the other forms of energy relevant to solar technology and introduce such fundamental topics as heat-transfer mechanisms and energy-conversion efficiency.

What *is* energy? Is it a flash of light? A burst of heat? Not really. These are just two among many forms of energy: electric, chemical, biochemical, nuclear, kinetic, angular, elastic, gravitational, magnetic, and forms we have not yet discovered. A good practical definition might be that energy is the ability to do work, that is, the ability to move matter. Of course this is not a precise definition and it evades the issue somewhat, but it at least provides us with a feel for what energy is in terms of what it does. Energy can be measured. Some of the units of energy and their conversion factors are included in Table 1.1.

Table 1.1 **Units for Measuring Energy**

Unit	Abbreviation
British thermal unit	Btu
Joule	J
Calorie	cal
Foot-pound	ft-lb
Kilowatt hour	kWh

Conversion Factors

1 Btu	= 1055 J	= 252.16 cal	= 778.16 ft-lb	= 0.0002931 kWh
1 J	= 0.0009479 Btu	= 0.2390 cal	= 0.7376 ft-lb	= 2.778 × 10^{-7} kWh
1 cal	= 0.0039657 Btu	= 4.184 J	= 3.0860 ft-lb	= 1.162 × 10^{10-6} kWh
1 ft-lb	= 0.0012851 Btu	= 1.356 J	= 0.32405 cal	= 3.767 × 10^{-7} kWh
1 kWh	= 3412 Btu	= 3.600 × 10^6 J	= 8.603 × 10^5 cal	= 2.655 × 10^6 ft-lb

Units for measuring energy

Energy varies in *quality*. If energy is the ability to do work, in principle 100 J of energy ought to be able to do 100 J of work. In practice, though, this is not true because some forms of energy have more *randomness* than others. Heat, in particular, can be thought of as a very random form of motion energy because it is the energy caused by the vibration of trillions of molecules in matter. Since there are so many small motions, each in a different direction, it is virtually impossible to get all of the energy lined up and in step, which would be necessary for any large-scale work. Thus heat is considered to be a rather low-quality form of energy. Electricity is a high-quality form of energy in that it can easily be made to do work and work can easily make electricity. But when electricity or some other form of energy is converted to heat, the process is called *irreversible* because it is impossible to convert all of the heat energy back to electricity. The irreversibility of some energy-conversion processes has large implications in solar technology, as will be seen in Chapter 9.

1.1 SOLAR ENERGY AND THE EARTH

1.1.1 The Sun and the Earth

In a sense, we already have a solar economy: the sun is responsible for most of our most easily accessible energy resources including oil, coal, hydroelectric power, and others. Photosynthesis in green plants and algae supported the prehistoric plant and animal life that later became fossil fuels. Hydroelectric power comes from snow and rain, which come from "sun-lifted" [evaporated] water.

Use of sunlight *directly* as an energy source has proved in the past to be less economical than use of these other sources of "concentrated sunshine." As technology advances, however, and the price of the more traditional energy sources continues to rise, perhaps it is only natural that we begin to look directly to the sun.

Direct use of solar energy means using *light*, which is a form of electromagnetic radiation. Electromagnetic radiation is a wavelike phenomenon that moves energy across distances. Not only light, but radio waves, microwaves, X-rays, and gamma rays are all forms of electromagnetic radiation. One thing that distinguishes each of these types of radiation from the others is its *wavelength*—the length of one complete wave. For example, a radio wave may be a meter or more in length, but a light wave may be less than a millionth of a meter long. Figure 1.1 gives the wavelengths of various colors of light, including infrared light and ultraviolet light.

Both ultraviolet and infrared light are invisible to the human eye, although they border on being visible. Ultraviolet light has wavelengths slightly shorter than violet light, and infrared light has wavelengths longer than red light. Ultraviolet light in small quantities is responsible

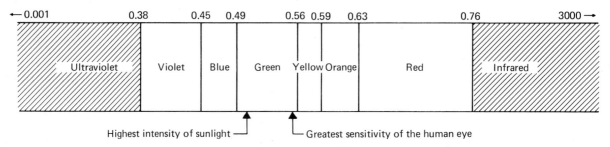

Fig. 1.1 Wavelengths of visible, ultraviolet, and infrared light.

for such beneficial things as suntans, but at high intensities it can be very dangerous. It can damage not only human skin and eyes, but plastics and other materials as well. Infrared light is sometimes called *radiant heat*, but is not really heat energy until it is absorbed by the skin or some other surface. Hot objects radiate infrared light. The sun mostly radiates visible light, but also substantial amounts of ultraviolet and infrared light.

Figure 1.2 is a simplified view of the sun and its relationship to the earth. Although the earth is shown very small, the drawing is still greatly out of proportion. In reality, the sun is 109 times larger in diameter than the earth and is separated from the earth by a distance of about 108 sun diameters. If the sun were represented by a 5-cm ball, the earth would be located 5.4 m away and would be about 0.5 mm in diameter.

From the earth the sun appears to move across the sky in an arc, roughly from east to west, owing to the rotation of the earth around its north–south axis. In the northern hemisphere sunrise and sunset occur farther toward the south during the winter, but move northward in the summer as the sun travels in a higher arc across the sky and days become longer. This seasonal motion is due to the slight tilt of the earth's axis and the earth's revolution around the sun. Using information presented in Table 1.2 and Fig. 1.3, those familiar with trigonometry (or who have access to a scientific calculator) can calculate the position of the sun in the sky at any time of the day and day of the year. Table 1.2 shows how to calculate *solar time*, or the time relative to the motion of the sun, where the sun is always highest in the sky at exactly noon. Owing to the time zones and depending on longitude, solar time may equal the actual time or it may be quite a bit different.

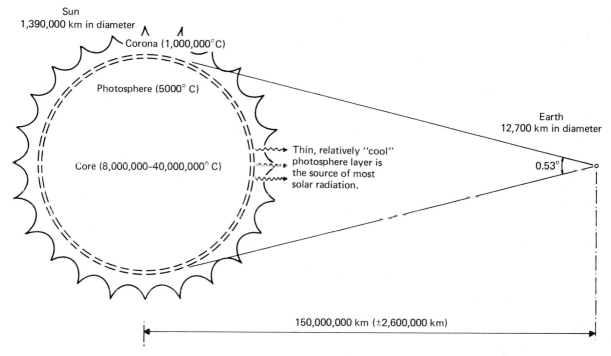

Fig. 1.2 The sun and its relationship to earth.

Fig. 1.3 Where (exactly) is the sun?

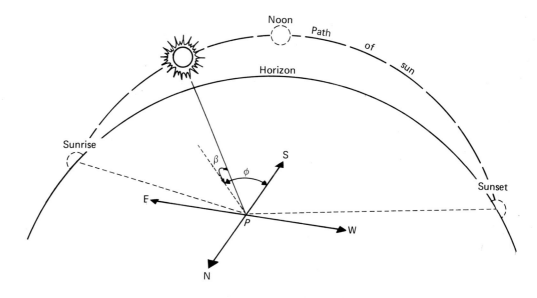

Table 1.2 How to Find Local Solar Time: Local solar time (in hours) = $GMT + [E - 4(\text{longitude})]/60$

U.S. Time Zone	To Find GMT (Greenwich Mean Time*) Add to Local Standard Time:
Eastern	5 hr
Central	6 hr
Mountain	7 hr
Pacific	8 hr
Yukon	9 hr
Alaska Hawaii	10 hr

	Equation of Time†, E, in Minutes			
Day Month	1st	8th	15th	22nd
January	−3.27	−6.43	−9.20	−11.45
February	−13.57	−14.23	−14.25	−13.68
March	−12.60	−11.07	−9.23	−7.20
April	−4.18	−2.12	−0.25	1.32
May	2.83	3.52	3.73	3.50
June	2.42	1.25	−0.15	−1.17
July	−3.55	−4.80	−5.75	−6.32
August	−6.28	−5.67	−4.58	−3.07
September	−0.25	2.05	4.48	6.97
October	10.03	12.18	13.98	15.33
November	16.33	16.27	15.48	14.03
December	11.23	8.43	5.22	1.78

*The local time in Greenwich, England, which happens to have a longitude of zero degrees.
†A correction factor that accounts for irregularities in the earth's orbit and rotation.

SOLAR ENERGY AND THE EARTH

The formula for determining the solar-altitude angle is

$$\beta = \arcsin(\cos L \cos \partial \cos H + \sin L \sin \partial),$$

and the formula for determining the solar-azimuth angle is

$$\phi = \arcsin\left[\frac{\cos \partial \sin H}{\cos \beta}\right],$$

where

L = latitude of point P,
∂ = declination at this time of year = $\partial \approx 23.45 \sin[360/365(n + 284)]$,
n = number of day in year (1–365),
H = hour angle = 15 × (number of hours from local solar noon).

The first formula gives the angle between the sun and horizon (called the *solar altitude*, β). The second formula gives the sun's direction relative to the south (called the *solar azimuth*, ϕ). The ability to calculate this information allows building designers to study shading effects and engineers to determine how the performance of a solar heating system changes throughout a day or a season. The formulas can be used to calculate the time of both sunrise and sunset or the length of a day. Note that

Azimuth angle, ϕ, is measured from the south (in this and most solar books).

ϕ and H are (customarily) positive numbers in the morning and negative in the afternoon.

Local solar noon is the time of day when the sun is directly over the south ($\phi = 0$).

The sun's altitude is highest at local solar noon. These equations differ from those that would be used south of the equator.

Local solar time can be calculated using the information presented in Table 1.2.

Example What is the local solar time in Phoenix, Arizona (112.4° west longitude) at 3 min past noon local mountain time on 15 September?

Solution Local solar time = GMT + [E − 4(longitude)]/60
 = (12.05 + 7) + (4.48 − 4 × 112.4)/60
 = 11.63 = 11:38 A.M.

Example What is the solar altitude angle at 3 min past noon local mountain time in Phoenix, Arizona (33.4° north latitude) on 15 September (day 258)?

Solution β = arc sin (cos L cos ∂ cos H + sin L sin ∂). L = 33.4.

Declination ∂ = 23.45 sin [360/365 (258 + 284)] = 2.22.

The local solar time from the example using Table 1.2 is 11.63 hr (11:38 A.M.), so the hour angle = 15° × (12 − 11.63) = 5.55°.

If we substitute these values into the equation for β,

β = arc sin [cos (33.4) cos (2.22) cos (5.55) + sin (33.4) sin (2.22)]
= arc sin (.852) = 58.4°.

1.1.2 Measuring Solar Radiation

It is obvious that sunlight is stronger than candlelight. Different sources of light have different *strengths*. Another word for the strength of light is *intensity*. We can measure the intensity of sunlight, but in solar technology a more useful quantity than intensity is *irradiance*. While intensity describes how powerful a ray or beam of light is, irradiance describes how much energy is striking a surface, or the total amount of energy that is passing through an area. Irradiance can be pictured as light rays striking or passing through an imaginary surface from one side only (Fig. 1.4). Irradiance is the amount of energy passing through this surface per unit of area and per unit of time. For example, if 7500 kJ of sunlight strike a 1 m² surface in 2 hrs, the irradiance at any point on the surface, assuming uniform distribution, is

$$\text{irradiance} = \frac{(7500 \text{ kJ})}{(1 \text{ m}^2)(2 \text{ hr})(3600 \text{ sec/hr})} = 1.04 \text{ kW/m}^2 \text{ (330 Btu/ft}^2 \text{ hr)}.$$

As Fig. 1.4 shows, the light that contributes to irradiance can come from any angle. It is useful to know the irradiance of sunlight because this measurement tells us how much solar energy is available for collection (of course we cannot collect it all).

Irradiance is a precise, technical term. There are two more commonly used terms for the irradiance of sunlight. One is *insolation*. Insolation is the solar irradiance received on a horizontal surface such as the ground. For decades and in hundreds of locations across the country, weather stations have been measuring insolation. As we will see in Chapter 6, these measurements can be used to predict how well a solar heating system will perform. Another term often used for solar irradiance is *solar flux*. Solar flux, solar irradiance, and insolation all mean

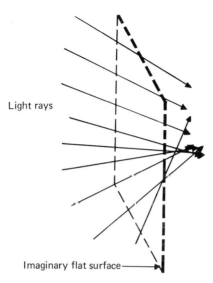

Fig. 1.4 Irradiance: the amount of solar radiation that passes through an imaginary surface, regardless of angle of incidence.

about the same thing. The only difference is that insolation refers to a horizontal surface, while the others may refer to surfaces at other angles.

The solar flux or irradiance that a surface receives depends on how directly it faces the sun. A surface that faces the sun squarely will receive the highest flux. If the surface is tilted away from the sun, the flux will drop (Fig. 1.5). A solar collector will receive the most energy when it faces the sun directly and will collect less energy when tilted away from the sun. The tilt of a solar collector must be taken into account if we want to be able to predict accurately how well it will perform (see Chapter 6).

Instruments for measuring solar radiation are called *pyranometers* or *pyrheliometers*. Pyrheliometers measure only direct or beam radiation directly from the sun. Pyranometers measure total radiation, both direct and diffuse. Solar radiation is measured in units of W/m^2, $Btu/hr\ ft^2$, or in langley/hr, where langley = cal/cm^2.

1.1.3 Solar Radiation at the Earth's Surface

Most solar radiation comes from a relatively thin, relatively cool (5000° C) layer near the sun's surface known as the *photosphere*. When the sun's energy reaches the earth's orbit, it contains a harmful percentage

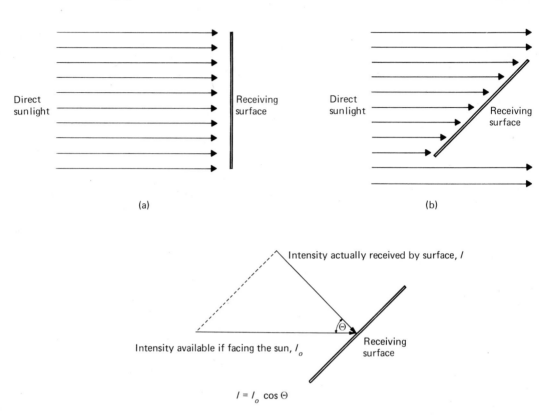

Fig. 1.5 Energy received by a tilted surface: (**a**) surface squarely facing the sun's rays receives maximum energy, (**b**) tilted surface is effectively smaller and receives less energy, (**c**) mathematical relationship.

of ultraviolet light and even a few gamma rays and X-rays. However, in passing through the earth's atmosphere, these harmful rays are largely filtered out along with some wavelengths of visible light. The graph in Fig. 1.6 is called the *solar spectrum*. It gives the relative strengths of the various wavelengths of sunlight and shows how strengths vary for different wavelengths. Two spectrums are shown. The upper curve is the spectrum of sunlight outside the atmosphere and the lower curve is the spectrum of sunlight inside the atmosphere at the earth's surface. Passing through the atmosphere weakens all wavelengths of light, some far more than others. There are several reasons why the atmosphere "dampens" sunlight and filters out certain wavelengths. Dust, water vapor, and absorption of light by ozone and air molecules are all partly responsible. In addition, the sun's brightness and spectrum as seen on earth depend on its position in the sky. This effect is obvious during a

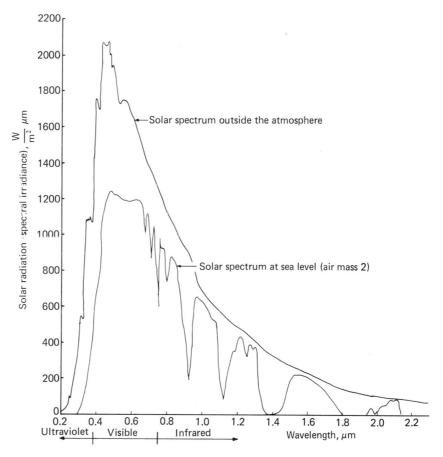

Fig. 1.6 The solar spectrum.

beautiful sunset, when the light has to pass a longer distance through the atmosphere and suspended dust and other particles filter out the shorter wavelengths, leaving a brilliant orange or red light.

Since the spectrum of sunlight depends on how much atmosphere it must pass through, the term *air mass* has been coined. Air mass is essentially a measurement of how much atmosphere sunlight must pass through before it reaches the earth. When the sun is directly overhead the amount of atmosphere that sunlight must pass through to reach sea level is called *air mass 1*. As Fig. 1.7 shows, *air mass 2* is the amount of atmosphere sunlight must pass through to reach sea level when the sun is about 30° above the horizon. Air mass 2 is two times greater than air mass 1.

The overall intensity of sunlight (averaged for all wavelengths) is never constant on earth. It varies from zero during the night to a maxi-

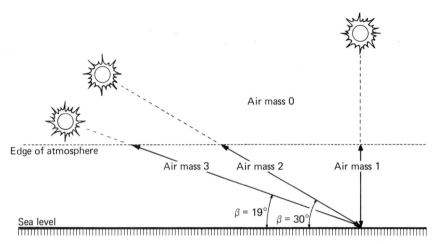

Fig. 1.7 Air masses.

For solar altitudes greater than about 15°, air mass is approximately equal to the cosecant of the solar altitude, β.

mum on a clear noon. If a pyranometer or pyroheliometer is set up and monitored during a typical day, the intensity will be seen to move up and down continuously as clouds pass over the sun and air conditions change. Outside the earth's atmosphere, however, sunlight almost never varies in intensity but stays remarkably constant, at 1353 W/m² or 429 Btu/hr ft². This means that any 1 m² surface facing the sun always receives about 1353 W of energy per hour (as long as it stays 150 million kilometers from the sun, which is the average sun–earth distance). This value is known as the *solar constant*. Actually, the radiation received outside the atmosphere is not constant but varies ±3% over the year owing to variations in the earth–sun distance. It is usually viewed as a constant, however, because of the greater uncertainty in atmospheric attenuation of solar radiation.

The solar constant can be thought of as a limit of sorts. Since sunlight inside the atmosphere is always less intense than the solar constant, it is possible to say with some conviction that no solar collector used on earth will ever collect as much as 1353 W/m² of collection area, and most will collect far less than this.

If we are interested in solar radiation for the purpose of capturing and using it, we must consider not only the direction of the sun, but the direction of sunlight as well—not all sunlight is received directly from the sun. The sunlight that strikes a solar collecting surface is *direct* (or *beam*), *diffuse*, or *reflected*. Direct sunlight comes directly from the sun. Since the sun is far from the earth, rays of direct sunlight are considered to be parallel. Diffuse sunlight does not come directly from the sun, but is first reflected from dust particles, air, clouds, and water vapor (Fig. 1.8). Diffuse sunlight comes from all areas of the sky and some solar collectors are not able to collect and use it. (Concentrating-type collectors require that all rays of light come in parallel to one another. See Chapter 9.) Even on a clear day approximately 15–20% of the sunlight is diffuse. When the sky is totally overcast, all of the sunlight is diffuse. The basic distinction between diffuse and reflected light is simply that diffuse sunlight comes in via scattering in the atmosphere, while reflected sunlight bounces from trees, snow, landscapes, mirrors, and other earthbound surfaces. The reason we distinguish among the three is that the direct and diffuse fractions of sunlight can be estimated for a given location based on weather bureau data, while the amount of reflected light reaching a surface depends entirely on what happens to be on the landscape.

One other source of energy in addition to direct, diffuse, and reflected sunlight is used by many solar collectors, namely *environmental radiation*, or *sky radiation*. Environmental radiation is infrared light that is radiated by the sky and surrounding landscape. An object tends to radiate infrared light in an amount related to its temperature. The hotter an object is, the more of its heat energy is converted to infrared light and radiated away. Since the sky and earth are warmed by the sun, they radiate their own energy in an amount large enough to aid measurably some solar collectors. To calculate the amount of radiation received from the environment an effective temperature, sometimes called

Fig. 1.8 The three components of sunlight.

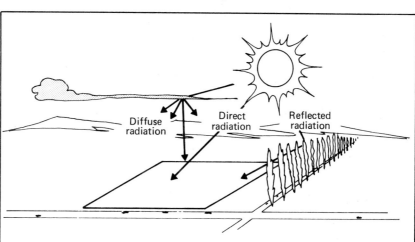

times called the *sky temperature*, is often used. Sky temperature is based on the actual air temperature but is usually several degrees lower.

Environmental radiation to the collector increases the total energy available for the collector to use. However, since the collector surface will also be warm, it radiates heat out to the environment. Usually the net effect of inward and outward environmental radiation is heat loss from the collector since the collector is warmer than the environment.

1.2 HEAT AND HEAT TRANSFER

1.2.1 Heat and Temperature

Heat and temperature are easy to observe but are rather difficult to describe and fully understand. For example, in the solar heating field we typically speak of *storing heat* and *stored heat* but, thermodynamically, heat cannot be stored. Rather, heat is an *interaction* and ceases to exist once an energy-transfer process stops. What is stored is *internal energy*, not heat.

Without being overly inaccurate, we can think of internal energy as the energy of motion and energy of potential motion of an object's molecules. Every moving object has energy of motion, or *kinetic energy*. In a solid the molecules move by vibrating back and forth as though they were all connected to one another by springs. In a gas or liquid molecules travel around freely. *Potential* energy of motion (one kind of potential energy) is like stored motion energy, for example, the energy stored in a compressed spring. There are springlike effects among moving molecules as well as motion energy that is stored in and among molecules mostly as a result of electrostatic forces. The amount of internal energy an object contains is a grand total of all the motion energy and potential motion energy of all of the molecules in the object.

Temperature is obviously related to internal energy, yet it is not measured in units of energy as heat is. Degrees Celsius, Fahrenheit, Kelvin, and Rankine are not the same as joules, Btus, calories, and other units of heat energy. And while temperature is a measure of how *energized* an object's molecules are, it is not really a measure of energy in the same sense that internal energy is. So what is it? The temperature of an object tells us something about what an average molecule is doing. Temperature is not a measure of energy, but the temperature of an object is related to the amount of motion energy an average molecule has. The faster the average molecule is moving, the higher the temperature of the object. This means that in some ways the temperature of an object is related to how much internal energy the object contains, since both temperature and internal energy are related to motion energy. But temperature is *average* energy and internal energy is *total* energy. Temperature indicates what the motion energy of an average molecule is, while internal energy indicates what the total motion energy is for all molecules. Temperature tells us nothing about the potential motion energy that might be stored in the object, whereas internal energy includes potential-energy effects.

Internal energy is usually related to temperature. When heat is added to an object, its temperature will usually rise. When an object loses some heat energy its temperature usually falls, *but not always.* Heat can be stored in an object in the form of potential energy, and no temperature rise will be seen. For example, to melt ice at 0° C (32° F), heat is added to it. But until all the ice has melted it will not rise in temperature even though heat is constantly being added. This is because the addition of heat does not make the molecules move any faster at first. Heat's first effect is to break some electrostatic bonds between molecules, turning the solid ice into liquid water. Breaking these bonds is like storing heat energy without raising the temperature.

Another way to see the difference between heat and temperature is to consider as an example a milliliter of water and an ocean of water, each at 10° C. Both have identical temperatures, but both do not contain the same amount of heat. The milliliter of water needs to contain a few joules of internal energy to have a temperature of 10° C, but the ocean needs to contain billions of joules of internal energy to be at 10° C because the ocean has billions of times as much mass. The average energy of a water molecule is the same in the milliliter and the ocean, but the ocean has many times more molecules so it has much more total energy. What is more, the addition of a few joules to the milliliter of water will cause a noticeable temperature rise, but will the addition of two or three joules of heat to the ocean cause any noticeable temperature rise?

The phenomenon in the ice example is known as *latent-heat* (hidden-heat) storage. As ice melts, it stores heat in a latent form. The example of the milliliter of water and the ocean illustrates the idea of *heat capacity*. The ocean has a much larger heat capacity than the milliliter of water so, to the ocean, a few more joules is a drop in the bucket. Both of these concepts, heat capacity and latent heat, are considered in Chapter 4 in relation to solar heat storage.

1.2.2 Introduction to Heat Transfer

We have spoken of adding heat to an object as though it were simply a matter of pouring it in. Needless to say, the process is a little more involved than this. Adding heat, removing heat, heat gain, and heat loss are the subjects of the theory of *heat transfer*. Heat transfer is a complex mathematical subject. It is also an extremely important part of the study of solar energy. It is beyond the scope (and space limitations) of this book to delve deeply into the quantitative aspects of heat-transfer theory, but to understand how solar energy can be used it is essential to acquire as much knowledge as possible about this all-important topic.

Heat-transfer theory is used in the study, design, and testing of solar heat collectors, heat storage units, heat exchangers, heating systems, industrial process heaters, solar air conditioning and power generating equipment, and all other solar applications that involve the use of heat.

There are three basic ways in which heat can be gained, lost, or transferred from one object to another. These are

1. Conduction
2. Convection
3. Radiation

We recommend that readers who are serious about pursuing a solar education read all the information they can find (and understand) concerning heat transfer [1, 2, 3, 4].

1.2.3 Conduction

Thermal conduction is simply the flow of heat through an object or material and it occurs as energized molecules transfer energy to their less-energized neighbors. Heat is conducted through a material when one area is more energized (at a higher temperature) than another area. Heat energy can only flow when one point in the material is at a higher temperature than some other point. The primary direction heat flows is always from the point of higher temperature to the point of lower temperature. Figure 1.9 illustrates heat conduction.

The *thermal conductivity*, k, of a material determines how fast heat flows through the material. The larger k is, the faster heat flows. For example, heat flows faster through aluminum than through wood, so alu-

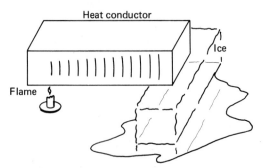

Fig. 1.9 Heat conduction: heat from the burning candle is conducted through the bar into the block of ice, causing the ice to melt.

HEAT AND HEAT TRANSFER

minum has a higher thermal conductivity. Table 1.3 shows the thermal conductivities of several common materials. The speed with which heat flows, however, depends on factors other than thermal conductivity, such as the size of the heat-flow path (Is the heat flowing along a short, wide metal bar or along a long, thin metal wire?), how far the heat has to flow, and how large a difference in temperature exists between the hot area and cool area. Figure 1.10 shows that the rate of heat flow, q, in watts or Btu/hr, is

$$q = \frac{kA}{d}(T_1 - T_2)$$

or

$$q = \frac{A}{R}(T_1 - T_2)$$

where

k = thermal conductivity in W/m°C or Btu/hr ft °F,
d = distance heat must flow,
A = cross-sectional area,
T_1 = temperature of hot side,
T_2 = temperature of cool side,
$R = d/k$ = conduction resistance in m²°C/W or hr ft²°F/Btu.

Table 1.3 **Thermal Conductivities of Several Materials at 0° C**

	k $\frac{W}{m\ °C}$	k $\frac{Btu}{hr\ ft\ °F}$
Copper	386	223
Aluminum	204	118
Steel	43	25
Concrete	1.7	1.0
Face bricks	1.3	0.75
Glass	0.73	0.42
Water	0.555	0.321
Hardwood	0.16	0.092
Glass-wool insulation	0.038	0.022

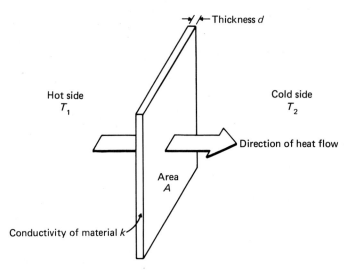

Fig. 1.10 Heat transfer by thermal conduction.

Example One end of a 10-cm-long copper wire 4 mm in diameter is maintained at 0°C while the other end is maintained at 100°C. If the thermal conductivity is 386 W/m^1°C, how much heat is conducted through the wire?

Solution We have seen that

$$q = \frac{kA}{d}(T_1 - T_2).$$

The cross-sectional area of the wire, A, is given by πr^2, or 1.26×10^{-5} m^2; q is then calculated as follows:

$$q = \frac{386\,(1.26 \times 10^{-5})}{0.1}(100 - 0)$$

$$q = 4.9\text{W} \,(17 \text{ Btu/hr.}).$$

A heat *insulator* is any material that has a low thermal conductivity such as plastic foam or glass wool. Insulation is used in building walls to reduce the speed of heat loss from the building; it is used around solar collector panels, around hot pipes, and in other situations when we wish to slow down the rate of heat conduction.

Often, to simplify heat-conduction problems, the term *thermal resistance* is introduced as

$$R = \frac{d}{k}$$

HEAT AND HEAT TRANSFER

where

R = thermal resistance,
d = distance heat must flow,
k = thermal conductivity.

Note that use of the heat-transfer resistance simplifies the heat transfer by conduction equation to

$$q = \frac{A}{R}(T_1 - T_2).$$

The resistance, or R value, of a material is also an indication of its insulating ability. The insulation with the higher R value is the better insulator.

Example One side of a sample of glass-wool insulation is maintained at $-20°$ C, while the other side is maintained at $20°$ C. If the thermal resistance of the insulation sample is 2.1 m² °C/W, how much heat is transferred through the sample?

Solution We have seen that

$$q = \frac{kA}{d}(T_1 - T_2) = \frac{A}{R}(T_1 - T_2).$$

Since the area is not given, the heat flow per unit area,

$$\frac{q}{A} = \frac{1}{R}(T_1 - T_2),$$

can be calculated as

$$\frac{q}{A} = \frac{1}{2.1}[20 - (-20)]$$

$$\frac{q}{A} = 19 \text{ W/m}^2 \ (6.0 \ \frac{\text{Btu}}{\text{hr ft}^2}).$$

1.2.4 Convection

The second important type of heat transfer is *convection*. Convection refers to the transfer of heat by a moving fluid. In thermodynamics and heat-transfer theory the word *fluid* is often used to describe both liquids and gases. For example, both air and water are fluids. Throughout this book, when we refer to a *heat-transfer fluid* we may mean air, water, or some other liquid or gas.

There are two types of convection, *natural* and *forced*. Natural convection occurs strictly because of natural density changes in the fluid, while forced convection is artificially caused, or forced.

In natural convection the fluid moves by itself because of gravity and because of changes in density. For example, when cool air comes in contact with a hot object the air picks up heat, becomes less dense, and tends to rise, taking some of the object's heat away with it. When warm air touches a cold object it gives up some of its heat to the object, becomes more dense, and tends to fall downward from the object. If a hot object is placed near a cooler object the air between tends to move in a circle, or a *convective loop* (Fig. 1.11). The result is heat transfer from the hot object to the cool object. Heat transfer by a convective loop is sometimes called *thermosiphoning*.

Natural convection may be both helpful and harmful. In solar heating, thermosiphoning can move heat throughout a room or throughout a building without the help of a fan. Natural convection is also an important cause of heat loss in flat-plate solar collectors because it transfers precious heat to the outside environment.

Forced convection is closely related to natural convection. Again, heat is transferred by flowing air or liquid. The difference is that in forced convection the fluid motion is not natural, but forced. For example, in a solar collector heat-transfer fluid may be forced over or through the solar absorbing surface with a fan or pump in order to remove the heat. Wind causes heat loss by forced convection because the flowing air picks up heat from the outer surfaces of the collector. Wind is a forced-air motion in the sense that it is not caused only by changes in density that happen at the collector surfaces. Forced convection implies that an outside force is responsible for the fluid's motion.

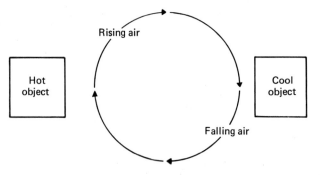

Fig. 1.11 Heat transfer by convective loop; thermosiphoning.

HEAT AND HEAT TRANSFER

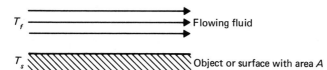

Fig. 1.12 Heat transfer by natural or forced convection.

The heat-transfer rate, q, in watts or Btu/hr, for convection (Fig. 1.12) is

$$q = hA(T_s - T_f),$$

or

$$q = \frac{A}{R}(T_s - T_f),$$

where

h = convective heat-transfer coefficient in W/m² °C or Btu/hr ft² °F,
A = area of flow surface,
T_s = temperature of surface,
T_f = temperature of fluid,
$R = 1/h$ = convective resistance in m² °C/W or hr ft² °F/Btu.

The convective heat-transfer coefficient, h, is not a constant, but depends on a large number of factors, including

The temperature difference, $T_s - T_f$
The geometry of the object or surface
The geometry and velocity of the flowing stream
The thermal conductivity of the fluid
The gravitational constant
The coefficient of thermal expansion and the constant-pressure specific heat of the fluid
The density of the fluid
The kinematic viscosity of the fluid

It also depends on whether the fluid flow is turbulent or laminar (even or steady) and whether the flow is forced. Often, h is given in the form of an equation that has been found by experimentation and applies only to a specific situation such as natural convection between plates [2].

Example Air at 50° C blows over a 0.5 m² hot plate maintained at 200° C. The convective heat-transfer coefficient is 16 W/m² °C. What is the heat-transfer rate?

Solution
$$q = hA(T_s - T_f)$$
$$= 16\,(0.5)\,(200 - 50)$$
$$= 1.2 \text{ kW (4100 Btu/hr)}.$$

Convection, whether natural or forced, is the result of conduction from the hot object into the individual fluid molecules or vice versa. Thus the convective heat-transfer coefficient, h, represents a combination of a conductive term and a true convective term.

The rate of heat transfer by natural or forced convection is usually no simple thing to calculate. It depends on (1) the temperatures of the fluid and the various surfaces; (2) the geometry of the situation; and (3) many properties of the fluid, such as its thermal conductivity and density. It also depends on how the liquid is flowing—whether the flow is turbulent or smooth. Usually, it is difficult or impossible to calculate the rate of heat transfer by simply using these properties. What is possible is to experiment and find simple empirical equations (equations based on experimentation) that apply to very specific cases. For example, experiments can be done with parallel sheets of glass to find how much heat is transferred by natural convection between the plates under various temperature conditions and with various plate spacings. The results can be put into equations that can be used by anyone interested in natural convection between glass plates. Many such equations have been written by various experimenters and are listed in heat-transfer handbooks and similar sources [2].

1.2.5 Radiation

Radiative heat transfer is the transfer between different bodies of heat energy by infrared or visible light due to differences in temperature, for example, radiation of heat by a hot stove to cooler objects in a room. Any hot surface tends to lose heat by giving off infrared light energy. The infrared cannot be seen, but it can be felt as it is absorbed by the skin. It is important to understand that infrared light is heat in the form of electromagnetic radiation that is emitted by warm objects. Heat is easily transmitted as infrared light by a warm surface and infrared light is easily accepted by many surfaces and transformed into heat. It is true that infrared light feels warm to the touch as it is absorbed by the skin, but so does visible light that is high enough in intensity. For example, the warmth of sunlight can be felt even through a window, but the light that comes through glass is almost all visible light—glass blocks most wavelengths of infrared light.

Why don't warm surfaces emit visible as well as infrared light? Why does an object have to be at several hundred degrees before it starts to glow with visible light? The reason has to do with the difference between infrared and visible light. Recall that infrared light has a longer wavelength than visible light. In addition, each photon of infrared light carries less energy than a photon of visible light. A photon is a *particle* of light, just as an electron is a particle of electricity. Actually light is neither particles nor waves, but at times it behaves like particles and at other times like waves. It may be considered as either, depending on the circumstances.

For a surface to emit light in the visible range, the surface has to be more highly energized with heat, that is, the surface has to have a higher temperature. Figure 1.13 shows how the wavelengths emitted by a high-temperature surface differ from the wavelengths emitted by a low-temperature surface. The graph shows the *spectral curve* of emitted radiation for a high-temperature object and a low-temperature object. (This spectral curve is simply a graph showing how the different wavelengths an object emits differ in intensity.) Note that a very-high-temperature surface (over about 3500° C) emits mostly visible light, while a lower-temperature surface emits mostly infrared light.

Figure 11.13 also shows that a high-temperature surface emits a lot more radiation (and thus loses a lot more heat) than a low-temperature surface. In fact, for any one surface the amount of light energy radiated depends only on the temperature. As the temperature rises, the amount of energy radiated goes up very quickly. Of course there are many dif-

Fig. 1.13 The relation of light radiated to temperature.

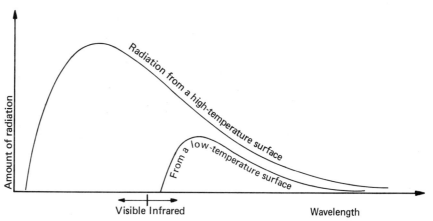

ferent kinds of surfaces. A painted metal surface at 100° C radiates heat several times faster than a polished metal surface at the same temperature.

If it is true that different materials radiate heat at different speeds, then is it possible to find some material that radiates heat infinitely fast no matter what its temperature? No. In fact, if such a material existed it would be the perfect refrigerator. No heat could stay in the material, so it would constantly absorb heat from everything around itself and radiate the energy away. No such material exists. An equation has been written that gives the maximum rate that any surface can lose energy at any temperature. By simply plugging a temperature into this formula, we can find out how fast a "perfect" radiating surface would emit radiation if it were at this temperature. This perfect heat radiator is called a *blackbody* (Fig. 1.14). The rate at which a blackbody radiates energy depends only on its temperature and its surface area:

$$q_{blackbody} = \sigma A T^4,$$

where

T = temperature of object, K or R,

A = area of radiating surface, m² or ft²,

σ = Stefan–Boltzmann constant

   = 5.67×10^{-8} W/m² K⁴ (1.71×10^{-9} Btu/hr ft² R⁴).

No surface in existence is truly a blackbody, since no surface has been found that emits as much energy as a true blackbody would. But there are surfaces (such as black-painted metal) that come close to being blackbodies, emitting as much as 98–99% of the radiation predicted by the formula.

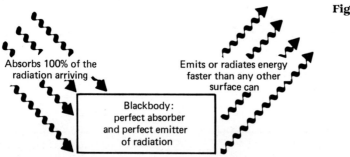

Fig. 1.14 What is a blackbody?

The fraction of blackbody radiation that a unit area of a surface emits is called the *emittance*, ϵ, of the surface.

Example The emittance for a black-painted surface is 0.98 at 20° C. What is the heat-transfer rate for the surface?

Solution The emittance is the ratio of the actual heat-radiation rate to the blackbody-radiation rate, or

$$\epsilon = \frac{(q/A)_{actual}}{(q/A)_{blackbody}}.$$

Rearranging, we see that

$$\left(\frac{q}{A}\right)_{actual} = \epsilon \left(\frac{q}{A}\right)_{blackbody}$$

and that

$$\left(\frac{q}{A}\right)_{blackbody} = \sigma T^4,$$

where T is in kelvin, and

$$T = 273° \text{ C} + 20° \text{ C} = 293° \text{ K}.$$

Therefore,

$$\left(\frac{q}{A}\right)_{actual} = 0.98 \, (5.67 \times 10^{-8})(293)^4$$
$$= 410 \text{ W/m}^2 \, (130 \text{ Btu/hr ft}^2).$$

The complete definition of a blackbody is that it is both a perfect emitter and a perfect absorber of radiation, absorbing all light, of any wavelength, that strikes it. That is why it is called a blackbody: since all radiation is absorbed and none is reflected, it would appear black to our eyes. A black-painted surface is almost a blackbody in that it absorbs nearly all the light that strikes it and emits nearly as much radiation as a blackbody would. Surprisingly, a white-painted surface meets half of the qualifications for being a blackbody because it is good radiator of heat. However, white paint reflects far too much visible light to qualify as a good absorber.

Absorption of radiation is what a solar heat collector is all about. A good solar collector should absorb and utilize a large fraction of the sunlight that strikes it (ideally, light of all wavelengths). The fraction of available light that a surface absorbs is called the *absorptance*, α, of the surface. The absorbing surface in a solar collector is usually made as black as possible so that it will have a high absorptance capacity.

Absorptance and emittance, while different properties, have the same value, or

$$\alpha_\lambda = \epsilon_\lambda,$$

where

α_λ = absorptance of radiation wavelength λ,

ϵ_λ = emittance of radiation at wavelength λ.

This equality is called *Kirchhoff's law*. It is desirable to make a solar collector with a high absorptance capacity for solar radiation but low emittance for infrared radiation (low heat loss). This type of surface obeys Kirchhoff's law because having a high absorptance and emittance for shorter-wavelength solar radiation is not in conflict with having a low emittance and absorptance for longer-wavelength infrared radiation. Since most solar radiation occurs at wavelengths shorter than 3 μm, this absorptance-emittance configuration can make for a very efficient collector. This type of surface is called a *selective* surface and is discussed in Chapter 3.

Heat transfer by radiation between two surfaces is shown in Fig. 1.15. Rate of heat transfer, q, in watts or Btu/hr, can be expressed as

$$q = F_\epsilon F_G \sigma A_1 (T_1^4 - T_2^4)$$

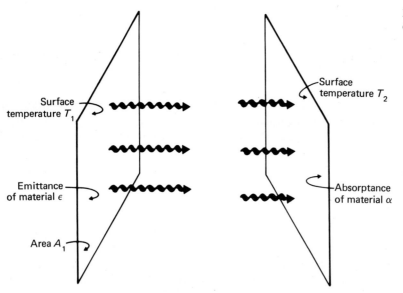

Fig. 1.15 Heat transfer by electromagnetic radiation.

in watts or Btu/hr, or

$$q = h_r A_1 (T_1 - T_2),$$

where

F_ϵ, F_G = factors concerning the emittance, ϵ, of the radiating surface and absorptance, α, of the receiving surface, and the geometry and areas of the surface,

σ = Stefan–Boltzmann constant
= 5.67×10^{-8} W/m² K⁴ (1.71×10^{-9} Btu/hr ft² R⁴),

A_1 = area of the radiating surface,

T_1 = temperature of hot surface in K or R,

T_2 = temperature of cool surface in K or R,

h_r = radiative heat-transfer coefficient
= $F_\epsilon F_G \sigma (T_1^2 + T_2^2)(T_1 - T_2)$ in W/m² °C or Btu/hr ft² °F (note that h_r is very temperature-dependent).

The radiative heat-transfer coefficient is usually a complex function of several variables and is very temperature-dependent. Note that the equation for q using h_r closely resembles the equations given earlier for heat transfer by conduction and convection. Temperatures in degrees Celsius or Fahrenheit may be used in this formula since it involves only a temperature difference. Absolute temperatures must be used, however, in the following version of this same equation:

$$q = F_\epsilon F_G \sigma A_1 (T_1^4 - T_2^4).$$

Example For the case of two infinite parallel planes $A_1 = A_2 = A$,

$$\frac{q}{A} = \frac{\sigma(T_1^4 - T_2^4)}{(1/\epsilon_1) + (1/\epsilon_2) - 1}.$$

One plane is maintained at 500° C and the other at 750° C and their emittances are 0.3 and 0.4 respectively. What is the heat-transfer rate by radiation between the planes only?

Solution

$$\frac{q}{A} = \frac{\sigma(T_1^4 - T_2^4)}{(1/\epsilon_1) + (1/\epsilon_2) - 1}$$

$$= \frac{(5.67 \times 10^{-8})[(273 + 750)^4 - (273 + 500)^4]}{1/0.3 + 1/0.4 - 1}$$

= 8.7 kW/m² (2800 Btu/hr ft²).

Note that absolute temperatures (K) are used.

1.2.6 The Overall Heat-Transfer Coefficient, U

As discussed with regard to Figs. 1.10 and 1.12, the heat-transfer rates by conduction and convection can be expressed in terms of thermal resistances as

$$q = \frac{A}{R}(T_1 - T_2)$$

where

$$R = \frac{d}{K}$$

for conduction, and

$$q = \frac{A}{R}(T_s - T_f)$$

where

$$R = \frac{1}{h}$$

for convection. Thermal resistance for conduction was discussed briefly in section 1.2.3. Heat-transfer resistance for thermal radiation can also be defined as $1/h_r$. The heat-transfer rate for radiation then becomes

$$q = \frac{A_1}{R}(T_1 - T_2),$$

where

$$R = 1/h_r.$$

Consider the heat transfer through the plane wall shown in Fig. 1.16. If conditions within the wall are steady, any heat transferred from one side of the wall must be transferred through the wall and then out the other side, or

$$q = q_{in} = q_{through} = q_{out}.$$

Heat transferred in and out of the wall can be considered to be transferred by convection only and heat transferred through the wall is transferred by conduction. If we use the equations for convection and conduction using thermal resistances, we see that

$$q = \frac{A}{R_{conv,\,a-1}}(T_1 - T_a) = \frac{A}{R_{cond,\,1-2}}(T_1 - T_2) = \frac{A}{R_{conv,\,2-b}}(T_2 - T_b),$$

HEAT AND HEAT TRANSFER

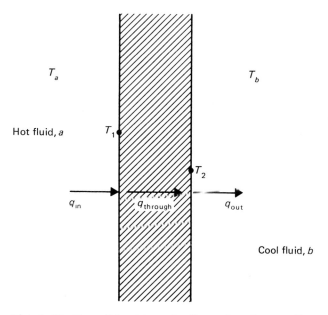

Fig. 1.16 Overall heat transfer through a plane wall.

which can be rewritten in terms of the two fluid temperatures T_a and T_b as

$$q = \frac{A(T_a - T_b)}{(R_{a-1}) + (R_{1-2}) + (R_{2-b})}.$$

It can similarly be shown that for a system of n layers

$$q = \frac{A(T_a - T_b)}{R_1 + R_2 + R_3 + \ldots + R_n}$$

or

$$q = \frac{A(T_a - T_b)}{R_{tot}}$$

$1/R_{tot}$ is defined as U, the overall heat-transfer coefficient. When we use U, the heat-transfer equation becomes

$$q = UA(T_a - T_b)$$

or

$$q = UA \, \Delta T.$$

Remember that the conduction, convection, and radiation of all materials and surfaces are included in the calculation of U. Often radiation resistance is combined with convection resistance to yield an effective convection resistance, which is then used in further calculations.

The U value, as it is sometimes called, is especially useful in heat-loss calculations for buildings. U values for different building constructions can easily be calculated and the heat loss through different areas of the building can then be determined. This method of heat-loss calculation is used in Chapter 6 in the solar-system sizing procedure.

Example One wall of your house consists of siding and sheathing ($R = 0.375$ m² °C/W), 10%; 5 cm × 10 cm wood studs ($R = 0.77$); 9 cm of glass-wool insulation ($R = 1.94$); and gypsum wallboard ($R = 0.079$). The convective resistance for the outside surface is given as 0.03, while the resistance for the inside surface is 0.12. What is the average U value for the wall? (Ignore radiation.)

Solution The calculation of the average U value for the wall must be broken down into two parts, at the framing and between the framing. The total resistance, R_{tot}, for each part is calculated as is shown in Table 1.4. The U value for each part of the wall can be estimated using relative areas of between-framing and at-framing sections as follows:

$$U_{avg} = 0.9 U_{(btw\ frm)} + 0.1 U_{(at\ frm)}$$

$$= 0.9 (0.394) + 0.1 (0.730)$$

$$= 0.43 \text{ W/m}^2 \text{ °C } (0.076 \text{ Btu/hr ft}^2 \text{ °F}).$$

1.2.7 Heat-Transfer Numbers

In the preceding sections we have introduced heat transfer. Heat transfer is not a simple subject, but it is very useful for predicting how well a new solar device is going to work or deciding what size equipment to install in a solar-heated building.

Table 1.4 **Thermal Resistance Values for Wall Parts**

	Between Framing	At Framing
Outside surface	0.03	0.03
Siding and sheathing	0.375	0.375
Studs		0.77
Insulation	1.94	
Wallboard	0.079	0.079
Inside surface	0.12	0.12
Total thermal resistance	2.54	1.37

There are many aspects of heat transfer that we did not mention but that may crop up in your future reading. Table 1.5 presents some dimensionless numbers that are often used to compare heat transfer in solar systems. Systems with identical heat-transfer numbers will exhibit similar behavior.

1.3 ENERGY CONVERSION

To make use of solar energy, it is necessary to convert it into a more useful form. To do this, we need an *energy-conversion device*. A solar cell is an example of an energy-conversion device because it accepts sunlight

Table 1.5 **Some Dimensionless Heat-Transfer Numbers Used in the Study of Solar Systems**

Name	Abbreviation	Value	Significance
Nusselt number	Nu	hL/k	Ratio of conductive to convective resistances; nondimensional heat-transfer coefficient
Reynolds number	Re	$\rho v L/\mu$	Ratio of internal to viscous forces for forced-convection systems; indicates transition from laminar to turbulent flow
Grashof number	Gr	$g\beta L^3(\Delta T)\rho^2/\mu^2$	Ratio of buoyancy to viscous forces; similar to Reynolds number but used for natural convection
Prandtl number	Pr	$\mu C_p/k$	Ratio of molecular diffusivities of momentum to heat; indicates thickness of boundary layers
Rayleigh number	Ra	$GrPr$	Ratio of thermal buoyancy to viscous inertia; minimum Rayleigh number must be exceeded for convection to occur

Definition

C_p = specific heat at constant pressure in (kJ/kg °C) (Btu/lb °F)
g = gravitational constant, 9.8 m/sec² (32.2 ft/sec²)
h = convective heat-transfer coefficient in (W/m² °C) (Btu/hr ft² °F)
k = thermal conductivity in (W/m °C) (Btu/hr ft °F)
L = distance, length, or diameter in m (ft)
ΔT = temperature difference in °C (°F)
V = fluid velocity in m/sec (ft/sec)
β = volume coefficient of thermal expansion in 1/K (1/R) (equals 1/T for an ideal gas)
μ = dynamic viscosity in (kg/m sec) (lb/ft sec)
ρ = density in (kg/m³) (lb/ft³)

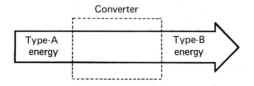

amount of energy in = amount of energy out

Fig. 1.17 An ideal energy converter.

and converts it to electricity. But there are many other examples of energy-conversion devices that are more familiar. An automobile engine is an energy-conversion device that takes the chemical energy in gasoline and converts it to mechanical energy. A light bulb is an energy-conversion device that takes electrical energy and converts it to radiant energy—light. An energy-conversion device takes in energy in one form and puts out energy in a different form.

No practical energy-conversion device is 100% perfect. Almost any energy converter takes some of the input energy and wastes it or converts it to a nonusable form. The law of *conservation of energy* tells us that all we can do is convert energy from one form to another. We cannot create energy out of nothing and we cannot destroy it or make it disappear. Therefore, whenever we waste input energy, we reduce the amount of output energy. An ideal energy converter, if one existed, could be represented by Fig. 1.17. All of the type-A energy we put in would be converted to type-B energy. If we put in 100 J of type-A energy, we would get out 100 J type-B energy. We could call this a 100% efficient conversion of energy.

What happens in a real energy converter is shown in Fig. 1.18. All real devices for converting energy take some of the input energy and turn it into unwanted forms of energy. Any such device would have to be described as less than 100% efficient since we get out less type-B energy than the amount of type-A energy we originally put in. Because the law of conservation of energy always holds true, we can say that the amount of type-B energy we get will always be equal to the amount of

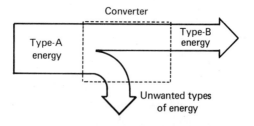

Fig. 1.18 A real energy converter.

ENERGY CONVERSION

type-A input energy minus the energy that was converted to undesirable forms.

We can define the efficiency of an energy-conversion device as the fraction of input energy that is converted to useful output form:

$$\text{energy-conversion efficiency} = \frac{\text{useful energy output}}{\text{energy input}}.$$

Efficiency can be expressed as a fraction or a percentage.

For an example of how this formula works, consider a solar cell. With a solar cell, we want to convert one type of energy (light) into a more useful kind (electricity). But even the best solar cells can only convert about 15 of every 100 J of incoming light energy into electricity. The other 85 J become unwanted forms of energy such as heat and reflected light. Here, the heat absorbed by solar cells is considered *waste heat* in the sense that a solar cell is supposed to make electricity, not heat. This waste heat can be used for other purposes, but what we are trying to measure here is how well a solar cell makes electricity (Fig. 1.19). Thus we can use the definition of efficiency to find that in such a solar cell

$$\text{energy-conversion efficiency} = \frac{15\,\text{J}}{100\,\text{J}} = 15\%.$$

The ideas of conversion efficiency, η, and conservation of energy, q_{in}, are summarized in Fig. 1.20 and can be expressed as follows:

$$\eta = \frac{q_{out}}{q_{in}} = \text{fraction of input energy that is converted to the desired output form,}$$

and

$$q_{in} = q_{out} + q_{lost}.$$

In the second equation, energy is neither created nor destroyed, but is changed in form. Table 1.6 gives the typical energy-conversion efficiency of some energy converters.

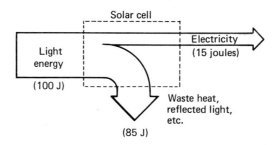

Fig. 1.19 Solar energy conversion by a solar cell.

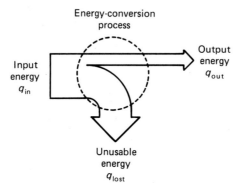

Fig. 1.20 Energy conversion.

The technology of solar energy is a technology of energy-conversion devices. In the following chapters we are primarily concerned with solar heating because the conversion of light to heat is such a simple process that solar heating is the most fully developed solar technology at this time. However, the future will see many developments in other energy-conversion areas; some of these are introduced in Chapter 9. We hope that solar energy will play a rapidly increasing part in humanity's energy usage as these technologies continue to be developed.

Table 1.6 **Performance of Typical Energy Converters**

Energy-Conversion Device	Type of Input Energy		Type of Output Energy	Typical Energy-Conversion Efficiency
Automobile engine	Chemical	→	Mechanical	15%
Steam power plant	Chemical	→	Electrical	40%
Electric motor	Electrical	→	Mechnical	90%
Light bulb	Electrical	→	Radiant	7%
Crop of wheat	Radiant	→	Biochemical	0.2%
Home furnace	Chemical	→	Thermal	70%
Solar cell	Radiant	→	Electrical	12%
Solar collector	Radiant	→	Thermal	40%

REFERENCES

1. Duffie, J. A., and W. A. Beckman (1974). *Solar Energy Thermal Processes*. New York: John Wiley.
2. ASHRAE (1977). *1977 Handbook of Fundamentals*. New York: ASHRAE.
3. Holman, J. P. (1976). *Heat Transfer*. New York: McGraw-Hill.
4. Meinel, A. B., and M. P. Meinel (1976). *Applied Solar Energy: An Introduction*. Reading, Mass.: Addison-Wesley.

BIBLIOGRAPHY

Threlkeld, J. L. (1970). *Thermal Environmental Engineering*. 2nd ed. Englewood Cliffs, N.J.: Prentice-Hall.

Yellot, J. I. (1977). Solar radiation measurement. In R. C. Jordan and B. Y. H. Liu (eds.), *Applications of Solar Energy for Heating and Cooling of Buildings*. New York: ASHRAE.

PROBLEMS

1.1 Using the conversion factors presented in Table 1.1, convert the following amounts of energy to the indicated units:

a) 10 ft-lb to _____ J

b) 45 cal to _____ kWh

c) 1050 Btu to _____ kWh

d) 30 cal to _____ ft-lb

e) 160 Btu to _____ kJ

f) 3 kWh to _____ cal

1.2 What color is light having a wavelength in micrometers of 0.57? 0.70? 0.40? 0.89?

1.3 How long is the day on 22 March at the equator? At 35° north latitude? At 57° north latitude? At 75° north latitude? At 30° south latitude?

1.4 Calculate the solar time and azimuth angle for sunrise at 45° north latitude on 15 January and 15 July (day 196).

1.5 Calculate the local time for sunrise for problem 1.3 if the longitude is 100° west.

1.6 At what angle should a solar collector (at 37° north latitude) be placed from horizontal if the sun's rays are to shine directly into the collector at solar noon on 31 May (day 151)?

1.7 Prove that at the time of the equinox the sun rises and sets due east and west for all locations on earth.

1.8 250 W/m² of solar radiation strikes a horizontal surface at 35° north latitude on 9 April (day 99) at solar noon. Assume that all radiation is direct and calculate how much radiation would strike a surface facing directly toward the sun.

1.9 A copper pipe with an inner diameter of 4 cm and an outer diameter of 4.4 cm is covered with 2 cm of insulation (k = 0.05 W/m °C). The inside wall of the pipe is maintained at 200° C and the outside of the insulation is at 35 °C. The area thermal resistance for a hollow cylinder is given by

$$\frac{R}{A} = \frac{\ln(r_2/r_1)}{2\pi kL},$$

where ln indicates natural log [2.303 log (x) = ln (x), r_1 is the inner radius and r_2 is the outer radius. Assume that the thermal conductance for copper is as given in Table 1.3 and calculate the heat loss per unit length of pipe.

1.10 The front side of a 5-mm-thick aluminum plate is exposed to a radiant heat flux of 1900 W/m² in a vacuum. If the front of the plate is maintained at 200° C and thermal conductivity of aluminum at that temperature is 228 W/m °C, what will the temperature of the other surface be if all the radiant energy striking the front surface is conducted through the plate? What is the significance of the vacuum? What would the temperature of the other surface be if the plate was made of wood ($k = 0.16$)?

1.11 The outside of a wall with thermal conductivity 0.021 Btu/hr ft^2 °F is exposed to the environment at 75° F. The inside surface of the wall is at 195° F, while the outside surface is maintained at 90° F by free convection alone to the environment. If the convective heat-transfer coefficient is 0.7 Btu/hr ft^2 °F, how thick is the wall? What is the heat-transfer rate?

1.12 Natural convection between vertical plates can be given [1] in terms of the Nusselt and Grashof numbers (Table 1.5) as Nu = 0.033 (Gr)$^{0.381}$. The plates are at 100° and 40° respectively and one each measures 2m^2. They are separated by $L = 0.025$ m. The air between the plates can be treated as an ideal gas $\beta = 1/T(K)$ and β should be evaluated at the average temperature of the two plates. ρ^2/μ^2 is given as 2.49×10^9 sec^2/m^4 and k = 0.0295 W/m °C. Calculate the natural convective heat transfer between the plates.

1.13 For the example on radiative heat transfer between two infinite parallel planes in section 1.2.5, what is the formula for the radiative heat-transfer coefficient, h_r, and what is its value?

1.14 For a collector radiating energy to the sky, the sky can be considered a blackbody and an effective sky temperature can be defined as $T_{sky} = 0.0552 \, (T_{air})^{1.5}$ [1]. The radiation exchange between the collector and the sky can then be written as

$$q = \epsilon \, A \, \sigma(T^4 - T_{sky}^4).$$

With the outside glass collector cover at 40° C, the radiative heat-transfer rate for the collector to the sky is 65 W/m^2. The radiative heat transfer for glass at 40° C to a black body is 479 W/m^2. What is the effective sky temperature? What is the outside-air temperature?

1.15 Solar irradiance of 900 W/m^2 strikes a plate ($\epsilon = 0.92$) that is perfectly insulated on the back side. The convective heat-transfer coefficient is 12 W/m^2 °C and the air temperature is 20° C. Assume that the sky temperature (see problem 1.14) equals the air temperature and calculate the temperature of the plate at equilibrium.

1.16 Assume that the sun is a blackbody and calculate its temperature using the solar constant.

1.17 A wall of your house is constructed as follows:

8 in. common brick ($R = 1.6$ hr ft^2 °F/Btu)

airspace ($R = 1.01$)

10% 2″ × 4″ furring ($R = 4.38$)

gypsum wallboard ($R = 0.45$)

The convection resistances for the outside and inside surfaces are 0.17 and 0.68 respectively. What is the U value for the wall? What is the overall U value for the wall if 3.5 in. of insulation ($R = 11$) is used between the two-by-fours?

2 SOLAR HEATING SYSTEMS

Putting solar energy to work involves converting it to another, more useful, form of energy. Without question, the simplest conversion process for solar energy is the conversion of sunlight to heat. This simple process goes on all around us in nature as sunlight strikes the earth and is converted to heat and as the earth warms the atmosphere. Near the equator this natural solar heating of the atmosphere is sufficient to maintain a comfortable temperature for the human population year around. Farther toward the poles days become too short and the sun is too low in the sky during the winter months to keep the air heated to a comfortable level; heat losses overtake heat gains and shelter becomes essential for survival. In recent decades it has been recognized that winter sunlight can be made adequate to provide a significant portion of the heat for a shelter. All that is required is that the shelter be constructed so as to let sunlight in and prevent as much heat loss as possible. The shortness of winter days can be compensated for by reducing the heat loss that occurs during long winter nights. In other words, the solar heating of buildings is accomplished by maximizing light gain and minimizing heat loss.

2.1 HOW SOLAR HEATING WORKS

Any dark surface that faces the sun is a *solar heat collector*; it is dark precisely because it has absorbed some light energy and converted it to heat instead of reflecting the light away. But a dark-colored object is a very inefficient solar collector during winter when the surrounding air is cold. Any heat the object gains is immediately stolen by the cold surrounding environment. Enter glazing.

Collector glazings, or transparent covers, can be made of glass, plastics, or fiberglass. Glass is chosen most often because of its long life, high solar transmittance (the fraction of sunlight it allows to pass through) and low transmittance for infrared radiation from the collector's absorber.

Glass is a rather miraculous substance. A pane of glass is a hard, solid object and yet visible light can pass through it almost as if it were nothing. In addition glass prevents the transmission of most radiant heat (infrared light). Thus if we take a dark surface and cover it with a sheet of glass, we have a fairly efficient solar collector—one that readily lets light in, converts the light to heat, and then slows down any radiant-heat losses. The glass cover has a couple of very helpful side effects as well. If it is placed within an inch or two of the dark collecting surface, it creates a *dead-air space*, or zone of fairly still air that acts as an insulator (nonmoving air is a good insulator). The glass also protects the surface from cold wind, rain, snow, and so on. The combination of a dark surface with a glass cover results in an "energy trap" of sorts. It lets light enter but allows neither light nor heat to escape in large amounts. This energy-trapping process is known as the *greenhouse effect* (Fig. 2.1).

Solar heating involves more than heat collection. It also involves distribution of heat to the rooms of a building and the storage of unneeded heat for later use (at night or during cloudy periods). An *active* solar heating system actively distributes the heat throughout the building: pumps, fans, or other equipment physically move the heat around. A *passive* system depends on natural forces to distribute the heat. For example, natural convection rather than fans move heated air in a passive system. How are collection, storage, and distribution of heat accomplished in both active and passive systems?

Figure 2.2 shows a simple passive solar heating scheme known as the *direct-gain* approach (see Chapter 7 for other schemes). Heat collection is accomplished via the greenhouse effect: large expanses of south-facing glass admit sunlight, dark-colored walls and floors absorb the solar energy, transforming it to internal energy, which may subsequently be given up as heat. The heat is distributed throughout the room by natural convection; as air near the floor and walls is heated, it rises and is replaced by cool, falling air. Heat is stored in the massive

HOW SOLAR HEATING WORKS

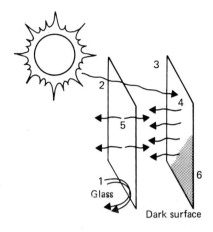

Number key:
1. Glass stops wind.
2. Glass lets sunlight through.
3. Dead-air space insulates.
4. Dark surface absorbs sunlight and radiates infrared.
5. Glass absorbs most infrared from the absorber and reradiates less infrared.
6. Part of hot surface's losses are made up for since the warm glass radiates some infrared back to absorber.

Fig. 2.1 How glazing improves the performance of a solar collector: the green-house effect.

Fig. 2.2 A direct-gain passive solar collection scheme.

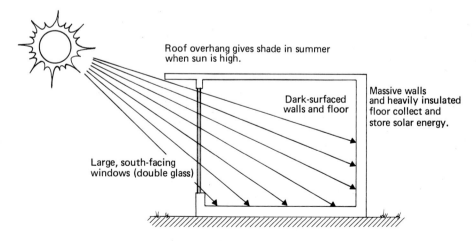

(e.g., concrete) walls and floors since heat collected at the surface slowly moves inside. During the night this absorbed heat energy is slowly released, heating the cool room air. A carefully designed direct-gain system can often provide 60–70% or more of the heat that buildings in many areas of the country require. (For best performance additional features are often included such as insulating shutters for the windows that can be closed at night to reduce heat loss.)

In active systems, solar energy is generally collected by a group (array) of several *flat-plate solar collector panels,* boxlike devices that are mounted on a south-facing wall or roof. Each *solar collector*, as it is often called, has tubes or channels that a *heat-transfer fluid* passes through. The heat-transfer fluid is the air, water, or other substance that removes the collected heat and carries it to the inside of the building. Figure 2.3 is a cross-sectional view of a simple air-type collector that shows how air enters at the bottom, blows across the hot, black *absorber plate*, and exits, heated, at the top. Figure 2.4 (p. 42) is a similar view of a liquid-type collector. The basic parts of any flat-plate collector are the absorber (or absorber plate), the cover, the insulation, and the box or frame that holds it all together. Flat-plate collectors are discussed extensively in Chapter 3.

Heat in active systems is stored in a separate *storage unit* located somewhere inside the building. In air systems the storage unit may be a container of small rocks (typically 2–4-cm [.75–1.5-in] rocks in a 2–3-m [6–10-ft] rectangular bin). For heat storage hot air is blown through the rocks. For heat removal the air to be heated is blown through the rocks. In a water system the heat-storage unit may simply be an insulated water tank (holding perhaps 3800 l or 1000 gal). Heat storage is discussed in Chapter 4.

In active systems heat is distributed with fans or pumps that move the air or water. Although flat-plate collectors are the most commonly used heat collectors, active systems often employ other devices. There are about four solar collection, distribution, and storage methods that can be considered practical today. Figure 2.5 (pp. 43, 44) describes them. The first two are the air- and liquid-type systems in which flat-plate collectors are used. These two systems are widely used for heating buildings. Figures 2.5(c) and (d) show liquid-type systems in which high-technology solar collectors are used. These two systems are used for large-scale installations where both solar heating and solar cooling are required; solar cooling requires high collection temperatures. Figure 2.5(c) shows the *evacuated-tube*-type collector, in which a vacuum is used for insulation instead of a dead-air space (see Chapter 3). Figure 2.5(d) shows a *linear-concentrator collector*, which is the only concentrating collector presently considered practical for solar heating and cooling systems. In

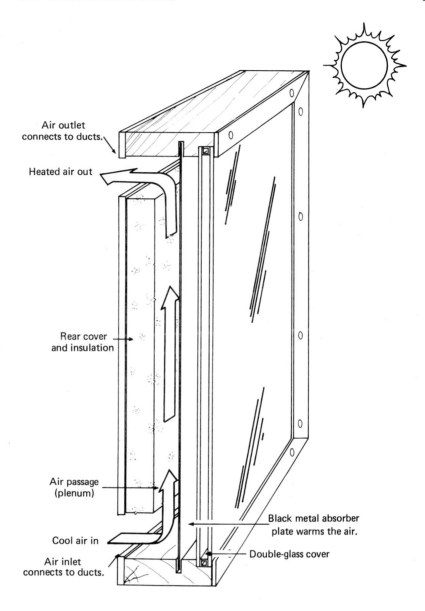

Fig. 2.3 A simple air-type solar collector.

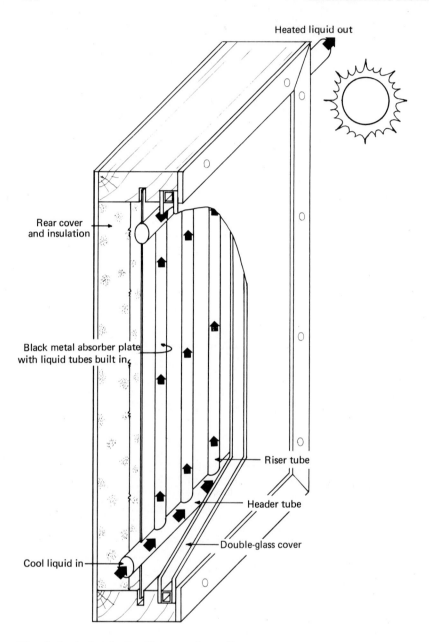

Fig. 2.4 A simple liquid-type solar collector.

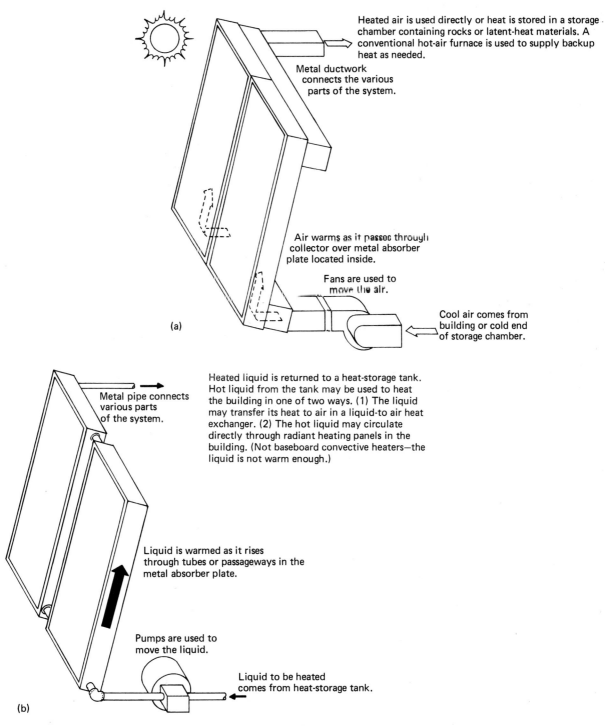

Fig. 2.5 Four practical active systems: **(a)** air type, **(b)** liquid type (continued)

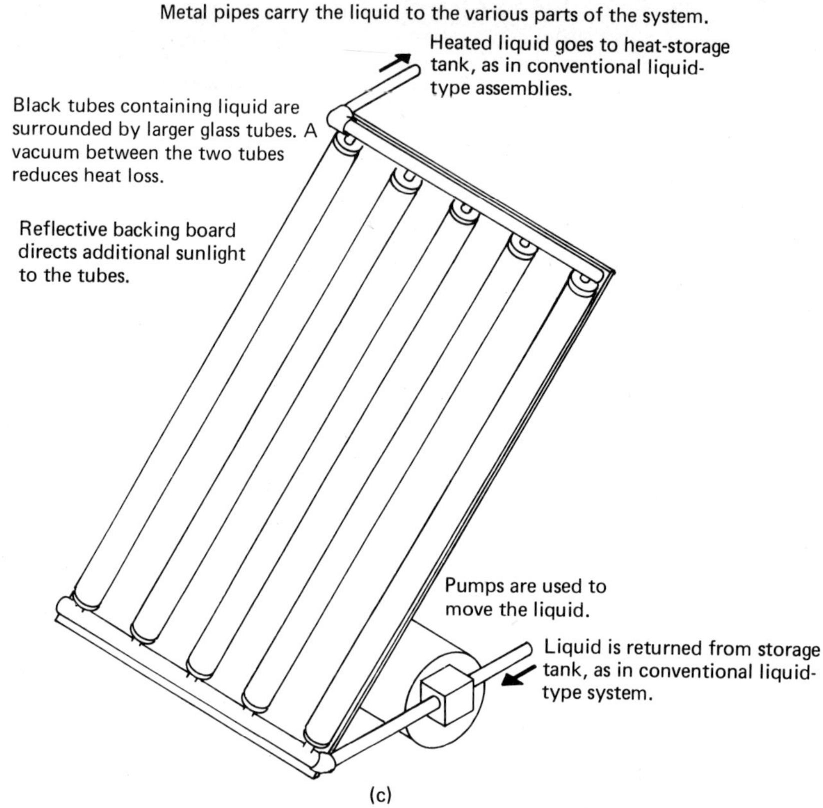

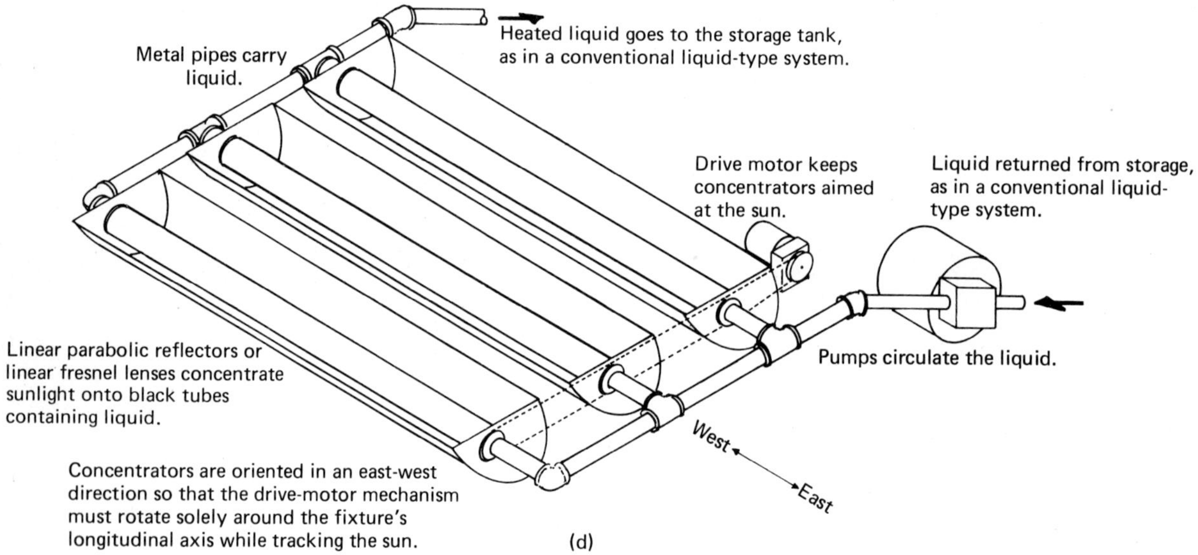

Fig. 2.5 (continued) (**c**) evacuated-tube type, (**d**) linear-concentrator type.

concentrating collectors, mirrors or lenses focus large areas of sunlight onto small collection areas (see Chapter 9).

One of the advantages of using an active system instead of a passive one is that having fans and other equipment gives the building occupants greater control over their environment. To gain a better understanding of how heat distribution works in an active solar heating system see Fig. 2.6, which shows a simple air-type system (the fans are not

Fig. 2.6 A typical solar heating system can (**a**) heat using energy from the sun, (**b**) store solar generated heat, and (**c**) deliver heat to the building from the storage unit and/or backup furnace.

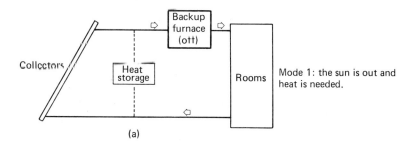

Mode 1: the sun is out and heat is needed.

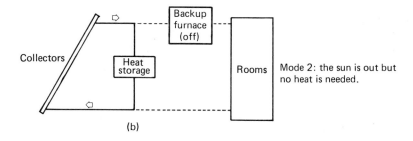

Mode 2: the sun is out but no heat is needed.

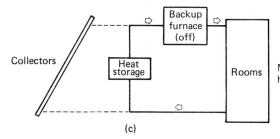

Mode 3: the sun is not out and heat is needed.

pictured). Three different heating *modes* are shown. When the sun is shining and the building needs heat, the system goes into mode 1. In this mode fans blow hot air directly from the collectors into the building and cool building air is returned to the collectors to be heated. Backup heat is not needed, so the backup heater stays off. Suppose now that the building becomes warm enough that no more heat is needed. The sun is still out, so what do we do with the unneeded solar heat? The system goes into mode 2 and stores the excess heat for later use. Later on, during the night, the system goes into mode 3 and heats the building with the stored solar heat. If this stored heat starts to die out toward morning, the backup furnace can be turned on. Almost any solar heating system requires some form of backup heat source, unless the system has an uneconomically large number of collectors and a large heat-storage unit. Besides the three functions illustrated in Fig. 2.6, the typical solar heating system might (1) exhaust excess heat in summer, (2) preheat household hot water, and (3) provide some air conditioning in summer.

2.2 ENERGY FLOW AND EFFICIENCY IN FLAT-PLATE COLLECTORS

2.2.1 What is Collector Efficiency?

As we have seen in Chapter 1, every energy-conversion device has an *energy-conversion efficiency* that indicates what fraction of the available input energy is converted to the desired-output form. Solar heat collectors are simply energy-conversion devices, so it should be possible to speak of the *efficiency* of a solar collector, and to use efficiency ratings to compare the performance of different designs. Also, if the collector efficiency is known, the solar heating system performance over an entire heating season can be estimated.

In a flat-plate collector, the *available input energy* can be defined as the sunlight energy that strikes the collector in some period of time. The *desired-output form* of energy is the useful heat energy the collector puts out during that same period. Thus we can define average collector efficiency as

$$\eta_{avg} = \frac{\text{useful heat output during period}}{\text{sunlight energy striking collector during period}}.$$

In many cases, though, it is more useful to think of rates of energy flow in and out of the collector at some precise instant. For this reason, it is helpful to define *instantaneous collector efficiency* as

$$\eta = \frac{\text{rate that useful heat is put out}}{\text{rate that sunlight strikes collector}} = \frac{q_u/A}{I_T}$$

where

q_u = rate that useful heat is put out by the collector,

A = gross collector area,

I_T = rate at which solar radiation is striking the collector surface measured in W/m² or Btu/hr ft² (T means that the radiation is measured at the same tilt as the collector is positioned).

Efficiency is a fraction and is commonly expressed as a percentage. For a collector to be 100% efficient, it would have to put out heat at the same rate that sunlight strikes it, which is extremely unlikely for two reasons. First, not all of the sunlight that strikes a solar collector makes it inside. Second, not all of the sunlight that makes it inside the collector is put out as useful heat.

Why doesn't all the radiation striking the collector make it inside? Basically, there are two reasons: some radiation is reflected or absorbed by the collector cover, and some is intercepted and blocked from the inside of the collector by the collector itself.

The fraction of sunlight that strikes the transparent cover and passes through is called the transmittance, τ. This is the fraction available for the collector to use. The fraction that is reflected from the cover is called the reflectance, ρ. The absorptance, α, was defined earlier. Since sunlight is reflected, absorbed, or transmitted, the sum of the reflectance, absorptance and transmittance must equal 1, or

$$1 = \rho + \alpha + \tau.$$

For glass, the reflectance at normal incidence, ρ, is 0.04 for each surface or 0.08 for one sheet, and the absorptance, α, varies from 0.01 to 0.14 depending on the type of glass and its thickness. Some of the radiation absorbed by the cover is reradiated and lost as heat to the outside air.

A cover that faces directly toward the sun has both a lower reflectance and lower absorptance and thus a higher transmittance for solar radiation than a cover that does not face the sun. Thus a collector that *tracks* the motion of the sun will put out more heat than an identical fixed collector since more solar energy is available to it. Different methods for increasing the cover transmittance have been used, including use of low-iron or water-white-crystal glass (low absorptance) and treatment of the surface of the glass to reduce reflectance (see section 3.2).

Efficiency calculations are based on the amount of sunlight striking the entire collector or on the collector's overall or gross area. Some of this area is often filled with glazing supports and insulation as well as structural connectors that attach the collector to the building. Obviously sunlight striking these areas has virtually no chance of being turned into useful heat by the collector. (A small fraction of this light may, however, be reflected into the collector and used, and the warmed collector supports can help prevent heat loss from the collector.)

Why isn't all the solar radiation that makes it into the collector turned into useful heat? First, some radiation is reflected away immediately by the absorber. The warmed absorber also loses heat to the cold outside environment because the hot absorber plate tends to transfer heat to the cover by radiation and convection, where it is easily lost to the outside, and because other heat is lost by conduction through the rear and sides of the collector.

Figure 2.7 summarizes the important energy gains and losses that occur in a single-glazed (one-cover) collector. The important energy gains are (1) absorption of visible light from the sun and (2) absorption of infrared radiation from warm glass. Important energy losses are (1) infrared radiation emission, (2) natural convection between absorber and glass (increased by cooling of glass due to wind, low outside temperature, and radiation from glass); and (3) conduction of heat through rear and sides of collector (depending on how it is insulated). In addition to the processes shown, heat is being put out to the building as well. This can be considered another "loss" from the absorber plate. For a double-glazed collector these same processes occur, but in a more complex way.

Fig. 2.7 Energy processes in a single-glazed collector.

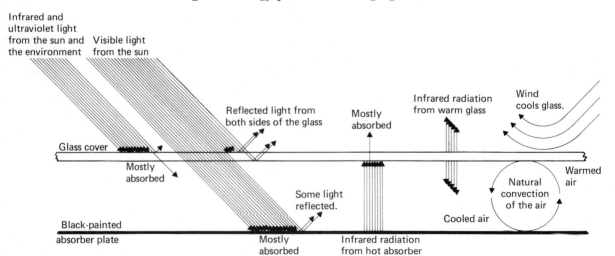

For example, natural convection still occurs between the inner glass and the absorber plate, but it also occurs between the inner glass and the outer glass since the inner glass will be warmer than the outer glass.

In a liquid-type collector, the liquid flows inside the absorber plate, meaning that all of the heat output comes from the absorber plate. In an air-type collector, however, air typically flows around the absorber plate. This means that the air is in contact with other surfaces in the collector and the situation is more complicated. For example, in some designs the flowing air touches the inside of the glass and may gain heat from it or lose heat to it depending on whether the air or the glass is warmer.

As the reader may have gathered by now, it is very difficult to calculate the efficiency of a particular collector. There are many subtleties that cannot be mentioned in a book of this kind due to the mathematics involved and space limitations. But while the subject is complex, it is also fascinating and not insurmountable. Readers who would like to know exactly what is involved in calculating an estimated efficiency should refer to works by Hottel and Woertz [1], Hottel and Willier [2], and Willier [3] for an analytical approach, and to the book by Meinel and Meinel [4] for an iterative (trial-and-error) approach.

To understand the concept of efficiency and how energy gains and energy losses are related in a flat-plate collector, we can think of a collector as a "black box" and study how energy goes in and comes out rather than what happens inside.

Figure 2.8 illustrates this way of thinking as it might be applied to a specific collector (the numbers are for illustration purposes only.) The efficiency of the collector in Fig. 2.8 is determined as

$$\text{Efficiency} = \frac{\text{rate of useful heat output}}{\text{rate sunlight strikes collector}}$$

$$= \frac{450 \text{ W/m}^2}{850 \text{ W/m}^2} = 53\%.$$

Total energy input, 1050 W/m², is made up of

Sunlight, 850 W/m²

Environmental infrared, 200 W/m²

Total energy output, 1050 W/m², is made up of

Heat output to building, 450 W/m²

Reflection loss, 150 W/m²

Infrared from glass, 350 W/m²

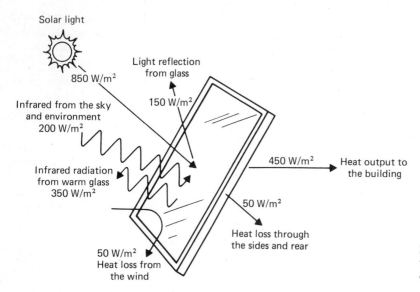

Fig. 2.8 Energy gains and losses that might occur with a flat-plate collector.

Loss to wind, 50 W/m²

Losses through sides and rear, 50 W/m²

Note that the sum of all energy inputs to the collector equals the sum of all energy outputs from the collector. This is always true when the collector is at *steady-state* conditions; that is, when wind and sunlight are steady, air temperature and collector temperatures are constant, and the heat-transfer fluid is moving through the collector at a steady rate.

2.2.2 Determining Collector Efficiency

The useful heat output in most flat-plate collectors can be summarized as

$$q_u = F_R A[I_T(\tau\alpha) - U_L(T_i - T_a)]$$

where

q_u = rate that useful heat is put out by the collector,

F_R = collector heat-removal efficiency factor,

I_T = rate that solar radiation strikes the collector surface measured at the collector tilt,

$\tau\alpha$ = transmittance absorptance (product of the transmittance of the cover and absorptance of the absorber),

U_L = overall collector heat-loss coefficient for the absorber plate,
T_i = inlet-fluid temperature,
T_a = ambient or outside-air temperature.

This equation is simply

$$q_{useful} = F_R (q_{\text{solar-in}} - q_{\text{solar-out}}).$$

Note that the far-right portion of this equation takes the form of

$$q_{out} = A U_L \Delta T$$

or

$$q_{out} = A U \Delta T$$

for heat loss, which we encountered in section 1.2.6 where U_L is the overall heat-transfer coefficient for the absorber plate. The $q_{\text{solar-in}}$ portion of the equation

$$q_{in} = A\, I_T (\tau\alpha)$$

evaluates the solar energy turned into heat by the collector as the area, A, times radiation, I_T, times the product $(\tau\alpha)$. $\tau\alpha$ is a proportionality factor relating the amount of radiation striking the collector to the amount that can be turned into heat. Only the fraction of radiation that both penetrates the cover and is absorbed by the absorber plate (thus $\tau \times \alpha$) can become useful heat. Since the values of both τ and α change with the radiation's angle of incidence, the value of their product, $\tau\alpha$, also changes with angle of incidence. $\tau\alpha$ is highest when the collector is facing directly toward the sun.

The temperature difference used in the q_u equation is the inlet-fluid temperature minus the air temperature, which does not really indicate the true case in the collector since the temperature increases through the collector and the outlet and average fluid temperatures are higher than the inlet temperature. The proportionality factor F_R, the collector heat-removal efficiency factor, is used to correct for this situation. F_R can be defined as

$$F_R = \frac{\text{actual useful energy collected}}{\substack{\text{useful energy collected if the entire collector} \\ \text{were at the inlet-fluid temperature}}}.$$

F_R is actually a complex function of several variables. Fortunately, it is not usually necessary to calculate F_R explicitly.

In the q_u equation for useful heat gain, only one area, A, is used. This is a simplification; actually, different areas should be used for different portions of the formula. The area of only the collector cover, or the aperture area, should be used for the $q_{\text{solar-in}}$ portion, and the area of the absorber (usually assumed to be equal to the aperture area) should be used for the $q_{\text{heat-loss-out}}$ portion. The efficiency of the collector is based on its overall or gross area. For simplicity, we do not make distinctions between these areas here.

If we put the formula for q_u into the equation for efficiency,

$$\eta = \frac{q_u/A}{I_T}$$

yields

$$\eta = F_R(\tau\alpha) - F_R(U_L)\frac{T_i - T_a}{I_T}.$$

Calculating the values of F_R, $(\tau\alpha)$, and U_L is difficult and often inaccurate. If there were a method of testing for collector performance, these variables could be experimentally determined.

In fact, testing collectors for thermal performance is the standard practice for evaluating the efficiency of a new collector design. With experimental efficiency results, values for the products $F_R U_L$ and $F_R(\tau\alpha)$ can be determined. It is not necessary to know the individual values of F_R, U_L, and $\tau\alpha$ since, as we have seen, only the products are used in the efficiency calculation.

Since efficiency is given by

$$\eta = \frac{q_u A}{I_T},$$

to evaluate the experimental efficiency only q_u or the rate of useful heat gain need be found experimentally, while the solar input, I_T, and the area, A, can be measured directly. The rate of useful heat gain can be determined if we know how much heat is picked up by the fluid stream as it passes through the collector. This is the *useful* heat gain. In equation form this becomes

$$q_u = AGC_p(T_o - T_i)$$

where

> G = fluid-mass flow rate per unit collector area (in kg/sec m² or lb/hr ft²)
>
> C_p = specific heat of the fluid (in J/kg °C or Btu/lb °F),
>
> T_o = outlet-fluid temperature.

A, G, C_p, and the inlet- and outlet-fluid temperatures can be measured and used to determine the instantaneous efficiency of the collector. The

ENERGY FLOW AND EFFICIENCY

efficiency is usually returned in the form of a graph (Fig. 2.9) of the efficiency versus $(T_i - T_a)/I_T$.

If we refer to our previous formula for collector efficiency,

$$\eta = F_R(\tau\alpha) - F_R U_L \left(\frac{T_i - T_a}{I_T}\right),$$

we see that if U_L can be considered a constant, this equation is in the form of a straight line $y = mx + b$ where the slope equals $-F_R U_L$ and the y intercept equals $F_R(\tau\alpha)$. Since the experimental thermal-performance tests are usually done near solar noon and the collector is aimed almost directly at the sun, this value is actually the $F_R(\tau\alpha)$ at normal incidence, or $F_R(\tau\alpha)_n$.

Without specifically knowing the $F_R(\tau\alpha)$ and the $F_R U_L$ values, we can sometimes use this equation to estimate the relative efficiencies of different collectors. For example, a collector with no insulation will have a high U_L and therefore higher heat loss and lower efficiency than one with insulation. A collector with water-white glass (high transmittance)

Fig. 2.9 Collector-efficiency graph.

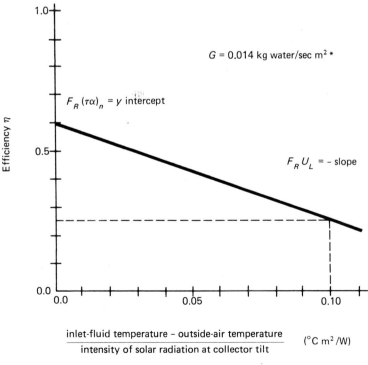

$$\frac{\text{inlet-fluid temperature} - \text{outside-air temperature}}{\text{intensity of solar radiation at collector tilt}} \quad (°C\ m^2/W)$$

*Fluid mass-flow rate G must be specified.

will have a higher $F_R(\tau\alpha)$ value than one with regular glass and therefore will be more efficient. A collector with two glass covers will have a lower transmittance and therefore lower $F_R(\tau\alpha)$, but it will also have a lower U_L and lower heat loss than a collector with only one cover. Since the y intercept, $F_R(\tau\alpha)$, will be lower for the two-cover collector but the slope, $F_R U_L$, will not be as steep, the efficiency lines for the two collectors can be expected to cross. Sometimes the one-cover collector will be more efficient and at other times the two-cover collector will be more efficient. Also, from this equation it can be seen that the collector will run most efficiently when the inlet-fluid temperature is at or below the ambient temperature.

Example What are the values of $F_R(\tau\alpha)_n$ and $F_R U_L$ for the hypothetical collector represented in Fig. 2.9?

Solution
$$F_R U_L = -\text{slope} = -\frac{0.25 - 0.60}{0.10 - 0} = 3.5 \frac{\text{W}}{\text{°C m}^2}$$

$$F_R(\tau\alpha)_n = y \text{ intercept} = 0.60.$$

Collector-efficiency graphs can be used to compare the relative merits of different collectors. Caution should be exercised in this regard, however. Efficiency graphs for different collectors with $T_i - T_a/I_T$ along the x axis cannot be compared directly *unless* the mass-flow rate, G, is the same. Therefore, G should always be specified. The same collector tested with different flow rates will have different values for $F_R U_L$ and $F_R(\tau\alpha)_n$.

Collector-efficiency curves can also be plotted against

$$\frac{T_o - T_a}{I_t},$$

the

$$\frac{\text{outlet-fluid} - \text{ambient temperature}}{I_T},$$

or

$$\frac{\text{average-fluid} - \text{ambient temperature}}{I_T}.$$

Values for $F_R U_L$ and $F_R(\tau\alpha)_n$ can also be obtained from these graphs, although not as directly.

The values of $F_R U_L$ and $F_R(\tau\alpha)_n$ obtained from any of these types of collector-efficiency graphs can be used to calculate estimated system performance for the entire heating season. These calculations are usually quite involved and tedious and so the task is usually assigned to a computer. One such method is explained by Beckman, Klein, and Duffie [5].

2.3 FACTORS AFFECTING THE PERFORMANCE OF SOLAR HEATING SYSTEMS

Many factors can affect the performance of a solar heating installation. In the succeeding chapters these factors will be considered in detail. By way of introduction, however, we will address the fundamental concerns that a solar heating system designer must keep in mind while designing or evaluating any solar installation. These are

1. Heat loss. This is perhaps the overriding concern of the solar-installation designer. If a solar collector had no heat loss, it could approach 80–100% efficiency (instead of 40–60%). If a building had no heat loss, it would require no heating system—heat it once and it would stay warm forever. There are many economical ways to reduce heat loss from collectors, distribution systems, and buildings, all of which allow the use of smaller, cheaper heating systems.

2. Light gain. Where sunlight exists, opportunity for light gain exists. The addition of (double-glazed) windows to the south wall of a building will often reduce the heating needs of the building, even when it has an active solar heating system on the roof. Reflective surfaces can be installed near collector arrays to increase energy collection by as much as 50%. Collection areas can be protected from shading. Covers for solar collectors can be made of antireflective or low-iron glass to increase light input.

3. Heat transfer. A heat source is not useful unless the heat can be effectively transferred to where it is needed. Careful attention to design is needed wherever heat must be passed from one medium to another; for example, from the absorber plate to the heat-transfer fluid or from hot air to the heat-storage material. Plenty of surface area is required; for example, fins installed on an air-type absorber plate increase the heat-transfer area and therefore heat transfer. Liquid-to-liquid heat exchangers must be large enough to transfer heat effectively from one liquid to the other.

4. Fluid friction. Large, power-hungry fans and pumps are needed if ducts, pipes, and equipment create a lot of resistance to fluid flow. Heat transfer in collectors, storage media, heat exchangers, and so forth is usually best enhanced by an increase in surface area, not by unnecessary fluid turbulence. Turbulence in the flowing air or liquid increases heat transfer at the cost of making the fluid harder to pump.

5. Simplicity. The simpler a system is, the fewer things there are that can go wrong. The less exotic a system is, the easier it is to obtain parts for and the simpler it is to service. Whenever a system can be changed to reduce its maintenance requirements, the economic break-even point is brought dramatically nearer in time. New ideas and new technology should be used whenever possible, but only after it can be shown that they will save on labor, maintenance, or cost. (Efficiency and thermal performance are not ends in themselves; they are means of saving money or increasing consumer satisfaction).

These are considerations that the system designer and product manufacturer had best keep in mind. They must also be among the basic criteria used by those evaluating systems—consumers, marketing organizations, and government agencies.

REFERENCES

1. Hottel, H. C., and B. B. Woertz (1942). Performance of flat-plate solar heat collectors. *Transactions, ASME* February, pp. 91–104.
2. Hottel, H. C., and A. Willier (1955). Evaluation of flat-plate solar collector performance. *Transactions of the Conference on the Use of Solar Energy*, vol. 3, Thermal Processes, part 2, pp. 74–104.
3. Willier, A. (1977). Prediction of performance of solar collectors. In R. C. Jordan and B. Y. H. Liu (eds.), *Applications of Solar Energy for Heating and Cooling of Buildings*. New York: ASHRAE.
4. Meinel, A. B., and M. P. Meinel (1976). *Applied Solar Energy: An Introduction.* Reading, Mass.: Addison-Wesley.
5. Beckman, W. A., S. A. Klein, and J. A. Duffie (1977). *Solar Heating Design by the f-Chart Method.* New York: John Wiley.

BIBLIOGRAPHY

Anderson, B. (1976). *The Solar Home Book; Heating, Cooling, and Designing with the Sun.* Church Hill, N.H.: Cheshire Books.

ASHRAE (1978). *Methods of Testing to Determine the Thermal Performance of Solar Collectors.* New York: ASHRAE, Standard 93-77.

SMACNA (1978). *Fundamentals of Solar Heating.* Washington, D.C.: U.S. Government Printing Office, stock no. 061-000-00043-7.

Sunset Books (ed.) (1978). *Sunset Homeowner's Guide to Solar Heating.* Menlo Park, Calif.: Lane.

PROBLEMS

2.1 Give two examples of the greenhouse effect experienced by many people as part of their everyday lives.

2.2 Why is providing 100% of your heating needs using solar energy almost always uneconomical?

2.3 A solar collector absorber ($\epsilon = 0.95$) at 80° C emits 130 W/m² of radiation to the glass cover. Assume the cover ($\epsilon = 0.88$) is at 26° C and the sky is at 10° C. Show how the greenhouse effect works in this example by calculating how much radiation is lost to the sky. How much would be lost if there were no cover?

2.4 The reflectance for radiation at normal incidence passing from a substance with index of refraction n_1 to a substance with index of refraction n_2 is given by

$$\rho = \left(\frac{n_1 - n_2}{n_1 + n_2}\right)^2.$$

What is the reflectance for light traveling from air ($n = 1$) into glass ($n = 1.52$)? If we neglect absorption and assume no multiple internal reflections, what is the transmittance of a sheet of glass? What is the overall reflectance?

2.5 The normal incidence transmittance (neglecting absorption) of a system of z covers, all of the same material, is given by

$$\tau = \frac{(1 - \rho)}{1 + (2z - 1)\rho}.$$

Using your single interface reflectance from problem 2.4, find the transmittance for one sheet of glass. Why is this answer different from the answer in problem 2.4? What is the transmittance of two glass covers? Of three?

2.6 The transmittance of radiation in partially transparent substances (neglecting reflection) is given as

$$\tau = e^{-KL}$$

where

$e = 2.7183$, base of natural logarithms,

K = extinction coefficient of the substance in 1/cm (high K = low transmittance),

L = actual path length of the radiation through the substance in cm (high L = low transmittance).

What is the transmittance for two sheets of 0.32-cm-thick glass with $K = 0.1$? If the transmittance (including both absorption and reflection) is the product of the two individual transmittances, what is the actual transmittance at normal incidence if you use your answer from problem 2.5? If the absorptance of the absorber plate is 0.95, what is the $(\tau\alpha)_n$ product for a double-glazed collector?

2.7 In this chapter, the rate of useful heat output by the collector was given as

$$q_u = F_R A \left[(I_T(\tau\alpha) - U_L(T_i - T_a)\right].$$

As was explained in the chapter, only one area, A, was used, when actually different areas should have been included. Derive the equation for instantaneous collector efficiency using the proper areas. Assume that the aperture area equals the absorber area. What are the values of $F_R(\tau\alpha)$ and $F_R U_L$ in this case?

2.8 Plot the instantaneous collector-efficiency graph for a liquid collector with $F_R(\tau\alpha)_n = 0.64$ and $F_R U_L = 4.7$ W/°C.

2.9 The following data points are from a theoretical collector thermal-performance test. Plot the instantaneous efficiency curve in terms of $T_i - T_a/I_T$ for this air collector and find the values of $F_R(\tau\alpha)_n$ and $F_R U_L$. The density of air is 0.001205 kg/l and its specific heat is 1012 J/kg °C.

Airflow rate = 15 l/sec m². Ambient temperature = 21.0 °C.

I_T (W/m²)	T_{in}(°C)	T_{out}(°C)
610	27.1	41.0
620	33.4	46.5
615	51.8	61.5
610	45.4	56.1
615	64.0	71.6

2.10 Collectors can easily be run at a variety of mass-flow rates. What is the effect on $F_R(\tau\alpha)_n$ and $F_R U_L$ of an increase in flow rate? What is the disadvantage associated with this?

2.11 Explain how the efficiency graphs of the indicated collectors will compare:

a) Single glazed with black absorber and double glazed with black absorber

b) Single glazed with black-painted absorber and single glazed with selective-surface absorber (selective surface reduces infrared radiation)

c) Single glazed with selective absorber and double glazed with black-painted absorber

2.12 Explain what effect the following changes have on collector efficiency. Assume that no other variables change.

a) Increase in ambient temperature

b) Decrease in inlet-fluid temperature

c) Increase in solar radiation, I_T

d) Increase in outlet-fluid temperature

e) Decrease in collector area

f) Increase in average fluid temperature

2.13 Instantaneous collector efficiency can also be plotted as a function of $(T_{avg} - T_a)/I_T$ where T_{avg} is the average collector fluid temperature, or as a function of $(T_o - T_a)/I_T$ where T_o is the exit temperature of the fluid. $F_R(\tau\alpha)_n$ and $F_R U_L$ may be found from these graphs as

$$F_R(\tau\alpha)_n = C \times (y \text{ intercept})$$
$$F_R U_L = -C \times (\text{slope})$$

where C is a correlation constant for the system. For the plot using T_{avg}, $C = GC_p/[GC_p - (\text{slope}/2)]$. Show that this is the correct constant for this plot. Using two different equations for collector efficiency, derive the value of C for the plot using T_o. (Hint: solve one equation for T_o and the other for T_a, then combine the results to get an efficiency equation in $(T_o - T_a)/I_T$.)

3 DESIGN FACTORS FOR FLAT-PLATE COLLECTORS

The process of acquiring flat-plate collectors for solar heating or cooling installations is complicated by several factors. Solar technology has not advanced to the point where one kind of collector has become dominant in the industry (for that matter, there is not much of a solar industry yet). In many cases it still pays for individuals to construct their own collectors. All things considered, when is it better to buy collectors and when is it better to construct them? What kinds of collectors are commercially available? What are the important design features of a flat-plate collector? How is each feature optimized? Which collector designs are most likely to prove effective in the long run? What kinds of innovations can be expected in the future?

Let us consider the last question for a moment. No one can say with certainty what the future holds, but every collector, even a future collector, must be

1. Efficient
2. Long lasting
3. Economical to buy, install, and maintain

By the very definition of efficiency,

$$\eta = \frac{\text{useful heat output}}{\text{light energy available}},$$

an efficient collector must

1. Absorb as much of the incoming light as possible (reject as little sunlight as possible)
2. Minimize heat loss to the environment
3. Effectively transfer its collected heat to the heat-transfer fluid flowing through it

To be long lasting, a collector must

1. Withstand occasional extremes of temperature (below zero in winter to 150–200° C [300–400° F] during stagnation, depending on its efficiency)
2. Withstand large daily temperature swings and the resulting thermal stresses
3. Withstand continuous bombardment by the ultraviolet radiation in sunlight
4. Resist corrosion from the heat-transfer fluid and other sources
5. Remain impervious to wind, rain, snow load, even hail storms

To be economical to buy, install, and maintain, a collector must

1. Contain a minimum of exotic materials and technology (unless it is *definitely* being mass-produced)
2. Be sturdy and relatively light in weight (for shipping, installation, and structural reasons)
3. Be simple to mount and simple to connect to air or liquid headers
4. Require cleaning and maintenance only at infrequent intervals
5. Be fail-safe in unusual conditions (for example, liquids in a collector must not be capable of freezing or boiling during a power outage)

The two fundamentally different types of flat-plate collectors are, of course, air-type and liquid-type collectors, depending on what type of heat-transfer fluid is used. It does not make sense to say that an air collector is better than a liquid collector (or vice versa) any more than it does to say that a furnace is better than a boiler. Each class of collector represents a different technology that is applicable to a different use. But because of the differences, it is likely that air-type collectors will be increasingly used for the heating of homes and small commercial buildings and liquid-type collectors will be used for heating large buildings, for industrial-process heating, and where solar-powered cooling is desired.

How do air and liquid systems compare in efficiency? If typical air and water collectors are set up and run side by side, with standard fluid-flow rates, identical sunlight input, and the same fluid-inlet temperatures, what will happen? In all probability, the water collector will put out more heat and operate at a higher efficiency. The reason is that the air collector will lose more heat. Why? Because the air collector runs with a hotter absorber plate so there is more heat to lose. The absorber in a water-type collector typically runs at a temperature that is only slightly higher than that of the liquid flowing through it, owing to the effectiveness of water in removing the heat. Air is not as effective at removing heat, so the absorber plate in an air collector may be 15° C (27° F) hotter than the air flowing through it, or even more. The hotter absorber results in more heat loss and less efficiency.

There is a problem with this type of comparison, though. In actual solar heating systems, the inlet-fluid temperatures will not be equal. The air entering air collectors will be 15–30° C (60–85° F). The water entering water collectors can be 45–55° C (110–130° F) or more, depending on the design. Other factors will be different too. While a water collector may be more efficient under certain conditions, a water system is not necessarily more efficient than an air system. Recent studies have shown air systems to be equal or even superior to water systems for residential heating [1]. Again, direct comparison of air to liquid collectors can lead to erroneous judgments.

The following sections explore the individual components and considerations that together make up efficient flat-plate collectors and their arrays. The collector design is discussed from the outside in, beginning with the frame, the glazing, and continuing in to the absorber plate, absorber coatings, and finally fluid-flow paths for the collector and entire array.

3.1 FRAMES, BOXES, AND INSULATION

The flat-plate solar collector has come to be thought of as a flat box with a window on the front. This, of course, may or may not be true. Active-type solar collection systems can arrive on a site as components to be installed rather than boxes to be mounted. They may have long, narrow sheets of plastic instead of glass "windows." Collectors may be round or wedge shaped. The important thing is that the end result be a device that efficiently traps solar energy. It just so happens that the modular approach—boxes or glass-sized panels—is convenient and economical. Glass is a good material that is cheap in certain sizes. Labor and frustration can be minimized during installation if premanufactured units are mounted.

Wood has been used in collectors for many years and may certainly be an acceptable material if properly used. A wood-framed collector is easy to build and has some natural insulating capability, but there is one inherent problem: wood and heat do not go together well. Initially, the danger of fire in wood collectors is low: temperatures of 175–200° C (350–400° F) are about the maximum that can be expected in a flat-plate collector (with strong sunlight and no fluid flow). However, as wood ages in the presence of heat it pyrolizes, becoming dry, weak, and more susceptible to fire. Furthermore, if a collector is constructed of wood that has not been properly seasoned it is likely to crack, check, or warp as it dries. Wood used in solar collectors should be fine grained, properly seasoned and, preferably, treated to resist aging and drying. The absorber plate (the hottest spot in a collector) should ideally not be in direct contact with wood surfaces. Wood can be used but because of its problems wood collectors are being restricted or banned in building codes in several localities.

Metal is strong and fire resistant but heavier than wood, more expensive, and an excellent conductor of heat. As a result, metal collector frames tend to be heavy, expensive, and heat losers. However, metal is often a more suitable material than wood, for example, in some commercial buildings. And metal frames can be thermally isolated from absorber plates. Aluminum can be used to mimimize the weight of metal frames.

Plastics are being considered for collector frames. New materials are becoming available that can withstand high temperatures and are as strong as metals. A plastic frame would possibly have to be painted to protect it from ultraviolet exposure, but it could be light in weight, fire resistant, and attractive.

Most plastics are made from fossil fuels, a nonrenewable resource. Metals are refined using large amounts of fossil fuels. Forests are being depleted for their wood. In fact, a lot of energy and other resources go into any solar collector, but at least these resources are being invested in the sense that a properly designed solar collector will eventually capture many times more energy than has gone into its construction.

The insulation in a solar collector separates the hot absorber plate from the cool outside environment. Chapter 8 discusses several types of building insulation. For a solar collector, the choices are more limited.

For instance, polystyrene foam cannot be used in collectors because it melts at a low temperature and is flammable. Some polyurethane foams may be used, and glass wool or rock wool may be used. Standard glass wool is not acceptable, however, because it is held together with a binder that will evaporate (*outgas*) at high temperatures, depositing a residue on the cover and absorber plate. Industrial-grade fiberglass board can be used instead (e.g., Johns-Manville No. 814) and is probably the best all-around insulation material for collectors. Asbestos, glass wool, and other potentially hazardous materials should never be used in the air passages of air-type collectors. All materials used inside collectors should be tested for temperature stability, preferably to the 175–200° C (350–400° F) stagnation temperatures that do occur rarely.

Allowance must always be made for the thermal expansion of various materials, particularly glass, which tends to self-destruct if not given sufficient room to expand. Table 3.1 gives a conservative estimate of the thermal expansion that can be expected for several substances. Note that an aluminum frame will expand faster than the glass it holds, while a wood frame will expand more slowly. (Beware of glass fitted tightly into a wood frame.) Note also that plastics expand about 10 times more than glass.

Table 3.1 **Thermal Expansion of Various Materials**

Material	Expansion in Length*
Aluminum	0.43%
Brass	0.33%
Concrete	0.20%
Copper	0.23%
Glass	0.14%
Iron, steel	0.18%
Plastic	0.80–3.3%
Rubber	1.3–2.2%
Wood (along grain)	0.05–0.11%
Wood (across grain)	0.55–0.80%

*Percentage of expansion in length of the material experiencing a 150° C (300° F) temperature rise.

The physical structure of a flat-plate collector frame is not overly critical. Many variations are possible, but certain features can be considered essential:

1. The frame must provide firm support for the collector and at the same time should insulate the absorber plate from the outer edges of the collector (e.g., silicone rubber gaskets or treated fiber supports instead of metal brackets).
2. A weatherproof and airtight seal for the cover, an allowance for thermal expansion, and firm physical support for the glazing (e.g., insulated glass unit supported by rubber gaskets) must be a feature of the frame.
3. Simple access through the frame into the collector, or a simple way to remove the cover to replace broken glazing, clean the absorber plate, or perform other maintenance should be a design feature (e.g., snap-on or screw-on glass stops).
4. The frame must also allow for watertight or airtight fluid passageways (e.g., liberal use should be made of silicone caulk at critical joints).

Figures 3.1 and 3.2 show the construction of two commercial-style flat-plate collectors, one an air type and the other a liquid type. Figure 3.3 shows a design for a simple air-type collecting wall (which may be of interest to a do-it-yourselfer, but is probably not practical for commercial use owing to its labor intensiveness). In evaluating the construction of these or any other flat-plate collection systems, it is helpful to examine them in light of the four conditions just listed as well as the conditions listed at the beginning of this chapter. The following list names actual problems that have been experienced in working flat-plate-collector systems over the past 20 years:

1. Materials Problems
   a) Foam insulation that melts, rots, or expands, breaking the glass
   b) Coatings that flake off absorber due to poor material or poor surface preparation
   c) Ultraviolet deterioration of sealants, gaskets
   d) Plastic covers that yellow, become brittle
   e) Condensation inside cover
   f) Wood frames that become soft, powdery
   g) Wood frames that crack or warp due to inadequate seasoning

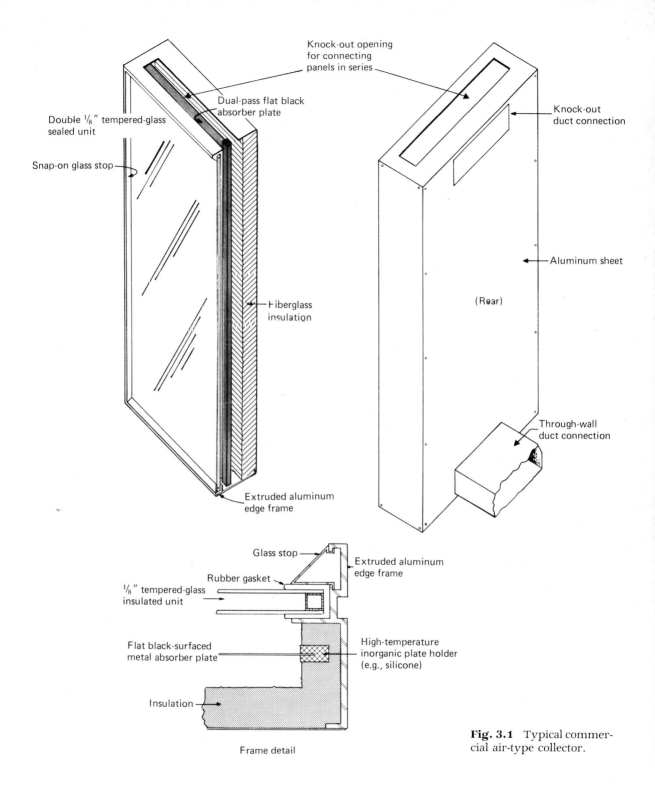

Fig. 3.1 Typical commercial air-type collector.

66   DESIGN FACTORS FOR FLAT-PLATE COLLECTORS

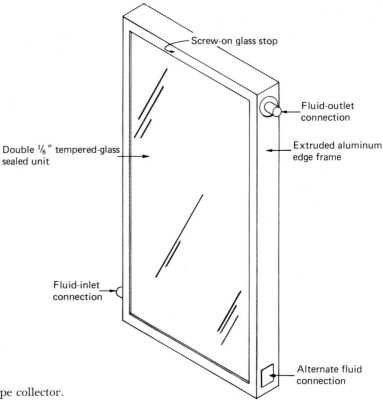

Fig. 3.2 Typical commercial liquid-type collector.

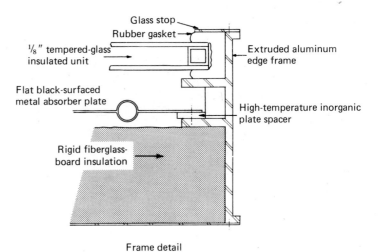

Frame detail

FRAMES, BOXES, AND INSULATION

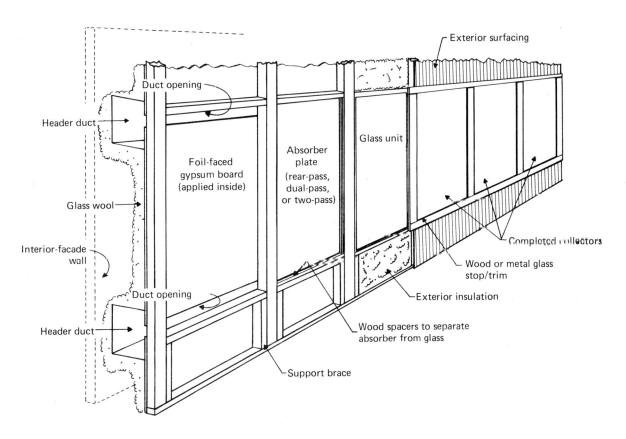

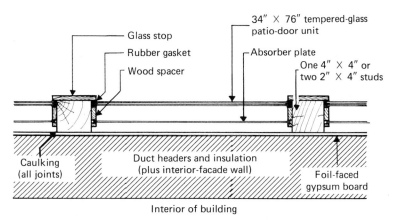

Fig. 3.3 A simple air-type solar collection wall.

h) Polyvinyl chloride (PVC) pipes and connectors that overheat and become deformed
   i) Leaking of water from corroded spots in absorber plates
   j) Erosion of outer surface of fiberglass cover due to weathering
2. Structural Problems
   a) Broken glass that was not tempered
   b) Broken glass due to inadequate space for thermal expansion
   c) Plastic covers that tear off in high winds
   d) Absorber plates that come loose, fall off, because stapled to wood frame
   e) Broken glass caused by hail storm
   f) Collectors that come loose from roof due to inadequate mounting
3. Performance Problems
   a) Poor heat collection because absorber plate connected directly to outer frame
   b) Excess heat loss due to 35% air leakage from air-type collector
   c) Reduction in light gain as plastic cover becomes clouded
   d) Reduction in light gain as absorber plate in air-type collector becomes dusty
4. Operational Problems
   a) Overheating and boiling in liquid-type collectors caused by restricted flow in some collectors in an array
   b) Leaking of absorber plate because freezing water left in low spots of collector when collectors drained for the night

3.2 GLAZING The transparent cover is a familiar part of most solar collectors. A glass or plastic cover is not always needed, however. A solar swimming pool heater works very well, sometimes even best, without covers—but only in summer when the outside air is warm. As soon as the outside air temperature begins to drop, putting a transparent cover on a solar collector inevitably results in a dramatic rise in heat collection. As mentioned in Chapter 2, transparent covers reduce heat loss in two basic ways. First, glass and some other materials have the ability to transmit visible light and intercept infrared light: they admit sunlight but reduce radiative heat loss. Second, a transparent cover keeps wind and natural convection from robbing heat from the absorber plate.

One of the most commonly used cover materials is glass—and for good reason. Glass is readily available, quite transparent to visible light, and opaque to infrared wavelengths beyond about 3 μm, meaning that it blocks heat loss from the absorber due to radiation. It is also, remarkably, one of the most weather- and sunlight-resistant materials on earth. Most transparent plastics have an effective outdoor lifetime of 1/2–3 years because of the destructive effects of ulraviolet light. A few, those more suitable for solar collectors, can last 15–20 years. Some glass windows, on the other hand, have been around for centuries without any apparent deterioration.

One sheet of 0.32-cm ($\frac{1}{8}$-in) glass transmits from about 82–90% of the visible light striking it. About 8% of the available light is reflected away, 4% from the front surface and 4% from the rear. (The fact that half of the reflection is from the back of the glass is surprising to some people, but the double image can be observed under the proper light conditions.) The remaining losses are due to absorption of light as it passes through the glass. The amount of light absorbed depends largely on how much iron oxide (Fe_2O_3) is in it. Ordinary glass has some iron in it and absorbs approximately 6–7% of the light passing through it. The more iron content it has, the more light is absorbed and the greener the glass appears when observed from its edge. High-quality glass is available (at fairly high cost) that can increase the efficiency of solar collectors. A 0.32-cm ($\frac{1}{8}$-in) sheet of *low-iron* glass absorbs 4–5% of the light passing through it and a 0.32-cm sheet of *water-white* glass may absorb as little as 1–2% of the light. The overall transmittance of various grades and thicknesses of glass is given in Table 3.2.

The actual transmittance of a glass cover depends on other factors as well. A ray of light that strikes it at a steep angle will have more of its energy reflected and absorbed than a ray that strikes the glass perpendicularly (Fig. 3.4). This is one reason why a collector is most efficient when aimed directly at the sun. Also, there is a relationship between how many covers a collector has and how much light is transmitted.

In general, increasing the number of glass covers on a flat-plate collector (1) reduces the heat loss from the collector (2) reduces the light input to the collector. Two glass covers is the optimum number for the northern United States. Three would result in higher output temperatures but lower efficiency. If this seems like a contradiction, remember that just because a collector is operating at a high temperature does not mean that it is operating more efficiently (see section 2.2.2), and any cover system that reduces the amount of light that comes in tends to reduce efficiency at all operating temperatures. A three-glass cover may be necessary in an extremely cold climate or where high temperatures are needed.

Special surface treatments are available that can increase the performance of glass. One quite effective treatment is a type of acid bath that

Table 3.2 **Properties of Various Glazing Materials (English Units Only)**

Material	Typical Trade Name*	Representative Thickness	Solar Transmission	Infrared Transmission	Maximum Recommended Size	Weight (Lb/Ft^2)	Flammability	Maximum Operating Temperature (F)	Estimated Lifetime for Solar Use (Years)
Clear glass (float lime)		1/8"	85%	Low	34" × 76"	1.63	Low	400°	Over 30
		3/16"	81%	Low	36" × 96"	2.51	Low	400°	Over 30
Low-iron glass	ASG Lo-iron®	1/8"	87%	Low	34" × 76"	1.03	Low	400°	Over 30
		3/16"	85%	Low	36" × 96"	2.51	Low	400°	Over 30
Water-white glass	ASG Sunadex®	1/8"	91%	Low	34" × 76"	1.61	Low	400°	Over 30
		3/16"	90%	Low	36" × 96"	2.41	Low	400°	Over 30
Stabilized fiberglass-reinforced polyester (FRP)	Kalwall Sunlite®	0.040"	87%	Low	24" (short dimension)	0.30	Low	250°	10–15
		0.060"	85%	Low	30" (short dimension)	0.45	Low	250°	10–15

Material	Trade name	Thickness			Max size			Max temp	
Acrylic sheet (polymethyl methacrylate)	Plexiglas®	1/8"	89%	Low	24" (short dimension)	0.75	High	190°	20
		3/16"	87%	Low	36" (short dimension)	1.10	High	190°	20
Stabilized Polycarbonate sheet	Lexan®	1/8"	81%	Low	24" (short dimension)	0.73	Moderate	250°	10
		3/16"	78%	Low	36" (short dimension)	1.17	Moderate	250°	10
Polyvinyl Fluoride (PVF) film	Tedlar®	0.004"	92%	Moderate		0.03	Low	250°	25
Fluorocarbon film (FEP)	Teflon®	0.001"	96%	High		0.01	Low	400°	25

*Note: The properties as specified in this table have been obtained from many sources and do not necessarily apply precisely to these trade-name products. Properties vary as formulations vary.

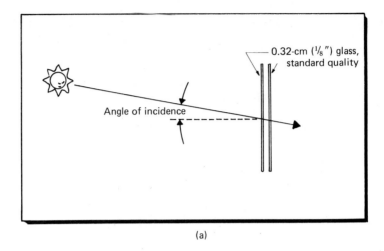

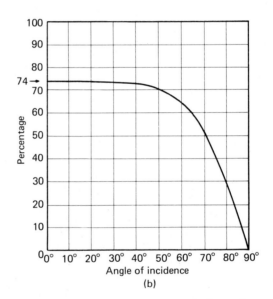

Fig. 3.4 Transmittance: (**a**) effects of sunlight angle on transmittance through a two-cover system, (**b**) percentage of available light transmitted.

greatly reduces the reflectivity of common glass. Glass that has been processed in this manner is known as *antireflective glass* or *AR glass* and may transmit 6–7% more light than ordinary glass. This can result in a 10–15% increase in collector efficiency at all operating temperatures. The process, known as Magicoat, was patented in the 1940s but the patent has expired. It involves dipping the glass in a fluosilicic (H_2SiF)

acid bath that has been saturated with silica [2]. This creates a microscopically porous, skeletonized surface that effectively has a low index of refraction (light-bending ability) and thus has low reflectivity. (A substance's reflectivity depends on its index of refraction.) One drawback is that glass is always costly to process because of its weight and fragility. Also, the porous antireflective surface tends to absorb grease readily and hold it, thus losing its properties. The treatment is so effective, though, that it might be considered for inaccessible surfaces, such as between two sheets of glass. Antireflective surfaces can be made in other ways [3], but the acid-dip process appears to be the most economical. Sandblasting or frosting does not result in an antireflective surface and sandblasted or frosted glass does not perform the same function as antireflective glass.

Another surface treatment for glass (and possibly plastics) is an *infrared reflective coating*. Ordinary glass absorbs infrared light that is radiated out from the absorber plate but does not reflect it back. Instead, it heats up and loses energy in two directions—inward and outward. The inward radiation from the warm glass is beneficial since the energy goes back toward the absorber plate. But part of the heat radiated from the glass goes to the outside. Heat loss would be reduced if infrared losses from the absorber plate were reflected back, which is what an infrared-reflective coating would enable the glass to do. One coating on the glass surface nearest the absorber plate would be sufficient.

Thus far it has not made economic sense to use infrared-reflective coatings since the only coatings known have had the unfortunate side effect of absorbing visible light, making them useless for solar collectors. In 1975 a more practical process was described [4]. This process, known as Pyrosol®, deposits a coating that transmits about 92% of visible light yet reflects about 85% of the infrared light. It has not received adequate study for its solar-collector potential, however. Perhaps the best way to reduce infrared losses will continue to be the selective-surface absorber plate (see Section 3.4).

We must consider structural factors when using glass in solar collectors. The glass must be tempered. Tempering is a common process that increases the strength of glass by setting up permanent stresses in it. Nontempered glass is not acceptable because it cannot withstand the thermal expansion and contraction that naturally occur in solar collectors. Unfortunately, glass cannot be cut after it is tempered, so it must be bought in standard sizes from tempering plants or made to order. A good if unsteady source of tempered glass is "scratch glass," glass that comes out of a tempering plant with minor imperfections such as scratches. The imperfections are usually too small to be seen, especially in solar collectors, and large cost savings are possible. The optimum glass thickness for solar collectors is 0.32 cm, which is thin for highest transmittance. When the width must be greater than 0.86 m (34 in), 0.48 cm (3/16 in) glass must be used for structural strength. There is a structural limit to how wide and how long a sheet of glass can be (Table 3.2).

Plastics are being used increasingly as glazing materials for solar installations. Although there are a large number of plastic sheets and films on the market, only a few have been found that are able to withstand the conditions that occur in solar collectors. Table 3.2 gives the properties of some of the more promising plastics, including FRP, Lexan, Plexiglass, Tedlar, and Teflon. The prices vary widely, but in quantity the prices for most plastics are in the same range as glass prices.

The basic problem with plastics is that they tend to deteriorate when exposed to the ultraviolet rays in sunlight. They turn yellow or brown, become brittle, and lose their ability to transmit light. The plastics used for solar applications are those that best resist ultraviolet light, but even most of these slowly lose their transmittance and mechanical properties over the years. The fluoroplastics such as Tedlar and Teflon have by far the longest lifetimes—25 years or more—but the others may lose 10–20% of their transmittance within 15 years. Their performance may be improved by adding chemical stabilizers to slow ultraviolet damage.

Other potential problems with plastics are fire safety and infrared transmission. Solar collectors are subject to local building codes, which may restrict the use of certain plastics. Fortunately, fiberglass and the fluoroplastics are quite low in flammability. Many plastics are not as effective as glass in capturing infrared radiation and must, therefore, be used with selective absorber plates to reduce heat loss. Fortunately, those listed in Table 3.2 have fairly good infrared characteristics.

The primary advantages of plastics are lightness and strength. Plastics are lighter than glass and often strong enough to be used in pieces of lesser thickness. And, of course, plastics don't have glass's tendency to shatter. The special plastics suitable for solar collectors are generally as expensive to buy as glass, or nearly so. Because they are stronger and lighter, however, the costs of transporting, installing, and supporting the collectors can be substantially reduced.

The cover of a solar collector can be considered a system in itself. It usually is made of two or more sheets of glazing material and must be tightly put together and carefully sealed. Not only the glazing but the sealants must resist sunlight, heat, and aging. If the seal on a cover system fails, unsightly condensation of water is very likely to occur. If it is an air-type collector, heat loss owing to air leakage can occur. If it is a roof-mounted collector, rain can find its way in.

The experience and techniques of window manufacturers should not be ignored when glass is to be the cover material. Double-glass windows are almost universally made with a *sealed unit* composed of two *lights* (panes) that are separated by a metal spacer around the edges. The sealed unit is manufactured separately from the window frame and carefully made airtight and watertight before it is installed in the frame. If it is not completely sealed it will someday "fail," and condensation will occur inside the unit (condensation is particularly ugly in solar collectors). The standard sealed unit for windows has a *single seal* of polysulfide, which resists ultraviolet well but does not hold up well under excessive heat. The recommended seal for solar collectors is a *double seal:* an inner coating of hot-melt butyl (isobutylene) around the metal spacer and a secondary seal of silicone caulking or metallized tape. If a collector overheats, the butyl may soften but it is likely to keep the unit airtight. The outer seal will provide mechanical stability and additional sealing action.

Silicone sealant is the universal and ideal sealant for use in solar collectors. It is flexible, good to 200° C (400° F) (very unusual among sealants), and impervious to sunlight, weather, and most chemicals.

Cover systems can be made from various combinations of plastics and glass (FRP and Tedlar, for example). It should be remembered, though, that different materials expand different amounts when heated and that a good, lasting seal is important no matter which material is used. Also, covers should be spaced 1–2.5 cm (⅜–1 in) apart for optimum insulation. If they are too close together, the air space conducts heat; if too far apart, natural convection begins to transfer too much heat.

3.3 ABSORBER PLATES AND HEAT-TRANSFER FLUIDS

The absorber plate is the element that performs the actual work of converting light to heat energy and transferring the heat to flowing air or liquid. It is a vital element of any collector, and several considerations affect its design. Of the many different absorber plates now in use, the two basic classes are air type and liquid type.

Pure air and pure water may be the simplest and most effective heat-transfer fluids available at this time, provided that measures are taken to guard against leaking, corrosion, freezing, and boiling. The type of heat-transfer fluid selected has a large bearing on the type of absorber plate used and its features.

3.3.1 Absorber Plates for Air Collectors

The use of air as the heat-transfer fluid leads to the simplest collector designs and has certain other advantages over a liquid fluid: air does not corrode absorber plates; it does not freeze or boil; and air leaks, while they do affect system performance, do not have the same destructive consequences that water leaks have. The disadvantages of air stem from its relatively poor heat-transfer characteristics: it does not accept or release heat as readily as do water and other liquids. Air absorbers, therefore, often contain many fins that increase the heated area and improve heat transfer. Heat transfer is also improved if large volumes of air are blown through the collectors to remove the heat. This large air volume requires large ducts (instead of small water pipes and pumps). Air collectors are generally less efficient than liquid collectors, although air systems, as discussed earlier, are a different story.

Most of the standard designs for air-type collectors are shown in Figs. 3.5 and 3.6. The designs in Fig. 3.5(a–e) are generally more efficient than those in Fig. 3.6(a–c). Rating the absolute efficiencies of different absorber designs, however, is very difficult and we do not attempt to do so here. A change in a small feature on the absorber plate can affect collector performance in many areas. For example, a change in the airflow pattern changes not only the heat gain but also the heat loss.

One of the most efficient air-type absorbers is shown in Fig. 3.5(a). This absorber has rear fins to improve heat transfer. The flat absorber front keeps radiation losses low.

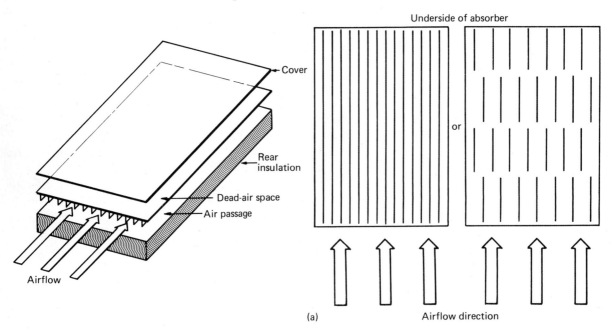

Fig. 3.5 The most effective air-type absorber-plate designs: **(a)** finned plate;

In the flat absorber with two-pass airflow (Fig. 3.5[b]), the first pass with cool building air keeps the glass and the front of the absorber at a relatively low temperature, thereby reducing outward heat loss. The second pass beneath the absorber continues to warm the air. This simple, flat absorber is inexpensive. Front airflow may necessitate occasional cleaning of glass and absorber, and/or air filtering.

In the flat absorber with dual-pass airflow (Fig. 3.5[c]), air flows on both sides of the absorber, providing double the heat-transfer area of the basic rear-flow collector. Some heat from the air in the outer passage is lost to the glass, but generally more heat is gained from the absorber than is lost to the glass. Front airflow may also necessitate occasional cleaning of the glass and absorber, and/or air filtering.

The absorber with Löf-type overlapping plates (Fig. 3.5[d]) is similar in design to the dual-flow absorber. Air flows on both sides of each glass plate. In addition, the clear portion of each plate captures infrared radiation from the plate below it, reducing heat loss. This absorber is a rather bulky, expensive, and difficult thing to construct. It requires occasional cleaning.

Figure 3.5(e) shows an absorber with a vee-corrugated plate and a selective surface. The corrugations perform the same function as the fins on the finned absorber, except that a selective black coating must be

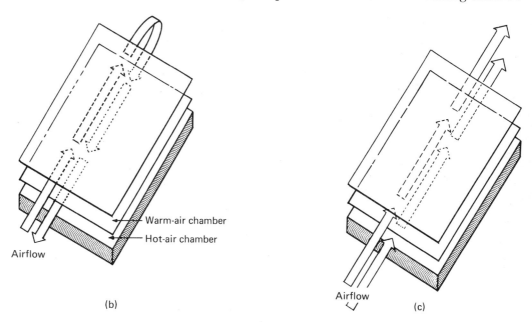

Fig. 3.5 (continued) (**b**) flat, two-pass airflow; (**c**) flat, dual-pass airflow;

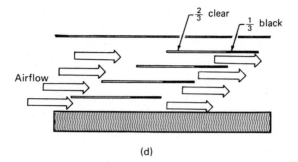

(d)

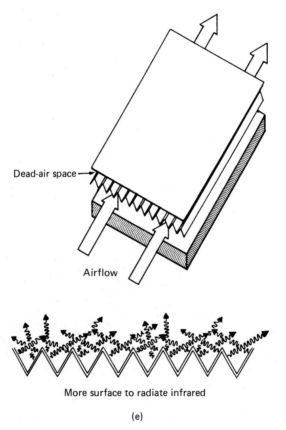

(e)

Fig. 3.5 (continued) (**d**) Löf-type overlapping glass plates; (**e**) vee-corrugated plate, selective surface.

used since the amount of area radiating infrared outward is greater than for a conventional flat absorber. This absorber is simpler to construct than the finned-type plate absorber.

Corrugated absorbers such as the one shown in Fig. 3.5(e) do not perform as well as flat-surfaced absorbers. The reason, as Tabor [5] points out, is that corrugating the absorber increases the surface area, which is exposed toward the outside by a certain percentage. This increases the amount of infrared radiated outward by a similar percentage since a surface radiates infrared in an amount proportional to its area. Heat loss is increased by a measurable amount, so the efficiency goes down. Corrugations, protrusions, and the like can only be justified when the absorber has a *selective* surface coating, which means that it absorbs visible light but does not radiate a lot of infrared. Since a selective absorber does not radiate much infrared to start with, a 50% increase in the amount of infrared radiated does not amount to much.

The slightly less effective absorber designs are illustrated in Fig. 3.6 (next page). The flat-plate, rear-flow-only absorber plate in Fig. 3.6(a) is a very basic absorber-plate and airflow design. It has less surface area than the designs in Fig. 3.5 and has no elements that cause air turbulence, so heat transfer and efficiency are lower.

With the matrix design (Fig. 3.6[b]), black mesh or a matrix of cloth, metal, or other material acts as the solar absorber. Air flows through the material from front to rear. The large amount of turbulence improves heat transfer, but at the expense of increasing air resistance and requiring greater fan horsepower. Collector efficiency may be large, but this system requires more nonsolar power for operation than do other systems.

The punched-fin absorber (Fig. 3.6[c]) is a rear-flow-type absorber. Fins are made by punching and bending the plate. This absorber is not as effective as regular finned-plate absorbers because of air motion through the holes, which disturbs the insulating air space. Depending on the orientation of the plate, the holes and fins may create a large amount of air turbulence, which increases collector efficiency but reduces energy efficiency since a larger fan is needed to overcome the resistance. Perforations reduce both advantages of the rear-flow configuration: they allow air motion above the plate and allow dust and dirt to collect on the glass and absorbing surface.

Perhaps one of the most effective air-type absorbers yet developed is the longitudinal finned type shown in Fig. 3.5(a). This design is relatively expensive to construct, but it exemplifies several of the important considerations that affect the performance of air-type absorber plates. In general:

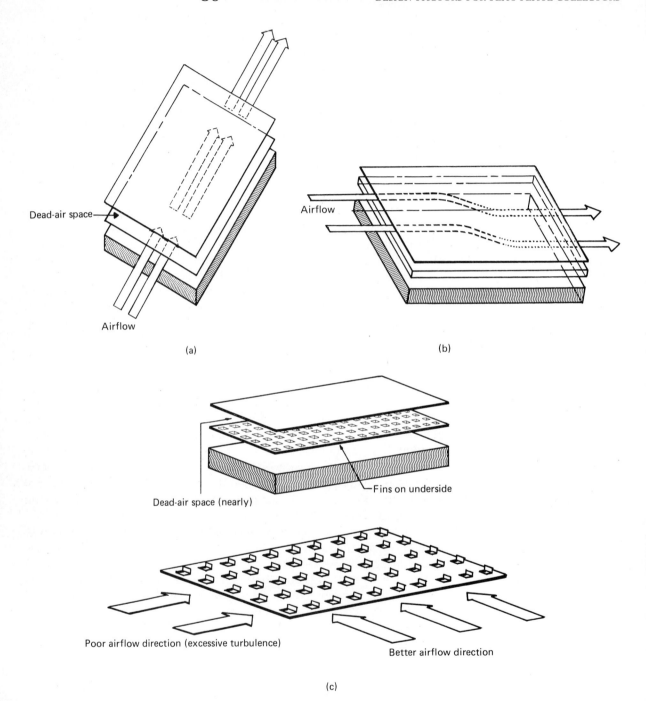

Fig. 3.6 Other air-type absorber-plate designs: (**a**) flat plate, rear flow only; (**b**) matrix type; (**c**) punched fin.

1. An air-type absorber should present a relatively flat surface to the sun.
2. An air-type absorber should present a large surface area to the flowing air.
3. An air-type absorber should not add a large amount of turbulence to the air flowing through the collector.
4. The air flow in an air-type absorber should generally occur at the rear. A *dead-air space* should be maintained in front.
5. The best material for air-type absorbers is metal. Aluminum and galvanized steel work the best.

First, an air-type absorber should present a relatively flat surface to the sun, especially if it has the standard, nonselective black coating. Fins on back of this absorber (Fig. 3.5[a]) provide a large surface area for good heat transfer. The fins offer low resistance to air flow because they are aligned in the direction of air flow. Rear air flow keeps dust and dirt away from the glass cover and the absorbing surface. A dead-air space in front provides insulation. Finned absorbers are more expensive than basic flat absorbers. A typical and effective fin spacing is one fin per inch. The staggering of fins, though not essential, can aid heat transfer and is common in heat exchangers.

Second, an air-type absorber should present a lot of surface to the flowing air. If we increase the heat-transfer area we increase the amount of heat transferred to the air. That is why a finned absorber performs better than a flat absorber: the fins add more surface area, the flowing air touches more hot metal, and more heat is transferred. It is unfortunate that corrugating the plate increases radiative heat loss since corrugations have an effect similar to fins as far as heat transfer is concerned and since corrugating a flat sheet of metal is substantially cheaper than welding fins on the back of it.

Third, an air-type absorber plate should not add a large amount of turbulence to the air flowing through the collector. It is true that turbulent air very effectively removes heat from the absorber, but increased turbulence causes increased resistance to airflow: this means that there is more pressure drop through the collector and a larger fan is needed to circulate the air. The types of absorbers that can lead to excess turbulence are the matrix type (Fig. 3.6 [b]) and the punched-fin type (Fig. 3.6 [c]), which tend to break up the air stream by swirling it or mixing it. Collectors having these types of absorbers can perform efficiently in tests, but this efficiency may be more than cancelled out by the larger blowers and increased power required to move the air through the collectors. Fan-power requirements are already quite large in air-type systems and there is no need to increase them further. The best way to improve heat transfer to the air is to increase the surface area without increasing air turbulence. This is exactly what happens with a finned absorber (the air flows along the fins, not across them).

Fourth, it generally pays to keep the airflow behind the absorber and maintain a dead-air space in front (as with a finned absorber). This reduces both heat loss and maintenance. If the absorber is adequately cooled from behind, the dead-air space reduces heat loss by insulating the absorber from the glass, and dust is prevented from collecting on the glass and black absorbing surface. For example, a dual-flow absorber (Fig. 3.5 [c]) may have exactly the same heat-transfer area as a certain finned absorber (since both sides are in contact with flowing air) yet may not perform as well simply because it does not have an insulating dead-air space in front of it, as the finned absorber does. Also, dust is likely to collect on its black surface with the passage of time because building air passes in front of the absorber.

The dust problem, while serious, is not crippling. An air filter or electronic air cleaner, which should be in the system anyway, can help. One air system using Löf-type collectors (Fig. 3.5[d]) has been in operation for over 25 years and the only maintenance required is to clean the glass plates occasionally.

The fifth performance factor we will consider concerns the choice of materials for air-type absorbers. In general, metals such as aluminum and galvanized steel work the best. Corrosion is not a problem, except over very long time spans, and metals conduct heat so well that localized hot and cool areas will not tend to develop—the temperature remains fairly uniform over the entire plate. Good heat conductivity is especially important for plates such as the finned absorber: the heat must be conducted from the front surface to the fins in back, where the air can take it away. In most cases, there are few if any advantages in using thick or heavy-gauge metals.

Despite the advantages of metals, with an air collector it is possible to use other materials in lieu of them. One very low-cost collector has an absorber made of a black, glass-foam-like material (and has front airflow). Cloth mesh has been used for matrix-type absorbers. High-temperature plastic has been suggested.

An air-type absorber plate, unless it is to be mass-produced, is best kept simple in design. Simple fabrication techniques should be used. For example, strengthening flat sheet metal by *cross-breaking* it (adding ridges) is usually much cheaper than making several cuts, welding it, or stamping it with a custom die. Thus the basic rear-flow, two-pass, and dual-flow absorbers are considerably less expensive to make, in small quantities, than finned absorbers or other complex designs.

3.3.2 Absorber Plates for Liquid Collectors

When water is used as the heat-transfer fluid, efficient if moderately expensive absorbers are possible. There are a large number of successful water-type systems in operation. It has been found, however, that care must be taken to avoid corrosion and freezing problems. Ordinary tap water is fairly corrosive, so precautions must be taken. For example, certain metals should only be used under certain conditions. All pure-water systems used in climates where freezing occurs even occasionally must include some method for preventing the collectors from freezing, such as a *draindown* mode (see Chapter 5).

A mixture of water, antifreeze, and other additives as the heat-transfer fluid eliminates the threat of frozen water and broken pipes, but it does cause some problems of its own. The first is expense. It is too costly to fill an entire system, including storage, with glycol antifreeze, so an extra heat exchanger is needed to separate the small collector loop from the storage loop and the rest of the system (an extra pump may be needed as well). This will slightly reduce system performance because it raises the cost. The other problem is that using water and antifreeze is not enough. A *buffering agent* must be added because glycol has a tendency to break down in the presence of high temperatures, forming the rather corrosive glycolic acid. Usually corrosion inhibitors are added as well. The result is a complex chemical balance that must be closely watched and maintained. Because careful maintenance is required, water-antifreeze systems are acceptable for large solar installations and industrial-process heating (where frequent maintenance is needed anyway), but should probably be avoided for average residential installations because it is unrealistic to expect owners to monitor their system or even to schedule an annual service call on time.

Nonaqueous, or water-free, heat-transfer fluids can also be used in liquid collectors. An example is Dow's Q2-1132 silicone-based heat-transfer fluid. The advantages of this fluid are its high flash point (fire safety), high boiling point, low freezing point, and noncorrosiveness. Unfortunately, there are disadvantages too. It is more viscous (thicker) than water and has a lower specific heat (holds less heat). It must flow two to three times faster than water through the system to collect the same amount of heat. Larger pumps that require more electricity must be used to achieve this increased flow rate. The liquid is also rather expensive. Other fluids, most of them organic, are available from various manufacturers. Each has advantages and disadvantages and almost all are expensive.

Figures 3.7 and 3.8 show most of the standard designs for liquid-type solar absorber plates. To construct the stamped absorber with tubular flow shown in Fig. 3.7(a). Two sheets of metal are stamped and

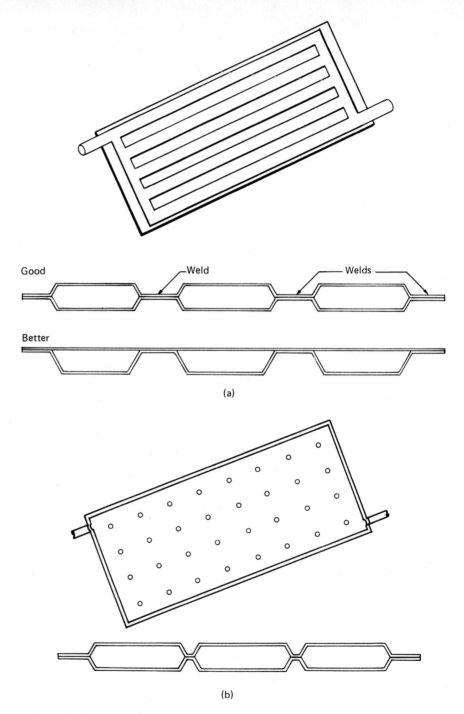

Fig. 3.7 The most effective liquid-type absorber-plate designs: (a) stamped absorber, tubular flow; (b) stamped absorber, full flow.

welded together. Water passages are formed out of the sheet itself, resulting in excellent heat transfer. A continuous weld around the perimeter can make for a reliable, leak-resistant unit. Ideally, the front surface is flat so that less surface area is exposed outward. This results in less infrared radiation and less heat loss. The stamped absorber with full flow shown in Fig. 3.7(b) is constructed like the one with tubular flow except that liquid is allowed to flow over the entire plate surface, resulting in excellent heat transfer. The flow should be uniform and there should be no hot spots or areas of restricted flow.

The linear-bonded or extruded-plate absorber illustrated in Fig. 3.8(a) is formed by a continuous roll-welding process or by extrusion. Plates may be cut to any length, but pipe connections are required on the ends. Owing to the many separate connections, some chance exists for leakage. Heat transfer to the fluid is very good, however. The bonded-tube absorber shown in Fig. 3.8(b) is one of the earliest designs. Liquid tubes are fastened to the sheet metal absorber by soldering, wiring or other methods. This absorber can be constructed by the do-it-yourselfer. Performance is not as good for bonded-tube absorbers as for the more unified design. Tubes must be continuously and firmly bonded to the plate for adequate heat transfer.

The designs shown in Fig. 3.7 are generally more efficient than those shown in Fig. 3.8. As with absorbers for air-type collectors, it is very difficult to rate absolute efficiencies of different absorbers. We intend the following discussion to provide only a qualitative understanding of the factors involved in absorber-plate design.

There are several important considerations that affect the performance of liquid absorber plates:

1. Balanced flow
2. The bond between the absorber plate and the fluid-flow passages
3. Elimination of corrugations and protrusions on the front of the absorber plate
4. Prevention of corrosion
5. Unitized construction

First, good performance is largely a matter of maintaining balanced flow throughout all of the liquid passages. Since fluid-flow patterns for collectors are similar to those for entire arrays, this subject is covered in section 3.5.

Second, performance is greatly affected by the bond between the absorber plate and fluid-flow passages. The first liquid-type absorbers were the bonded-tube type (Fig. 3.8[b]). Often, a pipe was simply bent

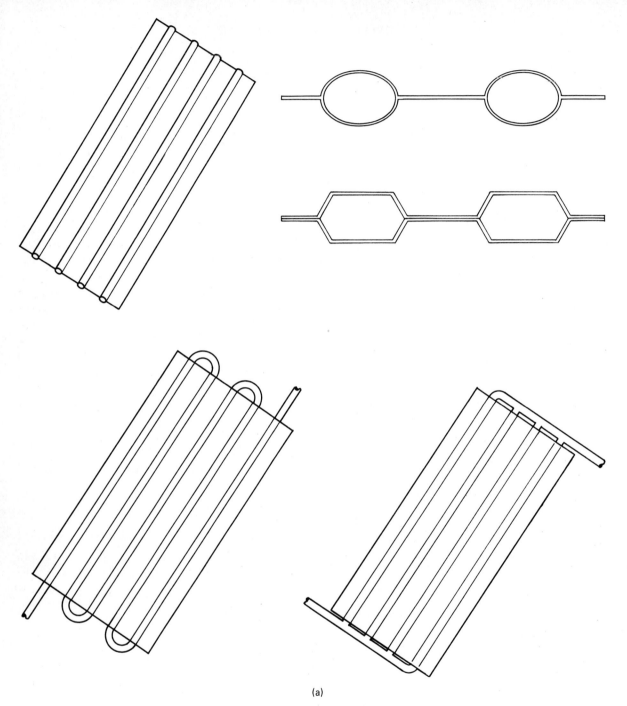

Fig. 3.8 Other liquid-type absorber-plate designs: (**a**) linear bonded, or extruded plate;

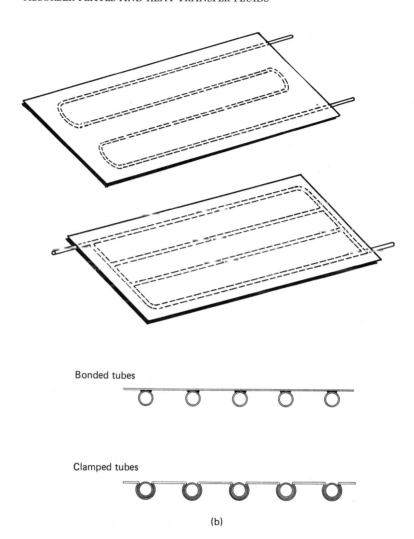

Fig. 3.8 (continued) (**b**) bonded tube.

into a serpentine shape and welded onto the back of a flat sheet of metal. With this design, any heat collected between the pipes must travel across the flat plate, through the weld, and through the pipe before it reaches the liquid. This reduces heat-transfer efficiency. If the tubes can be made an integral part of the plate (as in Figs. 3.7[a] and 3.8[a]) then heat will more easily enter the liquid and collector efficiency will be greater. If the tubes can be eliminated entirely and if liquid is allowed to flow over the entire absorber plate (as it does in the stamped absorber in Fig. 3.7[b]) performance will be better still. However, these full-flow-

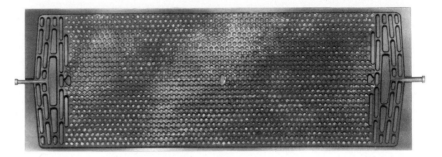

Fig. 3.9 SOLAR-BOND® absorber plate: Article 9171 by Olin Brass. (Photograph courtesy of Olin Brass, East Alton, IL 62024.)

type plates must be carefully engineered so that there are no back-eddies or areas where liquid is stagnant.

Third, as with air-type absorber plates, performance is reduced if there are any corrugations or protruding areas on the front surface of a liquid-type plate. For this reason the liquid passages ideally should bulge from the rear of the absorber, not from both front and rear as is usually the case. This often ignored consideration has a slight effect on performance.

Two commercially available liquid absorber plates of the type shown in Fig. 3.7(a) are shown in Figs. 3.9 and 3.10. The backplate shown in Fig. 3.9 is manufactured using the ROLL-BOND® process. The plate is created from two pieces of metal that are rolled together. Air is then forced into specially prepared areas that inflate to become the fluid-flow paths. This backplate is used for hot-water and space-heating systems. The backplate shown in Fig. 3.10 is also manufactured using the ROLL-BOND process. Many different and efficient absorber plates can be

Fig. 3.10 SOLAR-BOND® swimming pool absorber: Article 9154 by Olin Brass. (Photograph courtesy of Olin Brass, East Alton, IL 62024.)

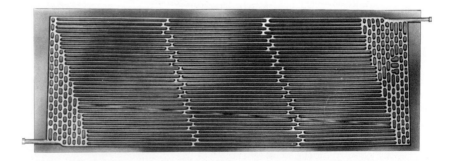

created using this process. The one pictured here was developed especially for use in swimming pool heating applications.

The fourth consideration involves the prevention of corrosion. Corrosion can be a very serious problem in liquid-type absorber plates. A poorly designed or improperly used liquid-type absorber can be full of leaks within weeks of its installation. The best way to prevent corrosion is to use only copper absorber plates and copper pipes in a liquid-type system. Under some circumstances any of the other commonly used materials—aluminum, stainless steel, carbon steel—can cause problems. Not even copper is totally trouble free, but it is by far the safest material to use.

In open-loop and draindown-type systems ordinary tap water is commonly used as the heat-transfer fluid. Tap water has the disturbing tendency to eat up rapidly both aluminum and carbon steel absorbers. Even stainless steel is susceptible to corrosion in areas where it has been welded. (Manufacturers of stainless steel plates should be carefully queried about their precautions along this line.) In closed-loop and water/antifreeze-type systems, copper, stainless steel, aluminum, or carbon steel can be used, but the latter two can only be used under the following conditions:

1. The fluid contains adequate amounts of corrosion inhibitors (special chemicals that slow corrosion).
2. If glycol-type antifreeze is used, buffering agents are added also. (These are chemicals that prevent glycolic acid from doing damage. Glycolic acid is an acid that tends to form when glycol is exposed to high temperatures.)
3. The chemical balance of the fluid in the system is monitored at at least six-month intervals, and optimum amounts of the inhibitors and buffers are maintained.
4. Other corrosion-prevention measures are taken.

Other corrosion-prevention measures can be taken. Wherever two dissimilar metals join (e.g., where a copper pipe joins a carbon steel absorber plate) electrically insulating connectors, usually ceramic or rubber, should be used to separate the metals. Corrosion is an electrochemical process that is aided by contact between electrochemically dissimilar metals.

The metal ions found in common tap water are responsible for much of its corrosiveness, so distilled or deionized water is often used in a closed-loop system. "Softened" water has not been deionized. The calcium ions in "softened" water have simply been exchanged for sodium ions, making the water even more corrosive.

A special device known as an *ion-getter* column can be installed in the system (see Chapter 5). None of these methods alone can prevent corrosion, but all can help.

Again, the best preventive measure is to use copper absorber plates or at least absorber plates that have copper tubes. Copper can last for decades. The only "corrosion" it is subject to is *erosion corrosion*, which can occur in areas of a copper absorber where local water turbulence gradually eats away at the copper. Prevention of unnecessary liquid turbulence through proper design will prevent erosion corrosion and also reduce the pumping requirements of the system since turbulence causes resistance to flow. After copper, the next best choice for an absorber is probably stainless steel.

The final performance factor we will consider for liquid absorber plates is the importance of construction. The sturdiest and most durable designs have proved to be those that are the most unitized and that have the fewest individual welds. For good performance and durability, the stamped/edge-bonded type (Fig. 3.7[a]) is very good. This absorber depends only on one continuous weld around the outside edge. Many of the older designs, which require the welding together of many tubes (Fig. 3.8), are more susceptible to failure and subsequent leakage. Solder joints should never be under much continuous mechanical stress because solder loses its structural strength as its temperature rises—even before it reaches its softening point.

The absorber plate is a vital element of any solar collector: it affects both collector efficiency and total system efficiency because it acts as an energy converter and heat exchanger and as a resistor to the flow of heat-transfer fluid, be it air or water. A good absorber plate not only absorbs sunlight well, but transfers as much heat as possible to the fluid and causes the least resistance possible to fluid flow.

3.4 SURFACE COATINGS

The most important function of an absorber plate is the conversion of the sun's radiant energy (light) to thermal energy (heat). The surface treatment on the absorber is what actually performs this function. The most commonly used surface treatments are black coatings, either flat black paint or a *selective surface* that is electroplated or vacuum-deposited onto the absorber. A solar absorber coating must meet several criteria if it is to work well in practice:

1. It must absorb nearly all of the visible light that strikes it; that is, it should be very black. Absorption of 98% is typical.

2. It must resist high temperatures because 150–200° C (300–400° F) can be reached in double-glazed collectors under unusual conditions.
3. It must have a long lifetime under widely varied conditions. In winter daily temperature swings of 65° C (150° F) are not uncommon. A coating must not fade or peel off over the years. It should be able to be cleaned without being destroyed.
4. It must not add excessively to the cost of the collector.

When paint is used as the surface coating it must be heat resistant and flat black. Heat-resistant paints often contain silicone-based resins. They can typically withstand temperatures up to 425° C (800° F)—far above the maximums reached in the average solar collector. Of course, black is best, but dark shades of other colors, such as green or brown, have occasionally been used for aesthetic reasons with acceptable results. It must be understood, though, that any slight improvement in solar absorptance tends to improve significantly collector efficiency.

The typical method of paint application is by spray gun. Most heat-resistant paints require priming and all require careful preparation of the surface prior to painting. The metal surface should be degreased with an industrial-grade degreaser and usually should be slightly roughened with a commercial etching compound (this is at least true for aluminum and galvanized steel). Primers should be applied as directed by the manufacturer.

The selective surfaces are coatings that absorb most visible light, as black paint does, but they emit far less infrared radiation than do painted surfaces. This results in lower heat loss from the collector which, in turn, means higher efficiency, especially at high collection temperatures. This last qualification is important. A solar collector can be operated at a low temperature or a high temperature, depending on how fast the heat-transfer fluid is forced through. If fluid moves through slowly it has more time to warm up. If it moves through quickly more fluid goes through in the same amount of time, so the absorber operates at a cooler temperature. In general, the higher the operating temperature, the more rapidly the collector loses heat so the less efficient it is.

This becomes obvious if we look at the standard efficiency curve of a collector. There are many ways to arrange the axes of this type of graph, but a rather clear way is to plot the efficiency of the collector as a function of absorber-plate temperature. Figure 3.11 shows this kind of graph for two different types of collectors. One type has a selective sur-

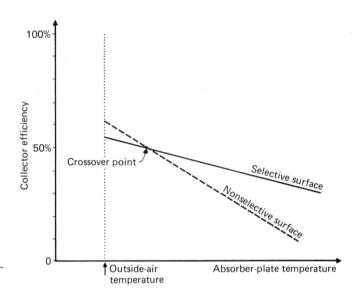

Fig. 3.11 How a selective surface affects collector efficiency.

face, the other does not. The data in Fig. 3.11 indicates that use of a selective surface results in higher efficiency, but only at higher operating temperatures. The efficiency curves for the selective and nonselective absorbers cross at some point on the graph. When the operating temperature of the absorber is below the crossover temperature, use of a nonselective absorber makes the collector more efficient. The reason has to do with the fact that black paints can be made blacker than selective surfaces; they absorb more light. The absorptance for flat-black paint may be as high as 0.98, while the absorptance for selective surface is usually in the 0.85–0.94 range.

The decision whether or not to use a selective surface in a given collector depends to a large extent on how hot the collectors are intended to be run. For solar-powered air conditioning, industrial-process heating, and other applications requiring high-temperature heat output, it is often advantageous to use selective surfaces. Heat loss must be minimized so that the absorber can operate at high temperatures without losing all of its heat. It may also be desirable to use selective surfaces in liquid-type heating systems since hotter liquid means that the heat exchangers will operate more effectively. In air systems, however, where house air is often directly heated in the collectors and returned to the house, extra-high temperatures are less advantageous and the collectors may be run cooler. The question then becomes whether to use the slightly more efficient selective surface or the substantially cheaper nonselective surface. In some cases, as we have seen, the nonselective surface can even lead to better performance.

There are no universal guidelines on this issue of selective versus nonselective surfaces. The decision must be made on an individual, case-by-case basis by the use of complex efficiency calculations or actual testing of the specific collector with two different absorbers. In general, it has been determined that a single-glazed collector (one glass cover) with a selective-surface absorber is approximately equivalent in performance to a double-glazed collector with a standard, nonselective-surface absorber. That is, adding a selective surface to a single-glazed collector is roughly equivalent to adding a second sheet of glass. The truth of this generalization depends on the design of the specific collector and the conditions under which it is to be used.

The most commonly used selective surfaces are very thin layers of metal oxides that are deposited by electrolysis on the polished-metal absorber plate. Most of the techniques for applying these surfaces have been published at one time or another [5], though some are patented. Figure 3.12 summarizes how this type of selective surface works.

Chrome black is a metal-oxide selective surface currently available in the United States as metal strips or as a custom-plating process for absorbers. A thin layer of nickel is plated onto the metal surface, then a chromium-oxide layer. It has proved to be one of the most durable and effective surfaces available and it can be applied to most metals.

Nickel black is similar to chrome black in many respects. It has been experimented with by several large research organizations and found to be similar in effectiveness to chrome black but slightly less moisture resistant. Nickel black has been in commercial use in Israel for many years, in solar hot-water heaters, with good results.

A third commonly used coating is copper oxide on copper or aluminum. Like nickel black, this coating has withstood the test of time, having been used in solar hot-water heaters in Australia for several years.

To obtain better performance than the metal-oxide surfaces can give, two approaches are possible. The first is to use a multilayer, vacuum-deposited coating. This coating is applied by vaporizing each coating material in a vacuum chamber and letting it collect in an ultra-thin layer on the metal surface, much as glass is coated with aluminum to make mirrors. This is an expensive, critical-tolerance process unsuitable for large-scale use. The other approach is to use an ordinary metal-oxide selective surface but use it on a sharply corrugated (vee-corrugated) absorber plate, as is shown in Fig. 3.13. As we have seen, corrugation of an absorber increases its infrared losses. Corrugation also in-

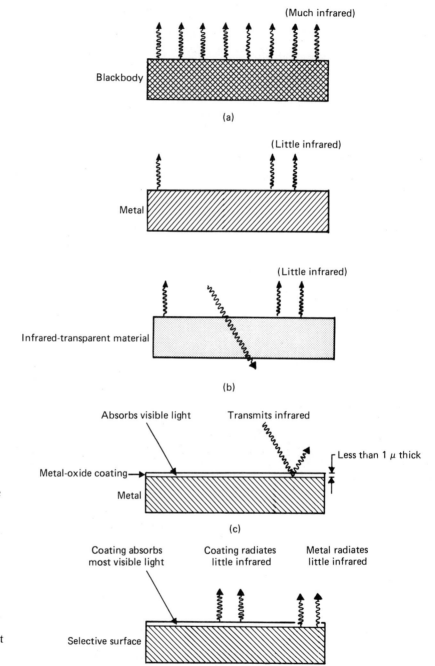

Fig. 3.12 How a metal-oxide selective surface works: (**a**) a blackbody, a perfect radiator of infrared; (**b**) bright reflective metals and materials that are transparent to infrared, not good radiators of infrared compared with a blackbody; (**c**) thin metal-oxide coatings on bright metal, transparent to infrared and absorb visible light; (**d**) selective surfaces, visible light absorbed but infrared not radiated well (even though coating is transparent to infrared) since little infrared is radiated from metal.

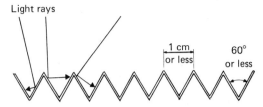

Most incoming rays bounce at least twice, each time releasing more energy.

Fig. 3.13 Increasing absorptance of a selective surface by corrugating the absorber plate.

creases losses caused by natural convection between the absorber and the glass cover. It increases the effective light absorption of the surface, however, since light will bounce more than once as it enters the grooves. The net effect is that corrugating a selective absorber will sometimes increase its efficiency. The vee grooves should be at a 60° or even sharper angle and should be 1 cm (0.4 in) or less wide to minimize the increase in convection losses.

Selective surfaces, because they are thin, are more delicate than painted surfaces. Special techniques are needed to clean them and some can be destroyed by water or even condensation. They should be tested for high-temperature stability since some can be damaged by temperatures as low as 175–200° C (350–400° F). The long-term performance of most selective surfaces is not adequately known.

3.5 FLUID-FLOW PATHS IN COLLECTORS AND ARRAYS

The efficiency of a solar collector or an array of collectors depends, in large measure, on how effectively the heat-transfer fluid performs its function of removing and transferring the heat. This in turn depends on what happens to the fluid during its passage through the collectors. For example, does the air or water split into several branches, or does each drop of fluid pass through all of the piping? Is the flow properly balanced among all of the collectors, or are some running hot owing to inadequate flow rate? The path in which fluid flows through a collector or an array has a considerable effect on performance.

Figure 3.14 summarizes the four basic flow-path designs found in liquid-type collectors. The parallel flow, reverse return (PFRR) pattern (Fig. 3.14[a]) has been found to cause the fewest problems in both collectors and arrays. Fluid tends to flow through each parallel tube at the same rate because each flow path is of the same length. It is better to split a large stream into smaller, parallel streams and rejoin them at the end than it is to take one stream and wind it around in a snakelike path

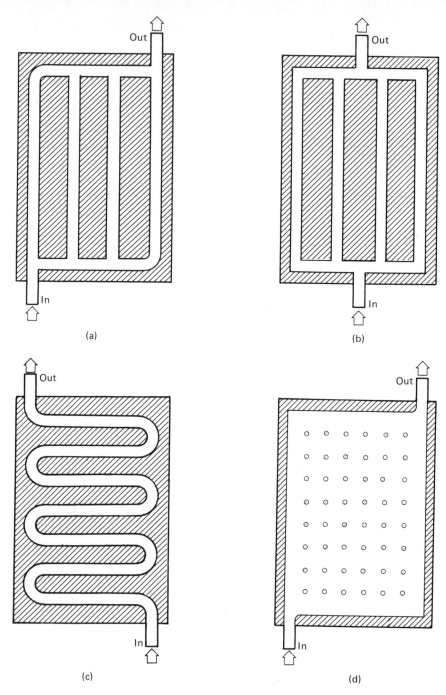

Fig. 3.14 Flow paths in liquid-type absorber plates: (**a**) parallel flow, reverse return; (**b**) parallel flow, direct return; (**c**) serpentine flow; (**d**) total-panel flow.

so that it goes to all collectors in series. The reason is that series flow causes substantially more resistance to fluid flow. It takes a lot less pump pressure to force a liquid through many parallel paths. The only time it might be advantageous to hook several collectors in series is when high temperatures are needed; for example, in solar-powered air conditioning machinery. As fluid passes through a series of collectors one right after the other, its temperature rises each time. This means that the collectors near the end of the array will be running hotter and at lower efficiency than those near the beginning.

Reverse return means that the fluid inlet is not located directly across from the outlet. Instead, the inlet and outlet are located on opposite corners of the collector or array. For example, the absorber plate shown in Fig. 3.14(a) has reverse return; the one in Fig. 3.14(b) does not. The parallel flow, direct return (PFDR) pattern shown in Fig. 3.14(b) is not a recommended one. When the entry tube is directly across from the exit tube, fluid flows faster in some tubes than others, owing to pressure drop along the headers. This unbalanced flow results in reduced efficiency and unnecessary thermal stress. (With careful design it is possible to equalize the flow rates by constricting some tubes or making some smaller than others.) Reverse return is superior to direct return because it equalizes the flow in each parallel branch. Without reverse return, fluid would flow more slowly in some tubes than in others. Reverse return is no guarantee that flow rates will be balanced (sometimes manual valves are needed in each branch of a large collector array to balance the flow) but it is helpful in most cases.

Serpentine flow (Fig. 3.14[c]) is not a recommended flow pattern. Though simple to construct, this design has inherently more resistance to fluid flow than parallel-type designs, and more pumping power is required. Total-panel flow (Fig. 3.14[d]) is a good flow pattern if carefully executed. Fluid flows over most of the absorber surface, so heat transfer is excellent. Fluid passage should be carefully designed so that flow is relatively uniform throughout the absorber.

Figure 3.15 shows the flow paths commonly used to connect liquid collectors into *arrays*. Liquid flow through one collector is similar to flow through an array of many collectors. The PFRR path in the array in Fig. 3.15(a) is the best method for ducting or plumbing a single row of collector panels. Fluid tends to flow through each panel at the same rate, so balancing of the fluid paths is usually not required. The PFDR flow path shown in the array in Fig. 3.15(b) should be avoided if possible. Because of pressure drop along the headers, fluid will flow faster in some collectors than in others. This reduces system performance and may cause serious problems because some collectors will be hotter than

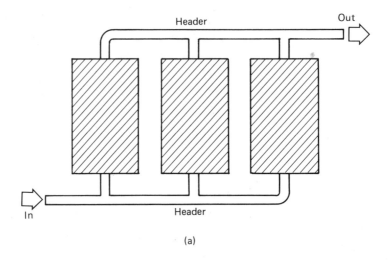

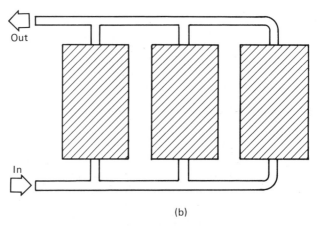

Fig. 3.15 Flow paths in liquid-type collector arrays: (**a**) parallel flow, reverse return; (**b**) parallel flow, direct return.

others. If this method must be used, headers should be as large as possible (to minimize pressure drop) and flow through each collector should be equalized with valves or constrictions in the lines. The series-parallel flow pattern shown in the array in Fig. 3.15(c) is good for double- or triple-height arrays. It does not require as much piping as it would if each row of collectors had its own headers. Generally this type of array does not require balancing as long as reverse return is used and each branch has the same number of panels. Connecting panels in series will

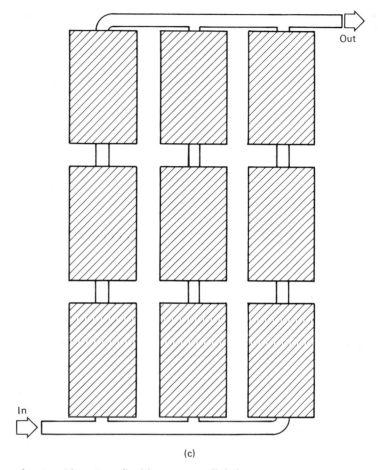

Fig. 3.15 (continued) (**c**) series-parallel flow.

increase the output temperature of the fluid but slightly reduce system performance since the hotter, upper panels will have greater heat loss.

The flow considerations mentioned for liquid collectors apply equally to air-type collectors. Reverse return is desirable for an air collector, though balancing the flow inside the collector is usually not a great problem because of the simplicity of air-type collectors. However, connecting air-type collectors into arrays is a subject unto itself owing to the bulkiness of air ducts. The ducts are usually so large that it is not practical to place them outside with the collectors (the extra insulation would be too costly), but putting the ductwork inside the building can mean using a large number of through-wall connections to get to each collector. Figure 3.16 illustrates the four basic ways of ducting an air-type collector array. Note that all are based on the PFRR principle.

In the first ducting method (Fig. 3.16[a]), collectors are mounted on the exterior of a wall or roof and *manifolds* (header ducts) are mounted on the inside of the wall or roof. Connecting ducts run through the wall to each collector. This is a commonly used and acceptable method, but requires the expense and labor of many connections.

In the second method (Fig. 3.16[b]), the header ducts are mounted on the exterior of the wall or roof along with the collector panels. This eliminates the through-wall connecting ducts and ductwork consumes less interior building space. A major disadvantage of the exterior-mount, exterior-manifold installation is the extra duct insulation required and the resulting large size of the ducts, which can mean undesirable appearance and possible shading of the collectors.

Exterior-mount array, internal-manifold collectors are designed to open directly into each other at the top and bottom on the sides (Fig. 3.16[c]) and thus, in effect, have header ducts built inside the collector array itself. Because there is a practical limit to how large the internal openings can be, this approach does not work well for long arrays having many collectors. Occasional through-wall connections must still be made to manifolds inside the building, but this method can substantially reduce the number of through-wall connections required. A disadvantage is that the collector panels must be deep and thus can be relatively expensive to construct.

Internal wall/roof mount, interior-manifold collectors (Fig. 3.16[d]) are designed to be set directly into the wall or roof structure between the studs or support members, which are set further apart than usual and made correspondingly sturdier). The collectors can open to the rear, directly into the header ducts. With proper design, open-sided ducts may be used to simplify connections to the collectors. The design, where applicable, has the advantage of eliminating through-wall connections since the collectors themselves extend through the wall or roof.

As with liquid-type collector arrays, design of air-type arrays should be centered around attempting to equalize the airflow to each collector while minimizing the overall resistance or pressure drop. Occasionally, adjustable dampers are needed to balance the airflow.

Proper flow-path design affects not only system performance but cost as well. For example, simplification of air-ducting methods can

Fig. 3.16 (opposite) Flow path and ducting of air-type collector arrays: (**a**) exterior mount, interior mainifold; (**b**) exterior mount, exterior manifold; (**c**) exterior-mount array, internal manifold; (**d**) internal wall/roof mount, interior manifold.

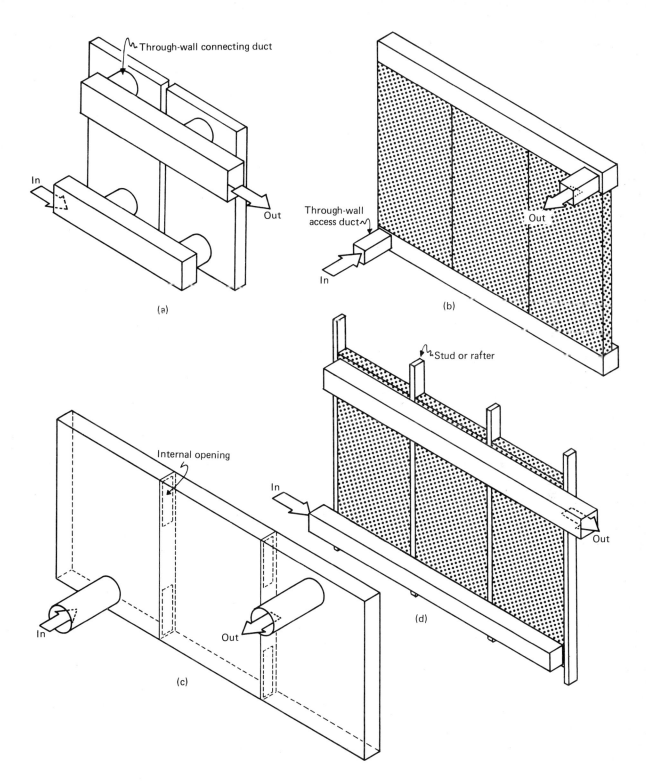

greatly reduce the amount of on-site labor needed to install air-type collectors, and labor is a large part of any solar installation.

The sizing of fans, ducts, pumps, pipes is complex and will not be covered here. For more information on these topics, information published by various professional organizations may be consulted.[1]

3.6 BREAKING THE RULES

Now that we have explained the rules it is time to demonstrate how they may be broken. Up to now, we have assumed that a solar heating panel can only be a flat-plate collector—a glazed, insulated box with an absorber plate inside. This is because almost all of the truly successful active solar heating systems to date have been constructed with flat-plate collectors or very similar arrangements. This is not to say that concentrating-type collectors and other exotic devices have not been used or that they have not provided adequate heat, but very few active-type solar collection devices have proved to be as economical as flat-plate collectors for the space heating of buildings. This situation could certainly change, however.

It is unlikely that concentrating-type collectors, in which mirrors or lenses focus light and motors follow the sun (see Chapter 9) will achieve wide usage for such applications as residential space heating. Their capability for generating high temperatures is not needed and their complexity means that they are expensive and require more maintenance than flat-plate collectors.

Other collector designs are being developed that could eventually compete with flat-plate collectors. One such device is the *evacuated-tube* collector. Figure 3.17 illustrates how this collector works. It breaks many of the rules for flat-plate collectors: it is shaped nothing like a pane of window glass; it needs no rear or side insulation material; it has only one glass cover. And yet some designs are so efficient that they may collect twice as much energy per square foot as does a conventional flat-plate collector. Having only one glass cover lets more light in, having a vacuum cuts convection losses to virtually nothing, and having a selective surface means small radiation losses. Thus far these tubular collectors have proved to be difficult to develop and expensive to make. Their success depends on how cheaply they can eventually be produced.

The installation shown in Fig. 3.18 contains 331 General Electric SOLARTRON™ evacuated-tube collectors. The system is designed to provide in excess of 25% of the hot water and heat for the inn. The collectors are tilted at 40° and their area totals some 490 m^2 (5280 ft^2).

---

1. Hydronics Institute, 35 Russo Pl., Berkeley Heights, NJ 07922; SMACNA, 8224 Old Courthouse Rd., Tysons Corner, Vienna, VA 22180; ASHRAE, 345 E. 47th St., New York, NY 10017; Air Conditioning Contractors of America, 1228 17th St. NW, Washington, D.C. 20036.

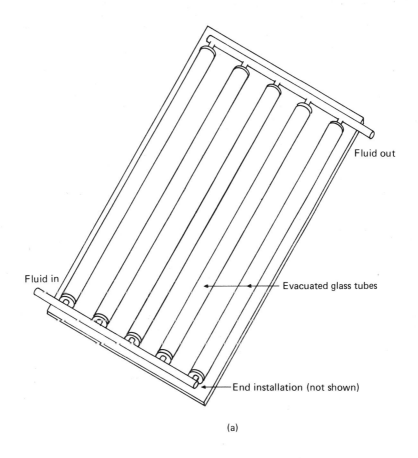

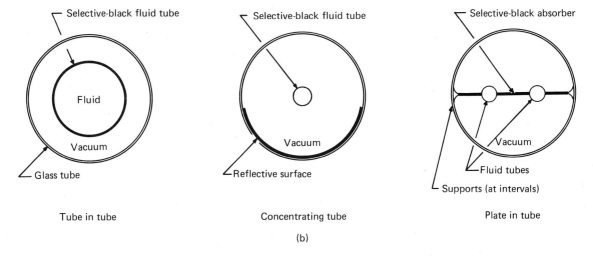

Fig. 3.17 The evacuated-tube collector: (**a**) the collector, (**b**) designs for evacuated tubes.

Fig. 3.18 Retrofitted solar system atop the Cherry Hill Inn near Philadelphia. (Photograph courtesy of General Electric, Valley Forge, PA.)

The system was dedicated in 1978 as one of the U.S. Department of Energy's (DOE) largest commercial installations of solar collectors.

Other solar collection technologies will emerge that will be radically different from anything we can imagine now. They will certainly appear to break the rules of good flat-plate-collector design. Yet neither the evacuated-tube collector nor any future designs can ignore the fundamental requirements for a solar collector listed at the beginning of this chapter: performance, durability, and economy.

REFERENCES

1. Löf, G. O. G. *et al.* (1976). Comparative performance of solar heating with air and liquid systems. In *Solar Heating and Cooling of Building*, vol. 3. Cape Canaveral, Fla.: American Section of the International Solar Energy Society, pp. 106–117.
2. Thomsen, S. M. (1951). Low-reflection films produced on glass in a liquid fluosilicic acid bath. *RCA Review* 12: 143.
3. Meinel, A. B., and M. P. Meinel (1976). *Applied Solar Energy: An Introduction.* Reading, Mass.: Addison-Wesley, pp. 249–252.
4. Blandenet, G., Y. Lagarde, and J. Spitz (1975). In J. Blocher *et al.* (eds.) *Proceedings of the Chemical Vapor Deposition Conference.* Princeton, N.J.: Electrochemical Society.
5. Tabor, H. (1967). Selective surfaces for solar collectors. In R. C. Jordan (ed.), *Low Temperature Engineering Applications of Solar Energy.* New York: ASHRAE.

BIBLIOGRAPHY

Duffie, J. A., and W. A. Beckman (1974). *Solar Energy Thermal Processes*. New York: John Wiley.

Jordan, R. C., and B. Y. H. Liu (eds.) (1977). *Applications of Solar Energy for Heating and Cooling of Buildings*. New York: ASHRAE.

Löf, G. O. G. (1977). Collectors. In *Solar Heating and Cooling of Residential Buildings: Design of Systems*. Solar Energy Applications Laboratory, Colorado State University, Washington, D.C.: U.S. Governent Printing Office, stock no. 003-011-00084-4.

Meinel, A. B., and M. P. Meinel (1976). *Applied Solar Energy: An Introduction*. Reading, Mass.: Addison-Wesley.

Ramsey, J. W. *et al.* (1975). *Development of Flat Plate Solar Collectors for the Heating and Cooling of Buildings*. Springfield, Va.: U.S. Department of Commerce.

PROBLEMS

3.1 Calculate the heat loss by conduction through a 4-mm-thick aluminum collector that is separated from the 2-mm-thick aluminum backplate by 5 cm of insulation. The backplate is inserted 1 cm into the insulation and is at 120° C. The thermal conductivity of aluminum is 204 W/m °C and that of the insulation is 0.036 W/m °C. There are no side losses and the outside temperature is 15° C. Compare your result with (1) the amount of heat that would be lost through a wood-framed collector (k = 0.16 W/m °C), where the wood holds the absorber plate and is 5.4 cm thick; and (2) with a plain metal frame, where the heat-loss path is also 5.4 cm.

3.2 A 2-m-long piece of plastic glazing is fitted into an aluminum frame. If expansion will be lengthwise only, how much longer than the glazing must the metal frame be to allow for the expansion of both?

3.3 A 0.32-cm sheet of regular window glass absorbs approximately 7% of the light passing through it. Use the formula for transmittance (neglect reflection) given in problem 2.6 to find the extinction coefficient for this type of glass. What are the extinction coefficients for 0.32-cm sheets of low-iron glass (5% absorptance)? For water-white crystal glass (2% absorptance)?

3.4 The reflectance of one side of a glass sheet (n = 1.52) at angles other than normal incidence can be estimated by the Fresnel equation:

$$\rho = 0.5 \left[\frac{\sin^2(\theta_2 - \theta_1)}{\sin^2(\theta_2 + \theta_1)} + \frac{\tan^2(\theta_2 - \theta_1)}{\tan^2(\theta_2 + \theta_1)} \right]$$

where θ_1 and θ_2 are the angles of incidence and refraction. Snells' law, n_1/n_2 = sin θ_2/sin θ_1, can be used to determine θ_2 if θ_1 and the refractive indices are known. What is the reflectance at 10° incidence? at 40°? At 70°? Use the formula for transmittance, $\tau = (1 - \rho)/[1 + (2z - 1)]$ for a system of z covers to find the two-cover-system transmittances for these angles.

3.5 Use your answers from problems 3.3 and 3.4 (assume the transmittance is a product of the reflective and absorptive transmittance) and find the transmittance for a two-cover system in which 0.32-cm low-iron glass is placed at 10°, 40°, and 70° angles of incidence.

3.6 An antireflective coating for glass can be made by adding a thin, evaporated coating (approximately $\frac{1}{4}$ of a wavelength of light thick) onto the surface. Light is reflected off two surfaces, that of the thin coating and that of the glass. Because of the $\frac{1}{4}$-wavelength thickness of the thin layer, the reflected rays are 180° out of phase and cancel themselves out to some extent. If the formula for reflectance given in problem 2.4,

$$\rho = \left[\frac{n_1 - n_2}{n_1 + n_2} \right]^2,$$

holds here, what must the theoretical index of refraction, n_1, of the layer be to eliminate all reflection?

3.7 What effect would you expect the following to have on a collector-efficiency graph (as discussed in Chapter 2)?

a) Antireflective glass used instead of regular glass
b) Infrared-selective coating on glazing
c) FEP used instead of glass
d) Use of a liquid heat-transfer fluid other than water in a collector previously tested with water

3.8 Discuss the advantages and disadvantages of the following segmented absorber plate for an air-type collector that allows airflow on both sides of the absorber.

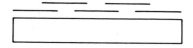

3.9 Discuss the advantages and disadvantages of the following absorber plate for a liquid-type collector.

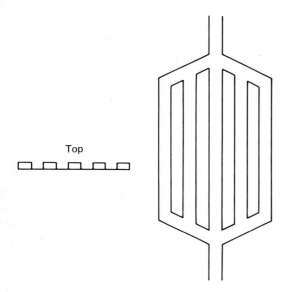

3.10 Invent your own absorber plate for an air- or liquid-type collector. Discuss its advantages and disadvantages.

3.11 A 1-kg magnesium (24.312 g/mole) getter column is attached to a steel liquid-type collector system. If the magnesium preferentially corrodes, forming MgO from oxygen (O_2), and 0.09 moles of oxygen can react daily with the column, how long will the getter column protect the steel from corrosion?

3.12 Two single-glazed collectors, identical except that one has a selective absorber, are run side by side so that the absorber plates are at the same temperature. For the two collectors, compare the temperatures of the glazing covers and the convection and radiation losses from the absorber to the glazing and from the glazing to the environment.

3.13 Show that the emittance of a low-emittance vee surface is approximately equal to P/a where P is the true area of the vee and a is the projected area of the surface. If other factors are the same, is the emittance actually lower or higher than this value? Why? (Hint: see problem 1.14.)

3.14 Study Figs. 3.15 and 3.16. What seems to be a general rule of PFRR systems that leads to balanced flow?

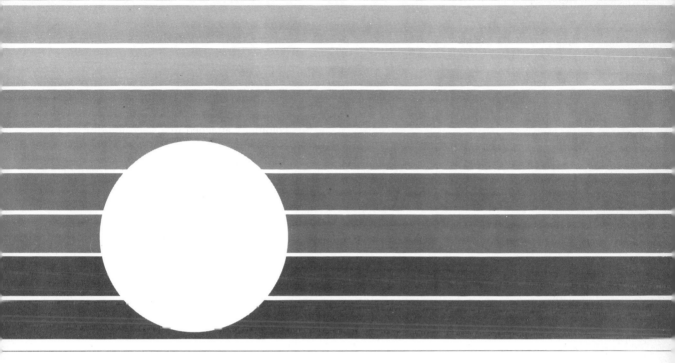

4 STORING SOLAR HEAT

Energy is constantly being stored and released around us. Just by lifting this book up from the table you have added energy to it. By moving the book higher into the air you have added *potential energy*. If it should fall from this height it would collide with the table and this extra energy would heat the book and the table slightly and could even force the table to move. If you lifted the book and placed it on a high shelf potential energy would be stored for later use.

For solar energy, however, different storage methods must be used. Why and how do solar heating systems store heat? Why does a solar heating system need to store heat?

Depending on weather, during the day solar collectors provide more heat than is necessary for hot-water and daytime heating. The collectors provide no heat at night or during stormy weather, however. Without heat storage, most solar heated buildings would become uncomfortably hot during the day and need to be vented, wasting precious solar heat. At night these same buildings would have to rely on expensive heat from backup furnaces. With a storage system, the extra heat

107

collected during sunny periods is available for use at night and during cloudy days.

How does a storage system work? For most solar heat-storage systems, the storage material is contained in an insulated compartment, or *storage unit*. Usually rocks or *phase-change materials*[1] in an insulated room are used for air-type collectors, while water in a tank is used as the storage medium for liquid-type collectors. The solar heated air or water is delivered to one end of the storage unit and cooler air or water is removed from the other end and returned to the collectors. The storage material continues to be heated in this fashion during the day. At night cooler house air is warmed by the hot storage material.

How much storage is needed? Winter overnight storage is usually all that is needed. An oversized storage system is less efficient, more expensive, and more difficult to maintain than a properly sized system. A large unit that would store several days' worth of solar heat would virtually never be filled in the winter, when efficiency is most important. The actual volume of space needed for heat storage depends on the size, type, and performance of the solar heating system. Guidelines for sizing storage units can be found in sections 4.2.2, 4.3.2 and 4.4.5.

Where is the storage unit located? The location and construction of the storage unit are important. Usually it is easiest and cheapest to build or install a storage unit while the building is in the early stages of construction. Often a separate or specially reinforced foundation is needed to support the weight of the storage unit to avoid the slab's becoming dished or cracked. Extra concrete installation and plumbing work can be completed as the foundation is laid and other plumbing work is done. Any large equipment needed for installing the unit or filling it with the storage material can easily be brought in at this time.

The closer the storage unit is to the collectors, the less ductwork or piping is necessary to connect the two installations, the less distance the air or liquid has to travel, and the less heat is lost while the fluid is in transit.

The storage unit can be located inside or outside the building and above or below grade. Ample access to the system must be provided for maintenance. A basement or crawl space is often chosen as the location for the storage unit. The unit can also be placed in an attic, but only if there is adequate structural support for the tremendous weights involved. Heat loss from outdoor storage units can be significant, so indoor storage is preferable.

The entire storage unit should be insulated, even the bottom. If the unit is located indoors, heat loss from an uninsulated bottom is not wasted during the winter—it helps heat the building. The heat-delivery

---

1. The phase change is usually from solid to liquid to solid.

rate from this leak cannot be controlled, however, and during the summer this heat leak adds to the cooling load for the air conditioning system. Heat lost through the bottom of an uninsulated outdoor storage unit is wasted.

How is the heat stored? Usually solar heat is stored as *sensible heat*, an upward change in the temperature of the storage material (generally rocks or water). It is also frequently stored in the phase-change process of some substances (as latent heat). Several new methods of storing heat are also being explored, including heat storage in salt ponds, annual (or summer-to-winter) storage, and storage in chemical reactions.

4.1 HOW SENSIBLE HEAT STORAGE WORKS

Every substance stores heat energy as its temperature rises. The heat is released later as the substance cools. The amount of heat that can be stored over a fixed temperature range depends on the amount of the substance and its ability to store heat. Over a fixed temperature range some substances absorb more heat than others. One cubic meter of water, for example, will absorb 4190 kJ of energy over a 1° C range. So it takes 4190 kJ of energy to raise the temperature of 1 m^3 of water 1° C. (It takes 62.4 Btu to raise 1 ft^3 1° F.) One cubic meter of rock pebbles will absorb only an average of 1410 kJ while increasing 1° C in temperature (21 Btu, 1 ft^3, 1° F). Water can store about three times more energy than the same volume of rock pebbles. Thus as a storage medium water is approximately three times more compact than rock pebbles.

Heat capacity is heat-storage ability. The heat capacity of water can be expressed in terms of volume as

$$4190 \frac{\text{kJ}}{\text{m}^3 \, °\text{C}} \left(\frac{62.4 \text{ Btu}}{\text{ft}^3 \, °\text{F}} \right)$$

and that of rock pebbles as

$$1410 \frac{\text{kJ}}{\text{m}^3 \, °\text{C}} \left(\frac{21 \text{ Btu}}{\text{ft}^3 \, °\text{F}} \right).$$

Heat capacity is also expressed in terms of mass by units of KJ/kg °C or Btu/lb °F. These two unit systems are equivalent. The heat capacities for several materials in terms of both mass and volume are shown in Table 4.1.

The heat capacity can be considered constant for all materials over the temperature ranges discussed in this book. Therefore raising the temperature of 1 m^3 of water 1° C takes 4190 kJ; raising it 2° C takes 8380 kJ (4190 kJ for the first degree Celsius and 4190 kJ for the second degree, or 2 × 4190); raising the temperature 30° C takes 83,800 kJ (30 × 4190).

Table 4.1 **Thermal Properties of Materials**

Materials	Heat Capacity kJ/kg °C	Density kg/m³	Heat Capacity by Volume kJ/m³ °C	Thermal Conductivity W/m °C	Heat Capacity Btu/lb °F	Density lb/ft³	Heat Capacity by Volume Btu/ft³ °F	Thermal Conductivity Btu/hr ft² (°F/in)
Adobe	1.0	1700	1700	0.52	0.24	106	25	3.6
Aluminum	0.896	2700	2420	204	0.214	169	36.2	1415
Brick	0.84	1920	1600	0.721	0.20	120	24	5.00
Concrete	0.92	2240	2100	0.9–1.7	0.22	140	31	6–12
Fiberglass-batt insulation	0.71–0.96	5–30	4–30	0.027–0.049	0.17–0.23	0.3–2	0.05–0.46	0.19–0.34
Polyurethane-board insulation	1.6	24	38	0.023	0.38	1.5	0.57	0.16
Rock pebbles	0.88	1600	1410		0.21	100	21	
Steel	0.48	7850	3800	45.0	0.11	490	54	312
Granite	0.88	2720	2400	1.7–4.0	0.21	170	36	13–28
Water	4.19	1000	4190	0.56	1.00	62.4	62.4	3.9
Wood (fir or pine)	2.5	510	1300	0.12	0.60	32	19	0.80

Rock pebbles and water are often used to store sensible solar heat. Concrete, steel, adobe, stone, and brick can be used for sensible solar heat storage also.

Another factor to consider when choosing the storage material is the speed with which it absorbs heat for storage. Heat capacity indicates how much heat a substance will store, not how long it will take to absorb that heat or to release it. The ability of a substance to transmit heat quickly (to have high thermal conductivity) is important. Insulating materials such as fiberglass batts do not transmit heat quickly. They exhibit low thermal conductivity. The thermal conductivities for several materials are listed in Table 4.1.

For solar heat storage in active systems, it is important that the limited amount of storage material be able to absorb the heat available to it quickly. Water does not have high thermal conductivity but it can conduct heat fairly well because convection currents occur inside containers and mix the heated and unheated liquid. Rock does not have high thermal conductivity; that is one reason why rock pebbles are used instead of solid rock. With pebbles, the surface area available for heat absorption is large compared to the small volume of each pebble. (With solid rock, the surface area available would be small compared to the rock's large volume.) The pebbles' large surface area makes up for rock's relatively poor thermal conductivity.

For some types of solar heating, however, the poor thermal conductivity of rock (also of concrete, adobe, and stone) can be used to advantage. Massive walls and floors of concrete, stone, or adobe are also used for the storage of sensible heat. These structures can be included in actively heated buildings, but are seen most often in buildings that are heated by passive systems. (See Chapter 7 for details on sizing and incorporation of mass in such buildings.) These massive structures must be large enough to make up for their low heat conductance and slow heat-absorption rates. If massive structures are placed indoors in direct sunlight they become solar collectors as well as heat-storage units. The poor thermal conductance of massive concrete structures is well suited to passively heated buildings because these structures release heat slowly. Several hours may elapse between absorption and release of the heat. This time lag makes some solar heat available during the evening hours when demand for heat is usually high.

4.2 ROCK STORAGE

4.2.1 How Rock Storage Works

Solar heat is often stored in pebble beds. A pebble bed is a large insulated container (cube, enclosed rectangle, or cylinder) filled with evenly sized rocks. Rock storage is used with air-type collectors. Airflow through the bed is usually vertical but can also be horizontal.

Rock pebbles are used for solar heat storage, despite their poor ability to store heat in a compact space, for several reasons. One is the fact that pebble-bed storage is well suited for air-type solar collection. Air is circulated through the collectors, through the storage unit, and through the house, making the system simple to operate and maintain. Hot water is obtained from a heat exchanger located in the system's ductwork. Figure 4.1 shows how a rock storage system works.

Another reason why pebble-bed storage works well is that usually the pebbles occupy only 50–60% of the bed's volume. The other 40–50% between and around the pebbles is occupied by air. The 40–50% of the

Fig. 4.1 Operation of the rock storage system: (**a**) excess heat stored in the pebble bed during the day, (**b**) stored heat released to the building the night and in cloudy weather.

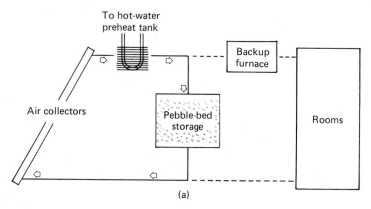

(a)

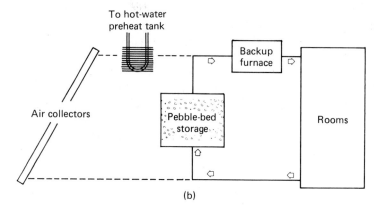

(b)

volume that is occupied by air is called the *void fraction.* The pebbles have poor thermal contact with one another and so heat is transmitted very slowly between them.

Rock beds operate efficiently because the void fraction leads to temperature stratification, or layering, in the storage unit. Usually hot air, often 70° C (160° F), from the collectors enters at the top of the storage unit and cool air is returned to the collectors from the bottom of the unit. This is desirable because collectors operate more efficiently with lower inlet temperatures. Cool air can be returned to the collectors because of temperature stratification in the storage unit. The unit is hottest near the top and coolest near the bottom.

In the morning the warm air from the collectors enters the top of the storage unit. The rocks at the top absorb the heat and become warm. As the sun becomes brighter during the morning, the collectors collect more solar energy and the temperature of the air coming from the collectors increases. This hotter air enters the top of the pebble bed and warms it even further. During this time some of the heat has migrated down into the bed. The temperature is warmest at the top and decreases to room temperature at a depth of about a meter. (Fig. 4.2).

To visualize this slow temperature migration and temperature stratification, we can think of the incoming air as "pushing" the temperature that is already stored in the top rocks down further into the pebble bed. The bottom rocks remain cool because this "pushing" action is too slow to warm the entire bed during the day.

In the afternoon the temperature of the air from the collectors will begin to decrease. This slightly cooler air will enter the pebble bed and "push" the warmer temperatures further down into the bed. In the late afternoon the hottest temperature layer in the pebble bed will be 0.6–1 m (2–3 ft) below the top. The top will cool down, although it will still be warmer than the bottom, which remains at near room temperature (Fig. 4.3).

The depth of the pebble bed is very important. A shallow bed with large top-surface area is desirable from the standpoint of fan operation. A smaller fan is required for a shallow bed because the pressure drop (pressure loss through a length of duct or piece of equipment due to friction, change in direction of fluid flow or path size, or change in velocity) through a shallow bed is less than that through a deep bed. It takes less power (and money) to pump the air through a shallow bed. However, if the bed were only 0.6–1m deep the entire bed could become heated during the day and hot air would be returned to the collectors in the afternoon, lowering their efficiency. Also, a shallow bed would take up a lot of floor space, which would be fine for a crawl space but probably not for a basement. A pebble-bed depth of approximately 1.5 m (5 ft) is

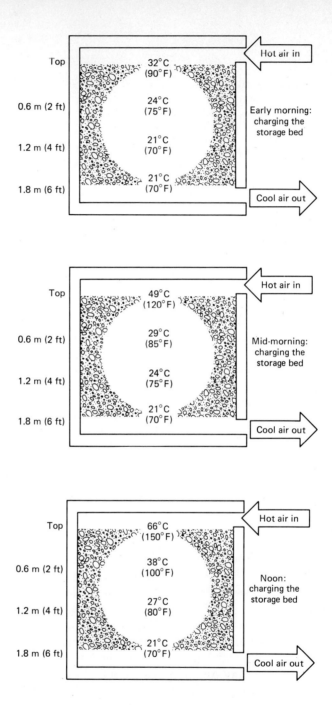

Fig. 4.2 Temperature stratification within a pebble bed (morning charging).

ROCK STORAGE

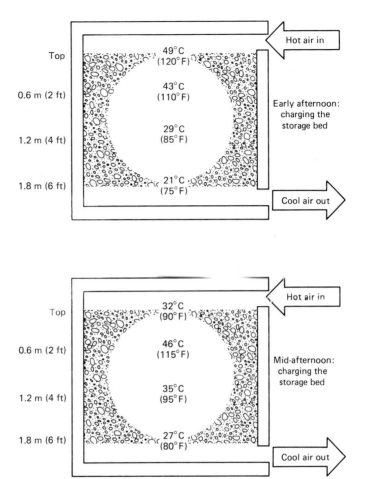

Fig. 4.3 Temperature stratification within a pebble bed (afternoon charging).

optimum and is a good compromise between a large shallow bed with a low pressure drop and a tall bed with high temperature stratification.

When the collectors are not operating and the stored heat is not being used, the "pushing" force of the moving air is not present. Then, temperature stratification in the pebble bed remains similar to what it was when the air circulation stopped.

When heat from the storage unit is needed, the airflow through the pebble bed is reversed. Cool air from the house enters at the bottom of the storage unit and hot air leaves at the top to heat the house. The entire "pushing" process is reversed also. The cool house air "pushes" the heat from the rocks out the top. As the storage bed discharges heat, the temperature migration through the pebble bed is reversed and the hottest temperature moves back up through the bed and out the top (Fig. 4.4).

This type of reverse-flow operation, with temperature stratification of the pebble bed, is very efficient. The advantages of temperature stratification may outweigh the disadvantage of the relatively large space needed for pebble-bed storage units. collector operation is kept at an efficient level because cool inlet temperatures are maintained: only cool air from the bottom of the storage unit is returned to the collectors. Even though the average temperature of the bed may be quite low, with temperature stratification and reversed flow for heat release from the

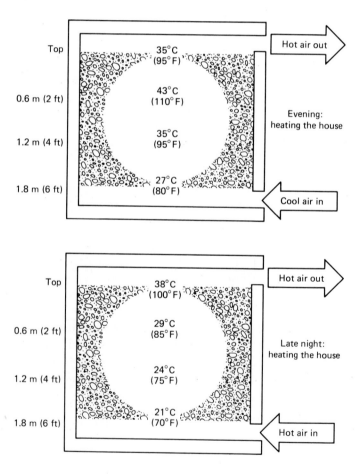

Fig. 4.4 Reverse flow through the pebble bed for house heating.

storage unit warm air from the hot upper layers of rocks is available immediately for heating the house.

Water-tank storage (section 4.3), though more compact than rock storage, is less efficient because the same high degree of temperature stratification cannot be achieved. The surging motion of pumped water mixes the water of the tank. Some heat exchangers do not physically mix the water, however; convection currents inside the tank distribute the heat. Elaborate baffle systems to preserve temperature stratification have had limited success. With water storage, warm liquid is returned to the collectors, and with higher inlet temperatures they operate at a relatively lower efficiency than air collectors with rock storage units. Also, the average temperature of the tank is the temperature available for heating. If it is too low to be used effectively, no stored heat can be used.

4.2.2 Sizing Rock Storage

It is generally agreed [1, 2, 3, 4] that approximately 0.25 m^3 (range 0.15–0.35) of 0.019–0.038 m pebbles should be provided for every square meter of optimized air-type collector area (0.8 ft^3 [range 0.5–1.15] of 0.75–1.5 in pebbles per ft^2 of collector). Optimizing the collector area is discussed in Chapter 6. This amount of pebbles, when charged, will supply approximately enough heat for one normal but sunless winter day's heating.

The pebble-bed depth should measure at least 0.76 m (2.5 ft) in the airflow direction to permit temperature stratification. A measurement of about 1.5 m (5 ft) is recommended because it permits temperature stratification but does not require a large cross-sectional area. Two and four tenths meters (8 ft) is the maximum recommended airflow distance through the bed. See Table 4.2 for a summary of pebble-bed sizing recommendations.

Example The optimized collector area for your house is 400 ft^2. How much storage should be provided? How much will the pebbles weigh? (The density of rock pebbles is approximately 100 lb/ft^3 and there are 2000 lb/ton.)

Solution Eight tenths of a cubic foot of pebbles is recommended for every square foot of collector area. Therefore,

$$(400 \text{ ft}^2 \text{ collector}) \times \frac{0.8 \text{ ft}^3 \text{ pebbles}}{\text{ft}^2 \text{ collector}} = 320 \text{ ft}^3 \text{ pebbles}.$$

Table 4.2 **Pebble-Bed Sizing**

Volume of pebbles required	0.15–0.35 m³/m² collector	0.5–1.15 ft³/ft² collector
Pebble size	0.019–0.038 m concrete aggregate	0.75–1.5 in concrete aggregate
Pebble-bed length (in airflow direction)	1.2–2.4 m	4–8 ft
Pressure drop through bed	25–75 Pa*	0.1–0.3 in of water

*Pa = Pascal (SI unit of pressure).

Three hundred and twenty cubic feet of pebbles should be provided. The density of rock pebbles is approximately 100 lb/ft³ and there are 2000 lb/ton, so

$$(320 \text{ ft}^3 \text{ pebbles}) \left(\frac{100 \text{ lb}}{\text{ft}^3}\right)\left(\frac{1 \text{ ton}}{2000 \text{ lb}}\right) = 16 \text{ tons.}$$

Approximately 16 tons of pebbles would be required. Since the recommended depth (if we assume vertical airflow) for a pebble bed is about 5 ft, these pebbles could be placed in a box measuring about 8′ × 8′ × 5′. The actual dimensions would be slightly larger to allow for bed construction and air passage along the top and bottom of the bed. If the entire pebble bed were allowed to store heat over a 60° F temperature range, 403,000 Btu (320 ft³ × 21 Btu/ft³ °F × 60°) of heat could be stored.

4.2.3 Construction of Rock Storage Units

The pebble bed can be enclosed by a variety of containers. These can be made of wood, concrete, or metal. The containers can be constructed at the site or hauled in, and are suitable for use with new construction and for retrofitting.

The storage unit should be well insulated on all sides and at the top (at least 9–15 cm [3.5–6 in.] of fiberglass batts). Styrofoam and some urethane insulations may not withstand the temperatures reached within the box and should not be used. If the hot air from the collectors passes through the bed from top to bottom, insulating the bottom of the pebble bed is not quite as important since its temperature remains fairly constant, near room temperature. It is recommended, however, that $7\frac{1}{2}$ cm (3 in) of rigid fiberglass insulation be used at the bottom of the pebble bed. If the container will be in an unheated space, at least twice that amount of insulation should be used.

Before rock is added to the container, all inside joints should be carefully sealed with a sealant that can withstand approximately 95° C (200° F) temperatures. The container must be airtight.

Wooden Container

A wooden container to hold the rocks can be constructed at the site (Figs. 4.5 and 4.6). The framing should be of studs on 0.3-m (1-ft) centers. About one third to one half of the way up the box a horizontal double-stud beam should be constructed on all sides. The ends of two steel tie rods (1.3 cm [0.5 in] in diameter) should be embedded in this beam, each connecting two opposite sides together. The tie rods help minimize bulging of the box caused by the weight of the rocks. Both sides of the stud walls should be covered with plywood and the spaces between the studs filled with fiberglass-batt insulation. If more insulation is desired, larger studs should be used.

Fig. 4.5 Structure of a wooden pebble bed

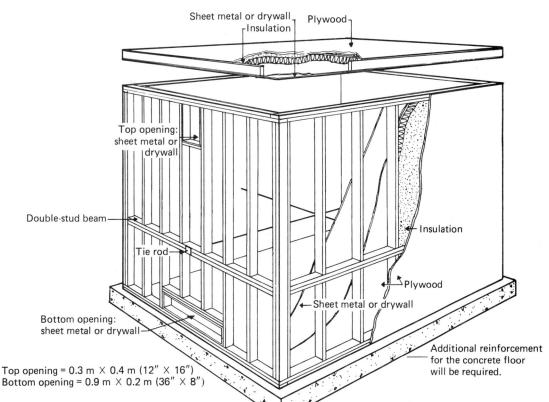

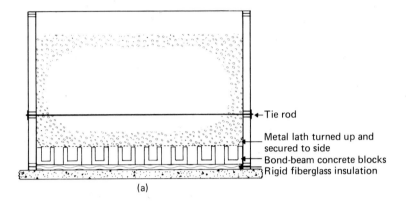

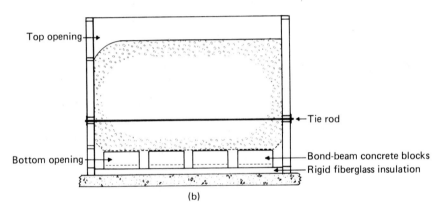

Fig. 4.6 Cross-sectional illustrations of a wooden pebble bed: (**a**) side view, (**b**) front view.

Openings for airflow must be provided at the top and bottom of one side of the box. The non-load-bearing lid of the box may be constructed of studs 0.6 m (24 in) on center. A butyl rubber gasket or other heat-resistant material can be used as a seal between the lid and the walls of the box.

All portions of the box and lid that come in contact with the hot air should be covered with sheet metal or moisture-resistant drywall. All inside seams and joints must be sealed with a sealant capable of withstanding 95° C (200° F).

The bottom *plenum* (air channel) is formed by bond-beam concrete blocks placed several centimeters apart on the floor of the box, perpendicular to the wall with the openings. The blocks should cover about 50% of the box floor. Heavy expanded-metal lath (screen) is placed over the blocks and bent up around all the walls. Rocks are placed on this screen to within about 0.3 m (1 ft) of the top of the box. The space above the rocks forms the top plenum when the lid is added.

Concrete Container

The pebble-bed container can also be constructed of concrete blocks (Fig. 4.7). The unit can be constructed relatively inexpensively if it is placed in a corner of the basement so that two of the basement's block walls become two walls of the box. Steel reinforcement rods (1.6 cm [$\frac{5}{8}$ in] in diameter) should be placed horizontally and vertically in the box's walls approximately every 0.5–0.6 m (1.5–2 ft). If the box is very large, it may be necessary to include *pilasters* (perpendicular concrete-block support pillars) in the walls or to use tie rods as in the wooden box.

Fig. 4.7 Concrete-block pebble bed.

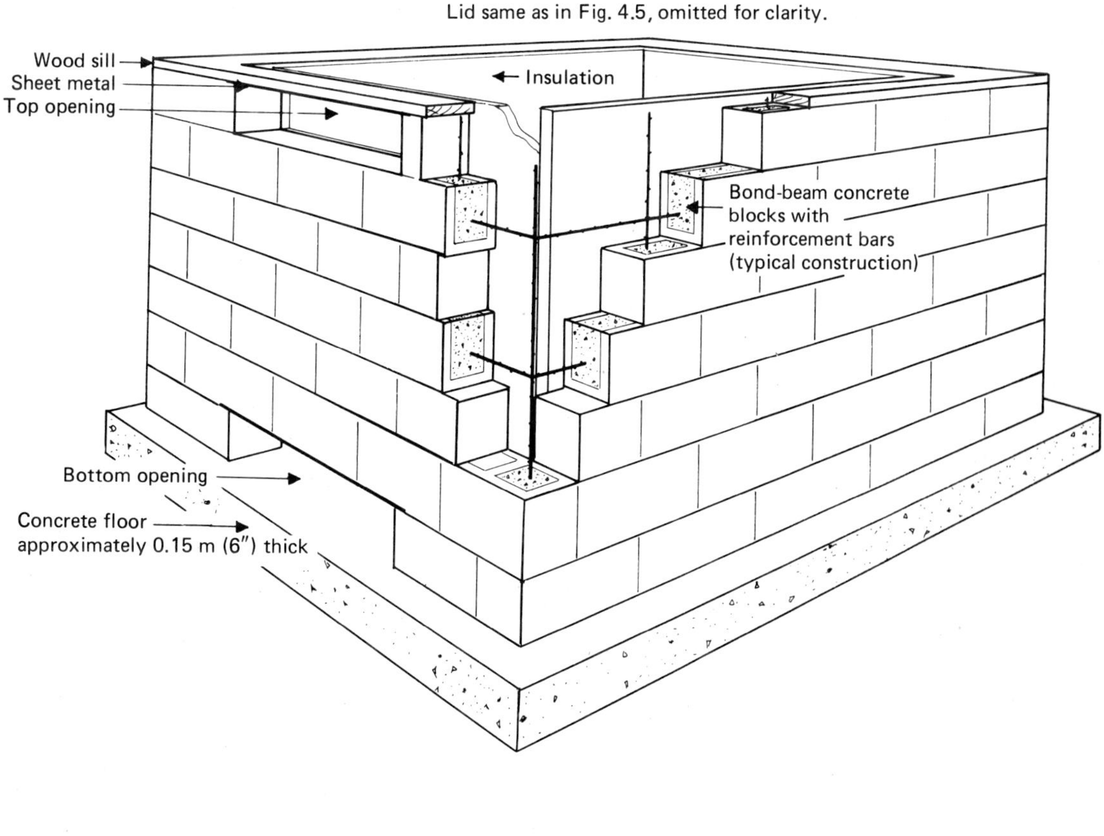

Lid same as in Fig. 4.5, omitted for clarity.

Openings for air circulation must be provided. The inside of the concrete box should be lined with a minimum of 5 cm (2 in) of rigid fiberglass insulation. The unit should be carefully sealed. The placement of the bond-beam blocks, metal lath, and rocks, and the construction of the lid are the same as for the wooden container.

Metal Tank

The storage pebbles can also be contained in a tank of steel or other metal (Fig. 4.8). Curved sections of the tank can be bolted together at the site. A heat-resistant sealant must be carefully applied to all the seams. The outside of the tank should be insulated with at least 9 cm (3.5 in) of fiberglass insulation.

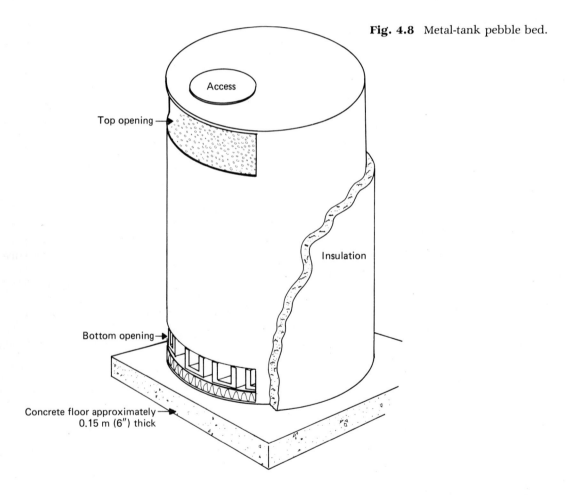

Fig. 4.8 Metal-tank pebble bed.

ROCK STORAGE

Air openings must be provided at the top and bottom of the tank. The placement of the bond-beam concrete blocks, metal lath and rocks, and the construction of the lid are the same as for the wooden container.

4.2.4 Airflow Direction in the Pebble Bed

In all the storage units described thus far airflow has been vertical, with air openings at the top and bottom of each unit. Vertical airflow generally results in better performance than horizontal flow and is therefore preferable.

With vertical airflow, the hot section of the pebble bed is near the top. Since hot air naturally rises, this arrangement is especially suitable. With horizontal airflow, the hot section of the bed is at one end. If the rocks have settled and do not completely fill the container, channeling of hot air across the top of the rocks will occur. Even without extensive channeling, the top portion of a horizontal storage unit is usually warmer than the bottom. Thus the full benefits of temperature stratification cannot be realized.

If horizontal airflow cannot be avoided (if a shallow crawl space is the only space available for storage, for example), vertical baffles that direct airflow through the rock should be provided (Fig. 4.9).

4.2.5 Selection of Rocks

The rock pebbles should be as uniform in size and as spherical as possible to allow proper airflow through the pebble bed. If the pebbles are too small, the airflow through the bed will be slow and the pressure drop through the unit will be large and require a large fan. If the rocks

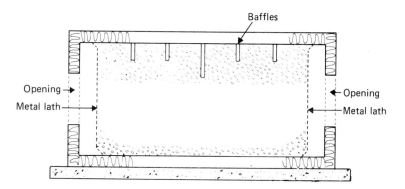

Fig. 4.9 Horizontal-airflow pebble bed.

are too big, only the exterior of individual rocks may be heated, lowering the effectiveness of temperature stratification in the bed. If the rocks are not sufficiently spherical, airflow through the bed will not be uniform and pressure drop through the bed will be too great.

Nineteen thousandths to 0.038 m (0.75–1.5 in) river rock is preferred, although any rock suitable for concrete aggregate can be used. Graded gravel cannot be used, however. The rock is sized by the seller by passage through a screen. Pebbles that fit through a 2-cm screen but not through a 4-cm screen are suitable. The pebbles should be as free of *fines* (pebbles less than 0.006 m [$\frac{1}{4}$ in] diameter) as possible and must be washed to minimize clogging by dust.

The bed should be filled by chute to minimize splitting of the pebbles and damage to the storage container.

4.2.6 Cooling with Rock Storage

In summer, pebble beds can be used to store *coolness* for air conditioning (Fig. 4.10). At night cool air from outside or outside air first cooled by an evaporative cooler is drawn through the storage unit. An evaporative cooler works by blowing hot dry air over moist pads. Heat from the air is absorbed by water in the pads (which evaporates), lowering the air temperature. During the hot day, outside air or air from the house can be circulated through and cooled by the cold rock and conditioned further by an evaporative cooler, if necessary, before being returned to the house.

The evaporative air conditioner is often operated at night to charge the rock storage unit. Outdoor temperatures at night may not be cool enough for comfort, but they may allow more efficient operation of the air conditioner than daytime temperatures. And if the cooler is operated primarily at night, it may be possible to take advantage of lower, off-peak electricity rates. Nighttime operation of the cooler is not wasteful because virtually all of the coolness stored at night is used the next day.

A potential problem with using an evaporative cooler in conjunction with rock storage is condensation on the rocks. The condition could lead to mildew and odors in the storage unit. It is advisable to include a drain at the bottom of a storage unit if it is to be used with an evaporative cooler.

4.3 WATER STORAGE

4.3.1 How Water Storage Works

Solar heat is often stored as an increase in the temperature of water. As we saw in section 4.1, water is a compact storage medium. Liquid-type solar collectors are usually used in conjunction with water storage.

WATER STORAGE

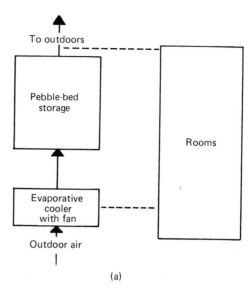

(a)

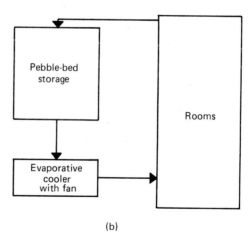

(b)

Fig. 4.10 Cooling with rock storage: (**a**) at night outside air is cooled by the evaporative cooler and is used to cool the pebble bed, (**b**) during the day hot building air is cooled by the pebble bed and the evaporative cooler.

Water can also be used as the storage medium with air-type collectors if the water is stored in small containers and the air allowed to blow among them.

Usually the water is kept in an insulated tank. If antifreeze or a heat-transfer fluid is used in the collectors, a heat exchanger (accessible for repairs) must be provided between the storage and the collector loop. If water is circulated through the collectors, no heat exchanger is necessary (Fig. 4.11).

Hot liquid from the collectors enters the top of the storage tank and cooler liquid is returned to the collectors either directly or through the heat-exchanger loop. Returning cool liquid to the collectors is desirable because it allows them to operate at higher efficiency.

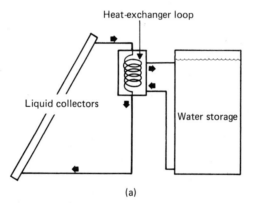

Fig. 4.11 Water as the storage medium: (**a**) heat exchanger between collectors and storage tank, (**b**) direct storage.

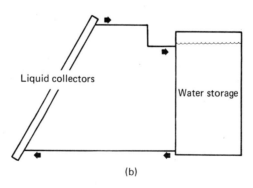

When heat for the building is needed, hot water from the top of the storage unit is pumped into the building's distribution system. The water may circulate through radiative-type heaters in each room or through a water-to-air heat exchanger in the ducting system of the building. In the water-to-air system air is warmed as it is blown across the heat exchanger by a fan and then is distributed to the building through the ductwork. A heat exchanger in the storage tank provides domestic hot water (Fig. 4.12). The cooled water is returned to the bottom of the tank.

Temperature stratification in water-storage units is not great, but some attempt is made to take advantage of the slight stratification that does exist. (Hot water goes in at the top and cool water goes out at the bottom in one collection mode.) Stratification is difficult to maintain because of the mixing motion caused as pumps move the water through the system and because convective currents in the tank mix the warm and cool water. Baffles can improve stratification, but the slightly increased performance that results is usually not worth the extra cost. In normal 4000-l (1060-gal) tanks, a temperature difference of approximately 5° C (9° F) between the top and bottom may be noticed. Rock storage, though bulkier, is considered more effective by some because of its greater stratification. In pebble beds, stratified layers may differ in temperature by as much as 45–50° C (80–90° F).

The water-storage tank can be pressurized or nonpressurized, though a nonpressurized tank is usually recommended because of its lower cost and simpler operation. With a nonpressurized tank, the upper temperature limit of the water is slightly below its boiling point of 100° C (212° F). The system is vented so that if boiling should ever occur the steam and possibly some water would escape. A float-controlled valve is needed to admit replacement water as is necessary.

Water-storage systems are especially susceptible to corrosion and scaling. Hard-water scaling can be reduced by the use of water softeners.

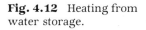

Fig. 4.12 Heating from water storage.

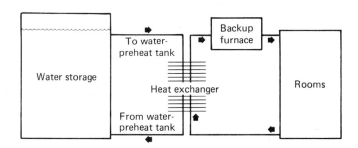

Other corrosion problems are not decreased by the use of softeners, however. Dissimilar metals must not come in direct contact with one another. Connecting dissimilar metal fittings with neoprene hose can help reduce corrosion. Special anodes (usually of magnesium or zinc) called *sacrificial anodes* can be included in the system and will corrode onto a metal tank, for example, thus protecting it from corrosion and pitting. Sacrificial anodes need periodic replacement. The most effective way to reduce corrosion of metal tanks is to line them. Lining materials able to withstand high temperatures include glass, neoprene, butyl rubber, and hypalon.

4.3.2 Sizing Water-Storage Tanks

It is generally agreed [1, 2, 3, 4] that approximately 75 l of storage water should be provided for every square meter of optimized liquid-type collector area with a range of 50–100 l (1.75 gal/ft^2 with a range of 1.25–2 gal). See Fig. 4.13. This amount of water, when heated, will supply approximately enough heat for one normal but sunless winter day's heating.

Example The optimized collector area for your house is 60 m^2. How much storage should be provided? What is the mass of water required?

Solution Seventy-five liters of water are recommended for every square meter of collector area:

60 m^2 × 75 l/m^2 = 4500 l.

The density of water is 1 kg/l:

4500 l × 1 kg/l = 4500 kg.

If we allow approximately 20% more room for water expansion, drainage from the collectors, and air space (see section 4.3.3), a tank that holds about 5400 l (4500 × 1.2 = 5400) is needed. A 5400-l cylindrical tank might be approximately 1.85 m in diameter and 2 m tall.

Fig. 4.13 Water-storage sizing.

Volume of water required	50–100 liters/m^2 collector	1.25–2 gallons/ft^2 collector

4.3.3 Construction of Water-Storage Units

A solar heated water-storage tank can be made of concrete, fiberglass, or metal. To minimize heat loss from the storage unit, a cylindrical or cubic tank constructed carefully to avoid excess stress on the corners) with a diameter-to-height ratio near 1:1 is preferred. Because the tank is large, its installation is easiest while the building is still under construction.

The tank must be protected from freezing temperatures, so it must be located inside a building or buried below the frost line. Buried outdoor tanks are more difficult to insulate properly and much more difficult to service than indoor tanks. Also, buried tanks must be strong enough to withstand water and soil pressures from the surrounding earth when empty and precautions must be taken in areas with high water tables to prevent an empty tank from floating out of position. Indoor tanks are usually best.

Since the entire tank will be at a high temperature, all surfaces of an indoor tank should be insulated with at least 18 cm (7 in) of fiberglass batt and of an outdoor tank with at least 25 cm (10 in). The bottom of the tank must also be insulated, preferably with moisture-resistant insulation. All pipes in the system should also be insulated. Before the insulation is applied, the entire system should be checked for leaks. It is very difficult to find a leak through wet insulation. Also, wet insulation does not insulate.

Pipe connections between dissimilar metals should be avoided. Carefully clamped neoprene rubber tubing may be used to separate metal fittings.

A drain from the bottom of the tank and an overflow-outlet pipe near the top of the tank should be provided and drain into a storm drain. A float valve can control the addition of *make-up* (replacement) water. A vent from the top of the tank, preferably to the outside, should also be provided to release any excess pressure.

Extra tank capacity must be provided to allow room for the expansion of heated water, possible drainage of water from the collectors into the tank, and an air space above the overflow outlet and below the make-up-water inlet.

The entire storage system should ideally be located in an insulated, vented room. In addition, each portion of the system should be able to be completely isolated by shut-off valves in case of failure.

Concrete Tank

Concrete tanks can be constructed of well-reinforced concrete blocks or of preformed utility vaults, septic tanks, or pieces of concrete pipe (Fig.

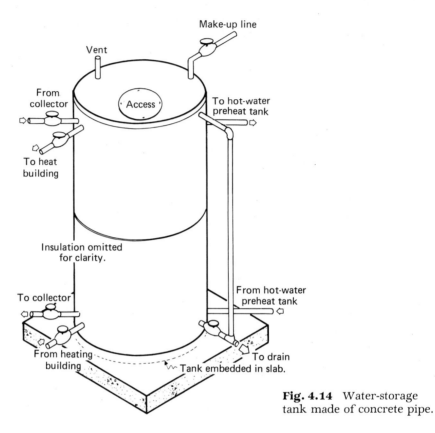

Fig. 4.14 Water-storage tank made of concrete pipe.

4.14). Precast concrete tanks may require a crane for installation and therefore must be installed while construction is in progress. Seams in concrete tanks can be very difficult to seal because of temperature stresses between the pieces. Usually a liner is required to prevent seepage. Pipe connections must also be carefully sealed.

Fiberglass Tanks

Most fiberglass tanks (Fig. 4.15) cannot withstand the high temperatures a solar storage unit is subjected to. Many fiberglass resins begin to flow above 70° C (160° F). Some resins capable of withstanding 100° C (212° F) have been developed and tanks made of these materials are satisfactory. Common fiberglass tanks include septic tanks and gasoline-storage tanks.

WATER STORAGE 131

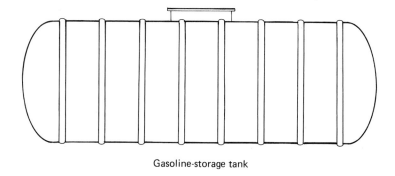

Gasoline-storage tank

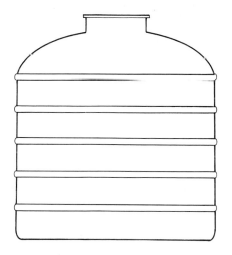

Septic tank

Insulation and piping
omitted for clarity.

Fig. 4.15 Fiberglass water-storage tanks.

Metal Tanks

Galvanized steel or glass- or stone-lined steel tanks are satisfactory for use as solar storage units (Fig. 4.16). Specially fabricated aluminum tanks, though quite susceptible to corrosion, may also be satisfactory. Eventually all metal tanks will corrode and need replacement. Care should be taken not to bump glass-lined tanks during installation. If the glass liner is fractured, the life of the tank will be reduced significantly. Glass- or stone-lined tanks or metal tanks lined with materials mentioned in section 4.3.1 are recommended.

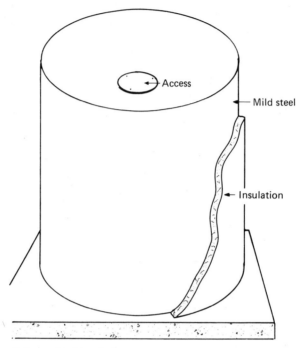

Fig. 4.16 Metal water-storage tank.

Piping omitted for clarity.

4.4 PHASE-CHANGE STORAGE

4.4.1 How Phase-Change Storage Works

Phase-change storage takes advantage of the energy that can be stored and then reclaimed during the phase change (from solid to liquid to solid) of certain substances. Phase-change storage materials are usually contained in small tubes, trays, or packets and stacked in the storage unit. They are used primarily with air-type collectors.

As with rock storage, air is blown through the storage unit to heat (in this case melt) the storage material. When heat is needed for the building, airflow through the storage unit is reversed and cool air is heated as the material cools and solidifies.

Some experimentation in the use of phase-change storage with liquid collectors has taken place. The feasibility of encapsulating the material and suspending it in water, making use of the greater heat-transfer ability of the water, has been shown [5]. Also, the use of an immiscible fluid-heat-exchanger system, in which an unmixable fluid is circulated through a solution containing the phase-change material, may prove practical [6].

PHASE-CHANGE STORAGE

The major advantage of phase change-storage is its compactness. It requires significantly less space than either rock or water storage. This compactness is usually accomplished at a high cost however, and when cost is considered the advantage often disappears.

How is energy stored in the phase-change process? Everyone knows that ice keeps a cold drink cold and that the drink warms up only after the ice has melted. This is a common example of the phase-change process. The water changes from its solid-water (ice) phase to its liquid phase.

It takes energy to turn ice into water. During that process the temperature of the ice-water mixture will stay near the melting temperature of ice, 0° C (32° F). The temperature of the mixture remains constant because all the available energy is being used to change phases. In our example the energy comes as heat transfer from the warm air surrounding the glass. The temperature cannot increase above the phase-change temperature until all the ice has melted.

Figure 4.17 is a phase-change diagram for melting. It shows graphically what happens to any substance undergoing a phase change. At point A on the diagram, an ice cube is in the freezer at −18° C (0° F). We bring it out and set it on the table. The relative warmth of the table and surrounding air heat the ice from −18° C along the curve to its melting point of 0° C (32° F), point B on the diagram. Now the ice begins to melt. All incoming energy is used in the melting process. The temperature of the ice-water mixture remains near 0° C until all the ice has melted, at point C on the diagram. Now incoming energy can finally be used to

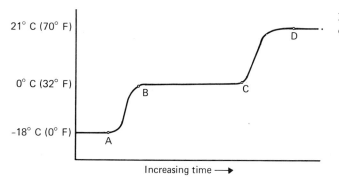

Fig. 4.17 Phase-change diagram: ice to water.

A = ice in freezer
B = ice beginning to melt
C = all ice melted
D = water at room temperature

heat the water. The water will continue to heat until it is the same temperature as the surroundings, at point D.

It takes energy to melt ice. When water freezes it gives up energy. As water freezes the temperature will, as we saw with the melting process, remain constant near 0° C (32° F) until all the water has changed—this time to ice. A phase-change diagram for freezing is shown in Fig. 4.18. As energy is removed from the water (in the freezer), the water temperature decreases, along the curve from point A, until the freezing point is reached, at point B. Energy continues to be removed as the water changes phase to ice. Only when all the water has become ice, at point C, can the temperature decrease below the freezing point of water.

The amount of heat energy that must be added to a substance to make it melt or that must be removed to make it solidify is called its *latent heat of fusion*. Different substances have different heats of fusion. A high heat of fusion is desirable because the more energy it takes to melt the substance, the more energy will be stored for later use.

Heats of fusion are usually given in units that relate energy to a given mass of the substance, for example, kJ/kg, Btu/lb, kcal/kg. For solar heating systems, it is more important to know the amount of energy that can be stored in a given volume than in a given weight. Therefore, when we select a substance for use as a phase-change storage material, we must consider its heat of fusion and density. A substance with a high heat of fusion and a high density will be able to store heat compactly. One with a low heat of fusion and low density will not be suitable

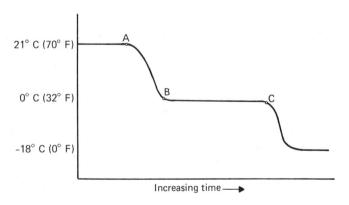

Fig. 4.18 Phase-change diagram: water to ice.

A = water at room temperature
B = water beginning to freeze
C = all water frozen

for solar heat storage. The heats of fusion for several phase-change energy-storage materials are listed in Table 4.3.

4.4.2 Substances Used for Phase-Change Storage

Water has a very high heat of fusion and, though it is not very dense, can store heat quite compactly. Solar heat must be stored at a temperature higher than the melting point of ice, however, so other substances are used. A comparison of the energy-storage abilities of several substances by volume, including water, is presented in Table 4.3. As we can see, aside from the water these substances can be divided into two groups, paraffin and salt hydrates (salts bonded to water molecules).

Paraffin wax has a high heat of fusion but because it is less dense than the salts more of it is required to store the same amount of heat. Paraffin does have one major advantage over the salt hydrates, however. Paraffin melts and resolidifies easily, always returning to its original form. Salt hydrates do not always resolidify at the right temperature and into the desired compound. Paraffin has several disadvantages also. Paraffin is more expensive than the hydrates and since it presents a fire hazard it must often be enclosed outside the building. In addition paraffin has a very low thermal conductance, so heat is not distributed to all of the wax quickly.

Salt hydrates (salts bound to water), like their water molecules, have high heats of fusion. Yet they are more dense than water and therefore make a more compact storage medium. Of the salt hydrates listed in Table 4.3, sodium sulfate decahydrate ($Na_2SO_4 \cdot 10H_2O$), or Glauber's salt, is the one used most often for phase-change storage.

Glauber's salt is inexpensive, is widely available, and stores heat very compactly. It is nontoxic (it was once used as a medicine), and is used today in the paper-making process, as an ingredient in mineral feed supplements for cattle, and in the manufacture of some detergents. Glauber's salt is mined and refined from salt lakes and sodium sulfate is produced as a by-product in industrial chemical reactions.

Maria Telkes is one of the foremost researchers into the uses of Glauber's salt and other salt hydrates for phase-change energy storage. She has developed several of the formulations for heat and coolness storage and has developed solutions to the problems faced by phase-change salts (see section 4.4.3).

Many of the salt hydrates used for phase-change solar heat storage have also been called *eutectic salts*. This term is used loosely to describe

Table 4.3 **Properties of Common Phase-Change Storage Materials**

Substance	Chemical Formula	Melting Point °C	Heat of Fusion kJ/kg	Density kg/m³	Heat of Fusion by Volume MJ/m³	Melting Point °F	Heat of Fusion Btu/lb	Density lb/ft³	Heat of Fusion by Volume Btu/ft³
Paraffin	C_{15}–C_{16} paraffin	44–48	209	786	160	112–118	90	49	4400
Calcium chloride hexahydrate	$CaCl_2 \cdot 6H_2O$	30	170	1670	280	86	75	104	7800
Calcium nitrate tetrahydrate	$Ca(NO_3)_2 \cdot 4H_2O$	40–43	140	1830	260	103–109	60	114	6800
Sodium carbonate decahydrate (washing soda)	$NaCO_3 \cdot 10H_2O$	32–35	247	1440	360	90–94	106	90	9500
Disodium phosphate dodecahydrate	$Na_2HPO_4 \cdot 12H_2O$	35	265	1520	400	95	114	95	10800
Sodium sulfate decahydrate (Glauber's salt, mirabilite)	$Na_2SO_4 \cdot 10H_2O$	32	251	1460	370	90	108	91	9800
Sodium thiosulfate pentahydrate (hypo)	$Na_2S_2O_3 \cdot 5H_2O$	48–49	210	1730	360	118–120	90	108	9700

eutectic mixtures as well as the salts that form them. (See section 4.4.7 for a definition of eutectic mixture and example of energy storage in such a mixture.) It is incorrect to refer to all phase-change storage media as *eutectic-salt storage*, and a reference to eutectic storage can lead to confusion between phase-change storage in a salt hydrate and storage in a eutectic mixture. To simplify this situation as much as possible, use of the ambiguous term *eutectic salt* is avoided. *Salt hydrates* and *eutectic mixtures* are referred to as such. Occasionally the term *phase-change salt* is used to describe behavior exhibited by both salt hydrates and their eutectic mixtures.

4.4.3 Problems of Phase-Change Salts

Two major problems are encountered when phase-change salts are used for solar heat storage: supercooling and stratification.

Supercooling

Most phase-change storage materials, including Glauber's salt, have a tendency to *supercool*, or cool below their freezing points, before starting to crystallize. As we can see on the phase-change diagram in Fig. 4.19 as heat is removed the temperature of the liquid falls below the freezing point (or supercools) before it begins to crystallize. Once the liquid begins to crystallize the temperature increases to the freezing point, where it remains until all the material has solidified.

It is believed that supercooling occurs because at the freezing point the molecules of cooling liquid are in random order. At this temperature they are supposed to line up and solidify in their crystalline form. It sometimes takes them a bit of time to line up correctly and during this

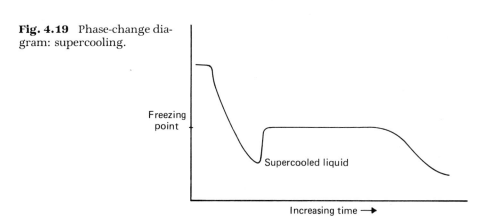

Fig. 4.19 Phase-change diagram: supercooling.

time additional sensible heat is being removed from the system, decreasing its temperature. As soon as the material starts crystallizing in the right pattern, the temperature increases to the normal freezing point of the material.

Substances that supercool to the extent that they become too cold to rearrange themselves into the proper crystal patterns may remain in the supercooled state as *solid-liquids*. These substances are called *glasses* after their most famous example, glass. Other common examples include many plastics such as polyethylene, vinyl polymers, and Teflon.

Supercooling to 10–15° C (18–27° F) below the freezing point can make the effective removal of heat from the storage unit quite a problem. To eliminate supercooling, a small amount of *seed crystals* (usually 3–5% by weight) is added to the salt. The seed crystals resemble the salt crystals in size and shape but remain as crystals in the liquid salt. When the liquid salt is cooling, crystallization begins at the freezing point because the proper crystal pattern is already present in the form of the seed crystals. The seed crystals also help to eliminate formation of lower-order hydrates that contain fewer than the desired number of water molecules and have lower heats of fusion and thus lower heat-storage abilities. The addition of seed crystals lowers the freezing point and the heat-storage ability of the salt slightly. Borax ($Na_2B_4O_7 \cdot 10H_2O$) is added to Glauber's salt as a seed crystal.

Stratification

Stratification, or separation of some of the components in the salt after it cycles through the phase-change process, severely limits the usefulness of salt hydrates as a solar heat-storage medium. Several satisfactory methods for preventing stratification are now available.

For example, as Glauber's salt ($Na_2SO_4 \cdot 10H_2O$) melts, the sodium sulfate (Na_2SO_4) separates from the 10 water molecules ($\cdot 10H_2O$) and partially dissolves in the water. Approximately 15% by weight of the *anhydrous* (waterless) salt cannot dissolve and because it is more dense than the surrounding solution it settles to the bottom. The highest solubility of sodium sulfate in the water occurs at the melting point. As the temperature increases, more and more anhydrous salt solidifies and settles to the bottom. When it is time for the salt to resolidify, some of the anhydrous salt on the bottom cannot find its 10 molecules of water to recombine with and crystallization is incomplete. With complete stratification, the storage ability of the salt decreases about 40%. Figure 4.20 illustrates stratification of Glauber's salt.

If the containers are shaken or stirred during crystallization, stratification does not occur. Of course, to shake every container in a heat-stor-

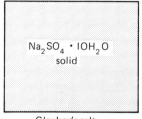
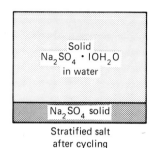

Fig. 4.20 Stratification of Glauber's salt.

age unit would be virtually impossible. Some research is being done with a heat-storage unit consisting of a large, constantly rotating drum. Rotation may provide the needed mixing action while also improving heat transfer to the salt [7]. Also, limiting the vertical height of the storage containers to 0.006 m (0.25 in) seems to permit recombination of the anhydrous salt and its waters.

The borax seed crystals are denser than the salt solution and they too tend to settle to the bottom of the containers over time. To prevent the settling of both borax and anhydrous sodium sulfate, thickeners are added to the salt. These thickeners (usually 6–10% by weight of the mixture) do not interfere with the melting or crystallization of the salt, but merely hold all the molecules in suspension during the cycle. The thickeners prevent the settling of the more dense particles, making complete crystallization possible.

Many substances have been tried as thickeners, including wood shavings, sawdust, paper pulp, silica gels, diatomatious earth, starch, and other chemicals. Some of these have been successful, though some decompose over time and others react partially with the solution. Addition of a thickener does reduce the heat-storage ability of the mixture, but without it the performance would deteriorate far below this slightly reduced level.

4.4.4 Phase-Change Storage versus Rock and Water Storage

As we stated before, phase-change solar heat storage is more compact and more expensive than either rock or water storage. Phase-change materials store energy primarily in the phase-change process. They can also store heat as sensible heat, just as rock and water do. The amount of heat stored in the phase-change process is fixed, but the amount of heat stored as sensible heat depends on the temperature range involved.

At the melting point of Glauber's salt with stabilizers, approximately 320 MJ/m^3 can be stored. If 50% of the space is allotted for air passage through the salt containers, salt storage is reduced to 160 MJ/m^3 (4250 Btu/ft^3). Molten salt in containers has a heat capacity of about 2.01 MJ/m^3 °C (30 Btu/ft^3 °F), which is higher than that of rock pebbles (1.41 MJ/m^3 °C, or 21 Btu/ft^3 °F) but lower than that of water (4.19 MJ/m^3 °C, or 62.4 Btu/ft^3 °F). If the temperature is allowed to rise to 2.8° C (5° F) above the salt's melting point, 160 MJ are stored per cubic meter of salt container when the salt melts, and another 6 MJ (2.8° C × 2.01 MJ/m^3 °C) are stored when the salt increases in temperature 2.8° C, for a total of 166 MJ (4400 Btu/ft^3). Within a 2.8° C range of the salt's melting point, it would take 14.1 m^3 of water (166 MJ × m^3 °C/4.19 MJ × 1/2.8° C) and 42 m^3 of rock pebbles (166 MJ × m^3 °C/1.41 MJ × 1/2.8° C) to store that same amount of heat.

If, however, we extend the temperature range to 16.7° C (30° F) from the salt's melting point, the advantage of salt storage decreases. The energy stored in the phase-change process at the melting point is still fixed at 160 MJ/m^3. Over the 16.7° C range, 1 m^3 of salt melts, storing 160 MJ via the phase change and 34 MJ as sensible heat (16.7° C × 2.01 MJ/m^3 °C), for a total storage of 194 MJ (5200 Btu/ft^3). To store the same amount of heat over that range would require 2.8 m^3 of water (194 MJ × m^3 °C/4.19 MJ × 1/16.7 °C), or 8.2 m^3 of rock pebbles (194 MJ × m^3 °C/1.41 MJ × 1/16.7° C).

Over a 33.3° (60° F) range, 1 m^3 of salt can store 160 + 67 MJ (33.3° C × 2.01 MJ/m^3 °C), for a total of 227 MJ (6100 Btu/ft^3). To store the same amount of heat over that temperature range would require 1.6 m^3 of water (227 MJ × m^3 °C/4.19 MJ × 1/33.3° C) or 4.8 m^3 of rock pebbles (227 MJ × m^3 °C/1.41 MJ × 1/33.3° C). The variation of storage-volume requirements with temperature range is illustrated in Fig. 4.21.

Results of a recent study [8] indicate that phase-change storage with air collectors, while more compact than pebble-bed storage, is less efficient. The high degree of temperature stratification present in a pebble bed, which permits very efficient operation of the collectors, cannot be obtained with a constant-temperature melting-phase-change system. The study shows the performance of paraffin wax to be much lower and that of Glauber's salt to be slightly lower than the performance of rock.

4.4.5 Sizing Phase-Change Storage

Incorporating phase-change storage into a solar heating system is generally not a do-it-yourself project. Recommendations from individual phase-change storage device manufacturers should be followed.

One study [8] has suggested that 7–20 kg of Glauber's salt or 7–25 kg of paraffin should be provided for every square meter of collector area

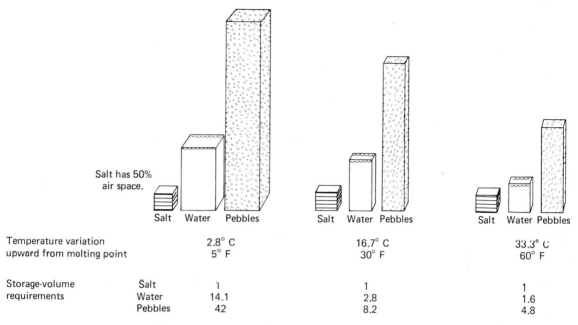

Fig. 4.21 Storage-volume requirements for containers of Glauber's salt, water, and rock pebbles.

(1.4–4.1 lb/ft² for Glauber's salt and 1.4–5.1 lb/ft² for paraffin). These recommendations translate to 0.0048–0.014 m³/² for pure Glauber's salt (0.015–0.045 ft³/ft²) and to 0.0086–0.031 m³/m² for paraffin (0.028 to 0.10 ft³/ft²)

The sizing recommendations are for pure Glauber's salt and paraffin, without allowances for additives or air space between containers. The addition of stabilizers to Glauber's salt will usually increase volume by 10–15%. The containers usually occupy only 50% of the total storage space needed.

These sizing recommendations lead to significantly lower heat-storage capabilities than do those for rock and water storage. Therefore, overall system performance may not be accurately predicted by the method explained in Chapter 6 if we use these storage-sizing figures. A method for deriving the amount of a phase-change material needed for equivalent performance has been suggested, however [9].

Example The collector area for your house is 65 m². How much Glauber's salt should be provided? (Assume that 0.014 m³/m² of salt will be used, that the addition of stabilizers increases the salt's volume by 14%, and that air space is 50% of the storage area.

Solution 0.014³ of Glauber's salt will be provided for every square meter of collector area. Additives increase the salt's original volume 1.14 times. Therefore,

$$\left[(65 \text{ m}^2 \text{ collector}) \times \frac{0.014 \text{ m}^3 \text{ salt}}{\text{m}^2 \text{ collector}}\right] \times 1.14 = 1.04 \text{ m}^3 \text{ salt}.$$

One and four one-hundredths cubic meters of Glauber's salt mixture must be provided. If we assume that 50% of the storage area is air space, the needed storage volume is (1.04 m³) (1/0.50) = 2.08 m³. If the storage temperature varied over a 30° C range,

$$458 \text{ MJ} \left[2.08 \text{ m}^3 \times \left(\frac{160 \text{ MJ}}{\text{m}^3} + \frac{2.01 \text{ MJ}}{\text{m}^3 \, °\text{F}} \times 30° \text{ C} \right) \right]$$

of heat could be stored.

4.4.6 Construction of Phase-Change Storage Units

Storage containers for phase-change materials must be durable and economical. Often the containers cost much more than the chemicals they hold. For salt hydrates, containers must be made of a material that allows very little water vapor loss.

Typical containers are *tubes*, *trays*, and *packets*. They are usually made of metal or plastic. Water vapor loss through plastic can be significant, so careful attention must be paid to container thickness and type of plastic. Usually walls 0.8–1.3 mm (30–50 mil) thick are required. Plastics with low water vapor permeabilities include polypropylene, polyethylene, and polycarbonate. Metal containers must be protected from internal corrosion and seal failure. Screw-cap seals are not adequate for preventing water vapor loss.

Generally the containers must be rather flat, with a large surface area compared to volume to make up for the poor heat-transfer abilities of the phase-change compounds. The storage room can be similar to a rock-storage bin (section 4.2.3), though supports that prevent bulging walls are usually not necessary. The phase-change containers may be stackable, with passageways between them to allow for good airflow. If the containers are not stackable, a supporting rack must be constructed for them inside the storage unit. Airflow through the unit is usually horizontal.

Figure 4.22 shows a phase-change storage unit in place in the Rainier Finance Building in Yakima, Washington. A portion of the insulated storage compartment lid has been removed to expose the self-stacking energy cells. Under normal operating conditions one 0.0027 m³ (17 in³)

PHASE-CHANGE STORAGE

Fig. 4.22 ENER-G-TRAY phase-change energy cells, manufactured by Saskatchewan Minerals. (Photograph courtesy of Solar Research Associates, Yakima, WA.)

"Concentrated Heat" tray will store 525 kJ (500 Btu) of energy, primarily in the phase-change process. The cells contain sodium sulfate decahydrate plus special substances to prevent supercooling and stratification.

4.4.7 Cooling with Phase-Change Storage

Coolness can also be stored in phase-change storage media. Substances that melt at low temperatures (7–20° C or 45–68° F) can be used to store coolness for air conditioning. At night a heat pump can be used to further cool outside air to the freezing point of the phase-change material, which then freezes. During the following day the material slowly melts as warm building air is blown through it and cooled. The temperature of the phase-change material remains constant (at its low melting point) until all of it has melted. With nighttime operation of the heat pump, building owners can take advantage of low, off-peak electricity rates.

A eutectic mixture has been developed by Maria Telkes [10] to store coolness. A eutectic mixture is made up of two or more substances that melt together at a constant temperature that is below the normal melting point of any of the substances. A eutectic mixture behaves like a pure substance (i.e., it changes phase at a constant temperature) and so such mixtures can be used for phase-change solar heat or coolness storage.

The eutectic mixture used for cooling contains sodium chloride (table salt), which melts at about 800° C (1500° F); ammonium chloride, which sublimes, or evaporates without melting, at about 340° C (650° F), and Glauber's salt, which melts at 32° C (90° F). (Seed crystals and thickeners must also be added in small amounts for reasons discussed earlier.) The mixture melts, as if it were a completely different substance, at a constant 13° C (55° F).

4.5 OTHER TYPES OF ENERGY STORAGE

The storage subsystems presented thus far have been limited to practical and commonly used systems. Research into new storage methods is constant, and new and different methods are being developed.

Salt-water ponds (*salt ponds*), which can effectively collect solar energy as well as store solar heat, are being researched. Large, centralized salt ponds hold promise for providing inexpensive solar heat to groups of houses.

Annual storage, or the practice of collecting solar heat during the summer for use during the winter, is economical now for some locations and applications. Eventually a method of storing solar energy indefinitely may be developed. *Chemical storage,* or energy stored in chemical reactions may be long term and even transportable.

4.5.1 Salt Ponds

Solar salt lakes, after which salt ponds are modeled, occur in nature and were first reported in the early 1900s. In normal fresh-water lakes the water temperature is highest near the surface and decreases with depth. In solar salt lakes and and ponds, however, water temperature increases with depth. Readers who have guessed that this reversal has something to do with the salt are correct. A salt pond has a high concentration of salt near the bottom that gradually decreases to plain water at the surface. High concentrations of salt mean high density, and high density leads to the unusual performance of salt ponds.

In normal lakes heat is dispersed at the surface by convective currents that form when heated water expands and its density decreases. In salt ponds increased density and salt concentration with depth prevent the formation of convective currents. Since water is a poor conductor of heat unless it can circulate, this characteristic effectively insulates the bottom of the pond from the surface, where heat is lost. The pond's temperature therefore increases with depth.

If the layer of the pond with the highest salt concentration is expanded it can be used to store solar heat that it and the dark bottom of the pond have collected. This lower section, which has constant high salt concentration and density, can form convective currents that distribute the heat evenly. It is called the *convective layer.* The upper section, where salt concentration, density, and temperature are increasing, is called the *insulating layer.* The solar energy absorbed by the insulating layer maintains the temperature gradient that prevents heat loss from the convective layer (Fig. 4.23).

Eventually the temperature in a salt pond will stabilize and not change with time or with day-to-day weather variations. Stabilization can take a year or more.

In artificial salt ponds sodium chloride (table salt, NaCl) or magnesium chloride (MgCl) are often used. Temperatures above the boiling point of water have been achieved in a NaCl pond [11].

Salt ponds will someday provide inexpensive solar heat for home heating. They are not as subject to daily weather variations as regular solar collectors and can provide solar heat even during extended periods of bad weather. Large, centralized solar ponds are advantageous because they have low heat losses through their sides, losses that can adversely affect the performance of smaller ponds. A large pond frees solar heated buildings from orientation and construction restraints imposed by solar collectors. There are several problems connected with their use, however.

Evaporation of water at the surface, which leads to higher salt concentrations there, and windstorms, which mix the water, can destroy the insulating layer and therefore the effectiveness of the pond. Concentration layering must therefore be maintained artificially.

Dust accumulation can also be a problem. Dust particles and leaves can remain suspended in the pond because of their small size or because their density corresponds to that of a layer of the pond. This reduces the effectiveness of the pond.

Finally, the removal of heat can be difficult. If hot brine is pumped out of the convective layer and cool brine returned to it, excessive mixing might occur that destroys layering. The operation of experimental salt ponds may soon provide solutions to these problems.

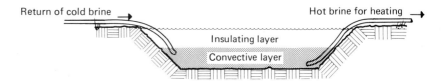

Fig. 4.23 Solar salt pond.

4.5.2 Annual Heat Storage

Annual heat storage, or the storage of heat during the summer for use during the winter, can be economical today. In the northern latitudes, where summer sunlight is significant but winter sunlight is nominal, annual heat storage can make sense. Annual storage may completely free buildings from restraints regarding solar orientation and construction. Also, annual storage systems are not seriously affected by extended periods of bad weather during the winter, as are regular solar heating systems. Annual storage is expensive, though, and only large systems may be economical.

In Studsvik, Sweden, plans exist for the construction of a large insulated water pit capable of supplying heat for 100 or more houses [12]. Collectors will be mounted on the floating, insulated lid of the pit. The lid can rotate completely, following exactly the motion of the summer sun.

In Alymer, Ontario, Canada, a 30-unit apartment house with 210 m² (2300 ft²) of collectors has water stored in a buried tank 6.7 m deep and 15 m in diameter (22 ft by 50 ft) [13]. Nearly 100% solar heating is expected.

Other methods of providing annual storage have also been proposed. At Oak Ridge National Laboratory research into annual-cycle energy storage (ACES) has shown it to be feasible [14]. In winter a heat pump is used to provide space heat while freezing a large tank of water. In summer the melting ice is used to provide air conditioning. This system does require the constant winter operation of a heat pump, but because the temperature of the heat source (freezing ice) is constant, the heat pump can be made very efficient. With the regular use of a heat pump for home heating, the outdoors is the heat source and a variety of outdoor temperatures must be used. If it gets too cold outside for effective use of the heat pump, expensive electric resistance heat is the only alternative available. A heat pump in an ACES system can be four times cheaper to operate than a regular heat pump.

4.5.3 Chemical Storage

Chemical storage, or the storage of energy by chemical reaction, is perhaps the best potential method of heat storage. It is the only true long-term storage method because the energy is stored as long as the chemical components remain separated. All other heat-storage methods depend on temperature, which means they are subject to heat losses.

Chemical storage can be very compact, requiring perhaps $\frac{1}{20}$ of the space needed for heat storage in water. No insulation is necessary for chemical storage and it can even be transported—energy collected at one location can be used at another. There are some problems connected with this type of storage, however.

Most chemical reactions used for chemical storage require very high temperatures, which would have to be provided by high-efficiency concentrating solar collectors. Also, regulation of the pressure within the reaction vessel is sometimes very important. Some of the chemicals used in these reactions are very costly and some are toxic. Until these and other problems are overcome, chemical storage of heat will remain largely experimental.

Some of the energy-storing reactions under study include (1) the distillation of water from sulfuric acid (H_2SO_4) and the recombination of the concentrated acid and water [15]; (2) high-temperature dehydration of magnesium hydroxide ($Mg[OH]_2$) to magnesium oxide (MgO) and calcium hydroxide ($Ca[OH]_2$) to calcium oxide, lime (CaO), and their rehydration [16, 17]; and (3) the decomposition of metal hydrides (MH) to the metal-plus-hydrogen gas on heating and their recombination [18, 19].

REFERENCES

1. Beckman, W. A., S. A. Klein, and J. A. Duffie (1977). *Solar Heating Design by the F-Chart Method*. New York: John Wiley.

2. Solar Energy Applications Laboratory, Colorado State University (1977). *Solar Heating and Cooling of Residential Buildings: Sizing, Installation, and Operation of Systems*. Washington, D.C.: U.S. Government Printing Office, stock no. 003-011-00085-2.

3. Solar Energy Applications Laboratory, Colorado State University (1977). *Solar Heating and Cooling of Residential Buildings: Design of Systems*. Washington, D.C.: U.S. Government Printing Office, stock no. 003-011-00084-4.

4. Kohler, J. (1978). Rules of thumb for sizing storage, *Solar Age* 3(4):23,

5. Mehalick, E. M., and A. T. Tweedie (1975). Two-component thermal energy storage material. In *Proceedings of the Workshop on Solar Energy Storage Subsystems for the Heating and Cooling of Buildings*, New York: ASHRAE, NSF-RA-N-75-041.

6. Edie, D. D., and S. S. Melsheimer (1976). An immiscible fluid heat of fusion energy storage system, *Proceedings of Sharing the Sun!* 8:262.

7. Anonymous (1978.) Efficient storage of solar heat. *Business Week* January 16.

8. Morrison, D. J., and S. I. Abdel-Khalik (1976). Performance of a solar heating system utilizing phase-change energy storage. *Proceedings of Sharing the Sun!* 4:244.

9. Jurinak, J. J., and S. I. Abdel-Khalik (1979). Sizing phase-change energy storage units for air-based solar heating systems. *Solar Energy* 22(4):355.

10. Telkes, M. (1974). Solar energy storage, *ASHRAE Journal* September, p. 38.

11. Business Publishers, Inc. (1980). *Solar Energy Intelligence Report* July 14, p. 274.

12. Margen, P. (1978). Central plants for annual heat storage. *Solar Age* 3(10):22.

13. Hooper, F. C. (1978). Annual storage. *Solar Age* 3(4): 16.

14. Fischer, H. C. (1975). Annual cycle energy system (ACES) for residential and commercial buildings. In *Solar Energy Storage Subsystems for the Heating and Cooling of Buildings.* New York: ASHRAE, p. 129.
15. Huxtable, D. D., and D. R. Poole (1976). Thermal energy storage by the sulfuric acid-water system. *Proceedings of Sharing the Sun!* 8:178.
16. Bauerle, G. *et al.* (1976). Storage of solar energy by inorganic oxide/hydroxides. *Proceedings of Sharing the Sun!* 8:192.
17. Ervin, G. (1975). Solar heat storage based on inorganic chemical reactions. In *Solar Energy Storage Subsystems for Heating and Cooling of Buildings.* New York: ASHRAE, p. 91.
18. Gruen, D. M., and I. Sheft (1975). Metal hydride systems for solar energy storage and conversion. In *Solar Energy Storage Subsystems for Heating and Cooling of Buildings*, New York: ASHRAE. p. 96.
19. Anonymous (1977). Solar heating: doing it the chemical way. *Chemical Week* October 5, p. 44.

BIBLIOGRAPHY

Allcut, E. A., and F. C. Hooper (1964). Solar energy in Canada. *Proceedings of the U.N. Conference on New Sources of Energy*, vol. 4, New York: The United Nations, p. 304.

Chahroudi, D. (1975). Suspension media for heat storage materials. In *Solar Energy Storage Subsystems for Heating and Cooling of Buildings.* New York: ASHRAE, p. 56.

Chahroudi, D., (1976). Thermocrete heat storage materials: applications and performance specifications. *Proceedings of Sharing the Sun!* 8:245.

Close, D. J. (1965). Rock pile thermal storage for comfort air conditioning. *Mechanical and Engineering Transactions of the Institution of Engineers, Australia* MCI(1):11.

Daniels, F. (1974). *Direct Use of the Sun's Energy.* New York: Ballantine.

Goldstein, M. (1964). Some physical chemical aspects of heat storage. *Proceedings of the U.N. Conference on New Sources of Energy*, vol. 5. New York: The United Nations, p. 411.

Grodzka, P. G. (1975). Some practical aspects of thermal energy storage. In *Solar Energy Storage Subsystems for Heating and Cooling of Buildings.* New York: ASHRAE, p. 68.

Leshuk, J. P. *et al.* (1978). Solar pond stability experiments. *Solar Energy* 21(3):237.

Löf, G. O. G. (1977). Systems for space heating with solar energy. In *Applications of Solar Energy for Heating and Cooling of Buildings.* New York: ASHRAE.

Löf, G. O. G., and R. W. Hawley (1948). Unsteady-state heat transfer between air and loose solids. *Industrial and Engineering Chemistry* 40(6):1061.

Lorsch, H. G. (1975). Thermal energy storage for solar heating. *ASHRAE Journal* November, p. 47.

MACNA (1978). *Fundamentals of Solar Heating.* Washington, D.C.: U.S. Department of Energy, EG-77-C-01-4038.

McCormick, P. O. (1975). Analytical modeling group report. In *Solar Energy Storage Subsystems for Heating and Cooling of Buildings.* New York: ASHRAE, p. 174.

Nathan, L. (1976). Pebble bed heat storage. *Solar Age* 1(12):30.

Offenhartz, P. O. (1976). Chemical methods of storing thermal energy. *Proceedings of Sharing the Sun!* 8:48.

Pickering, E. E. (1975). Residential hot water solar energy storage. In *Solar Energy Storage Subsystems for Heating and Cooling of Buildings.* New York: ASHRAE, p. 25.

Rueffel, P. G. (1976). Saskatchewan's sodium sulphate industry. *Canadian Geographical Journal* 93(2):44.

Sienko, M. J., and R. A. Plane (1966). *Chemistry: Principles and Properties.* New York: McGraw-Hill.

Stanford Research Institute (1976). *Residential Hot Water Solar Energy Storage Subsystems.* Washington, D.C.: U.S. Department of Commerce, PB-252-685.

Telkes, M. (1975). Thermal storage for solar heating and cooling. In *Solar Energy Storage Subsystems for the Heating and Cooling of Buildings.* New York: ASHRAE, p. 17.

Telkes, M. (1977). Solar energy storage. In *Applications of Solar Energy for Heating and Cooling of Buildings.* New York: ASHRAE, p. 17.

PROBLEMS

4.1 Compare heat storage in adobe with heat storage in rock pebbles. Why would heat storage in adobe be impractical for an active solar heating system? For which type of solar system would adobe be suitable?

4.2 Compare the volume of material needed to store 100 MJ of heat energy in aluminum, polyurethane insulation, wood, and water over 30° C temperature range.

4.3 How much space must be allowed for a rock storage unit for a 37.5 m² collector? Assume that the horizontal cross section of the storage bed is square and that the vertical dimension of the pebble bed is 1.5 m. Plenums and construction materials add 20 cm to each side of the bed and 80 cm to its height.

4.4 If the rock bed in problem 4.3 were built, what would be the load in kg/m² on the floor beneath the bed? Assume that the construction materials add 10 kg/m². Approximately how many lb/ft² is this?

4.5 What size should the water storage tank for a 400 ft² collector array be? Tanks with capacities of 500, 750, and 1000 gal are available. Which size tank would you use for this system and why?

4.6 Compare the absolute energy-storage potential of the recommended 1.75 gal/ft² for water with the recommended 0.8 ft³/ft² for rock pebbles over a 60° F temperature range. Suggest at least two reasons for this difference.

4.7 Draw a phase-change diagram for a substance that supercools to form a glass.

4.8 Why is supercooling a problem for salt-hydrate storage systems even though the substance always eventually crystallizes?

4.9 Why do solar researchers go to all the fuss and bother of preparing mixtures to avoid stratification of salt hydrates? Compare the storage of heat in stratified $Na_2SO_4 \cdot 10H_2O$ with that in water and rocks, as in Fig. 4.21, for temperature ranges of 5° F, 30° F and 60° F upward from the melting point of the salt. (Use 2900 Btu/ft³ as the volume heat of fusion for stratified salt in containers and 30 Btu/ft³ °F as the heat capacity of the molten salt mixture.)

4.10 Compare the efficiency of a solar salt pond with that of a flat-plate collector. Try to make qualitative assessments of the relative $F_R(\tau\alpha)_n$ and $F_R U_L$ values for the two solar collectors.

5 DESIGN OF ACTIVE SYSTEMS

A typical active solar heating system involves more than just an array of solar collectors and a heat-storage unit. Pipes or air ducts are needed to carry the heat-transfer fluid (the water or air). Pumps or fans are required to circulate the fluid. Some provision is needed for backup or auxiliary heat when sunlight alone is not enough to keep the building warm. Automatic valves or dampers are needed to switch the path of fluid flow. Heat exchangers, filters, vents, and other miscellaneous parts are required. To coordinate all of this equipment and make the decisions as to which device to turn on and when, a control circuit is needed, along with thermostats and other sensing devices. All of this sounds rather complex—and it is. Yet active solar systems are no more complex than many other devices and systems that are useful or necessary parts of our lives. This chapter introduces some of the basic concepts used in solar-system design and some techniques that can help to make the design process manageable. Of course, persons designing a system for the first time should seek the advice of qualified professionals at every phase of the design process.

5.1 BASIC CONCEPTS

5.1.1 Conventional Heating Systems

Active solar heating is an extension of conventional heating principles. Heat is produced in one place and moved to where it is needed by a pump-propelled heat-transfer fluid, such as air or water, through pipes or ducts. There are four major types of conventional heating systems:

1. Warm-air systems have a furnace as heat source. The furnace has a fan that pumps warm air through air ducts to the rooms of the building.
2. Hot-water heating systems have a boiler to heat water and pipes to transfer the heated water throughout the building. In each room the water releases heat through radiators or radiant-heating panels. In some systems the water circulates by natural convection, but most hot-water systems are *hydronic*, or forced-hot-water, systems, in which the water is artificially circulated.
3. Steam-heating systems are similar to water systems, except that steam is circulated through the piping system. In the past steam heating was the most common system, but it has since been replaced by other systems in residential and small commercial buildings.
4. Electric room heating makes use of individual heating units in each room. These may be radiant-heating panels, baseboard-type heaters (natural convection), or forced-air heaters.

The first three types of heating systems have a wide variety of heat sources. Gas burners, oil burners, coal burners, electric resistance elements, heat pumps, and even wood burners are all widely used as heat sources. Solar collectors can be added to this list.

5.1.2 Air-Type Solar Heating Systems

An air-type solar heating system is a warm-air system that has air-type solar collector panels as its heat source. Air is heated as it circulates through an array of solar collectors.

Solar heating systems are usually *multiple-source* systems. That is, more than one source of heat is used. Generally, three sources are used: a solar collector array, a heat-storage unit, and a backup heating unit of some kind. Because more than one heat source is used, active solar systems can be very efficient and can effectively maintain a building at a comfortable temperature. Active solar systems are also complex since the heat-transfer fluid must flow to different places at different times. Each type of system function is called a *mode.* The key to understanding solar-system design is to develop a good grasp of what modes are and how to select them.

Essentially, a mode is the specific *setting* of the system to perform some specific task. One of the most basic modes required in an air-type system is the *collectors-to-rooms* mode, or the *direct solar heating* mode (Fig. 5.1). In this mode, solar heated air is drawn from the collector array by a fan and blown into the building. To complete the circle, cool building air is returned to the collectors for heating. All modes involve moving the heat transfer-fluid (the air or water) in some kind of continuous circle to take heat from one place and put it somewhere else. For example, in the mode shown in Fig. 5.1 heat is taken from the solar collectors and moved to the rooms.

When the sun is out and the building becomes too warm to require more heat, we may wish to continue collecting solar heat anyway and store it for later use. To do this, we use the *collectors-to-storage* mode (Fig. 5.2).

Even a simple air system ordinarily needs at least two additional modes. A *storage-to-rooms* mode is needed at night or during periods when there is not enough sunlight. A *backup-to-rooms* mode is needed when the sun disappears for a long time and the heat-storage chamber cools down (the backup heat source must be used) (Fig. 5.3).

The heat-storage unit will usually perform best if we put heat into and take heat out of the same end (compare Figs. 5.2 and 5.3[a]). This is true for any rock-type or phase-change-type storage unit that has temperature stratification (see Chapter 4). One end of the storage chamber is always warmer than the other and is called the *hot end* of storage. In a rock-pebble bed the top of the chamber is usually the hot end.

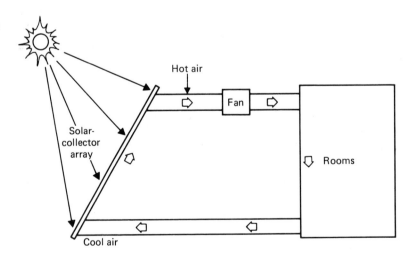

Fig. 5.1 Collectors-to-rooms mode for air-type system.

BASIC CONCEPTS 153

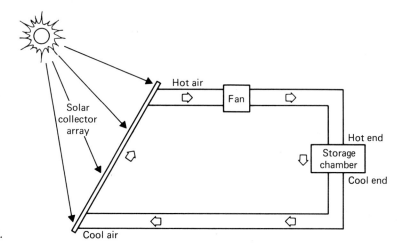

Fig. 5.2 Collectors-to-storage mode for air-type system.

Most systems have more than the four modes already mentioned; for example, a mode to cool the collectors in summer or a mode to air condition the building.

If a separate loop of ductwork and a separate fan had to be included for each mode of the system, much of the building would be filled

Fig. 5.3 Two air-type-system modes: (**a**) storage-to-rooms mode, (**b**) backup-to-rooms mode.

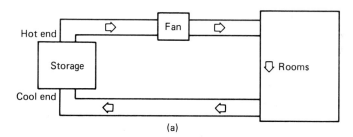

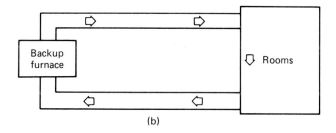

hardware and costs would soon get out of hand. System design involves determining all of the modes that will be required and then coming up with the simplest, cheapest equipment layout that will provide for all of these modes. Wherever possible a single air duct is used in two or more modes. Automatic, motorized dampers are used to switch airflow among different ducts, perhaps eliminating the need for an extra fan. Figure 5.4 shows a basic system that has each of the modes shown in the first three figures.

The system shown in Fig. 5.4 may appear to be rather complex, but it is actually one of the simpler designs. The right half of the top draw-

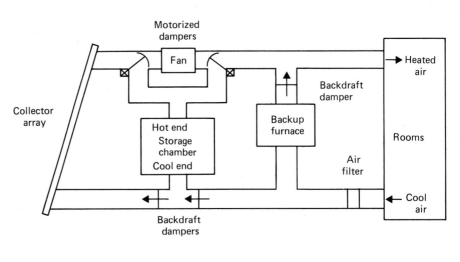

Fig. 5.4 A basic air-type system.

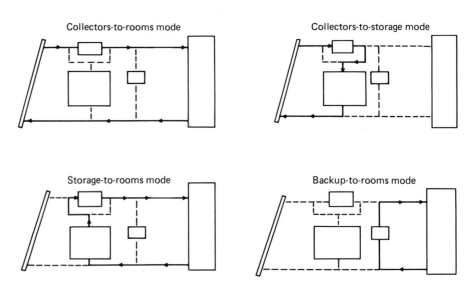

ing represents a conventional warm-air system. The solar system simply joins to this standard system. Two damper motors are required (see section 5.2.2) and one solar fan is needed, but these are about the only active components required. Additional examples of air systems are given in section 5.3.2.

5.1.3 Liquid-Type Solar Heating Systems

A liquid-type solar heating system is an extension of the hot-water heating system concept. Liquid-type solar collector panels are used as a heat source. Water in the collectors would be subject to freezing on a cold night, so other heat-transfer fluids, such as a water/antifreeze solution, are often used in the collector loop of the system.

It might be assumed that a liquid system would operate in a collectors-to-rooms mode similar to the one in air-type systems (Fig. 5.5). Liquid systems are almost never set up like this, however. Though the solar-heated liquid may be used in large, radiant-heat panel heaters, it will generally not be hot enough to pump directly into radiators or baseboard convective heaters in the rooms. Instead, the heat is often transferred to air by a *heat exchanger* and the air is blown through the building with a fan (Fig. 5.6). In liquid-type systems heat from the collectors is seldom used directly to heat the rooms (though it can be). Solar heat is usually stored immediately in a heat-storage tank, and storage-tank water is used to heat the building (Fig. 5.7). One reason for this is simply that it eliminates a few automatic valves (section 5.2.2). What it means in practice is that the storage tank must be heated before the building can receive any solar heat. There is a slight performance penalty for this, but not a serious one.

Another complicating factor for liquid-type systems in freezing climates is that the antifreeze solution used as a heat-transfer fluid is expensive and cannot be used to fill the entire system, storage tank and all. Because of this, a *closed-loop* design is often used for the collection loop. Figure 5.8 shows a basic liquid-type system that includes this closed-loop

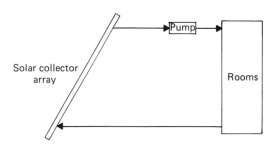

Fig. 5.5 Seldom used collectors-to-rooms mode for liquid-type system.

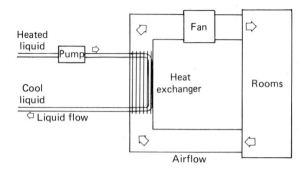

Fig. 5.6 Use of a heat exchanger for heat transfer from a heated liquid to cool air.

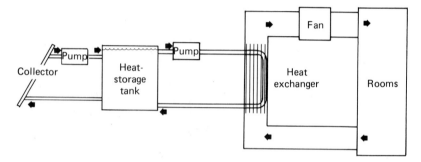

Fig. 5.7 Collectors-to-rooms mode for liquid-type system.

feature. A heat exchanger is used to separate the collector loop from the rest of the system. This liquid system has three pumps, but they can be small ones and they are the only active components.

Another way to solve the problem of freezing in liquid-type collectors is to design the system so that all of the water drains out of the collectors as they approach the freezing point. The water may simply drain back into the heat storage tank. A system such as this is known as a *draindown*-type water system. A draindown system and other liquid systems are pictured in section 5.3.2.

5.2 ELEMENTS OF SOLAR HEATING AND COOLING SYSTEMS

To understand and design systems, we must become familiar with their parts—all of the large and small pieces of equipment that are available. This is a formidable task because there are so many items to consider. It is not possible to discuss each of the system elements in any adequate detail, so in the sections that follow we merely introduce many of the more

ELEMENTS OF SOLAR SYSTEMS 157

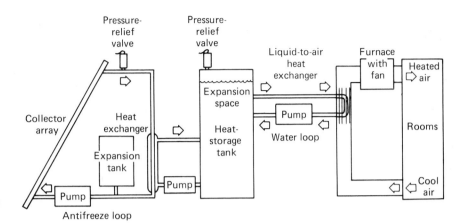

Fig. 5.8 A basic closed-loop liquid system.

important elements, briefly describe their function, and give pictorial symbols that can be used to represent them in schematic drawings such as those in this chapter. The symbols are not necessarily accurate representations of the individual items; they were chosen to make the system sketches clear and uncluttered and give some idea of the appearance or function of the items. Various standards exist for the symbols used in heating-system schematics. These symbols are not used in this text because they vary depending on the source and because they are occasionally more confusing than enlightening. Readers who are heavily engaged in system engineering would do well to learn the basic symbols used in current practice. These, as well as more specific information about equipment, can be found in the books listed at the end of this chapter. Manufacturers' catalogs and product literature are often extremely valuable sources of information, particularly about products that represent new concepts.

5.2.1 Basic System Elements

The following components are the basic items or major pieces of equipment found within solar heating systems.

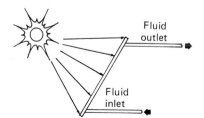

Solar Collector Panel

A *solar collector panel* is a heat-gathering device (normally a flat-plate collector) that is mounted on a south-facing wall or roof of a building. It traps sunlight and transfers its energy to moving air or liquid, which enters at one end and exits (heated) from the other end.

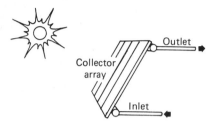

Collector Array

A *collector array* is a group of several solar collector panels. The ducting or piping of all panels is usually connected together, so from a systems standpoint a collector array may be thought of as one large solar collector with one fluid inlet and one fluid outlet.

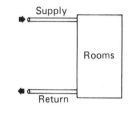

Conditioned Space

A *conditioned space* is the building area that is to be heated or cooled. Usually the conditioned space consists of several rooms, each with its own ducts or heating pipes. Since all of these *branch lines* feed into *trunk lines*, essentially the entire conditioned space has one *supply line* (where fluid enters), and one *return line* (that returns cool fluid to the heating system).

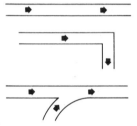

Air Ductwork

Air ductwork is the tubing of sheet metal that carries air throughout a forced-air-type system. Air ducts are usually rectangular or round and may measure from several inches to several feet across, depending on how much air must flow. Ducts are made of galvanized steel (or sometimes aluminum) sheet metal and are available with many different types of joints and seams, some able to withstand more air pressure than others. Most building heating and cooling systems operate at fairly low pressures, but even so air leakage must be minimized. Any duct system must be carefully sized so that the flow is adequate and balanced throughout the system [1, 2, 3, 4].

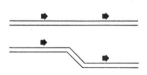

Liquid Pipes

Liquid pipes are the tubes that carry liquid throughout a liquid-type system. Copper, polyvinyl chloride (PVC), and other pipes have been used in the past, but copper pipes with soldered connections best withstand high temperatures and corrosion. Sizing information is available elsewhere [2, 3, 4].

Heat-Storage Chamber

The *heat-storage chamber* is a small, insulated room containing rocks, phase-change salts, cans of water, or other storage materials. This chamber stores excess

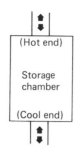

solar heat for use at night or during cloudy periods. Charging the storage with heat is accomplished by blowing hot air into the hot end. The air gives up its heat as it passes through the chamber and then exits from the *cool end.* Later, when stored heat is needed, the airflow is generally reversed: building air is drawn into the cool end and heated air is returned to the building from the hot end. In most cases the top of the chamber is the hot end and the bottom is the cool end.

Heat-Storage Tank

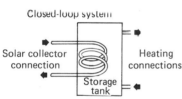

The *heat-storage tank* is a tank containing water. It stores solar heat in hydronic (liquid-type) systems. In an open-loop system water from the tank heats as it is circulated directly through the collectors. In a closed-loop system the water/antifreeze mixture is circulated to the collectors; this mixture transfers its heat to the tank via a heat exchanger, which is commonly just a coil of pipe inside the tank. Thus in a closed-loop system tank water never flows to the collectors, where it would be susceptible to freezing at night. In both types of systems stored heat is generally removed when tank water is circulated to the building.

Furnace

The *furnace* is used to supply auxiliary, or backup heat in air-type and some liquid-type systems when the sun is not shining and the temperature of the storage medium is not high enough to keep the building warm. Any type of conventional fuel may be used, including electricity, gas, and oil. Backup heat may also be supplied without a furnace in some cases; for example, from a wood stove or electric baseboard heaters. Occasionally in air-type systems the furnace fan may be used alone (with heat off) to eliminate another fan from the system. Most furnaces have a control box that allows fan-control capability with little or no rewiring.

Boiler

A *boiler* is a heater used in conventional (nonsolar) hot-water heating systems. It is used in some liquid-type solar heating systems to supply backup heat when the temperature of the storage medium becomes too low.

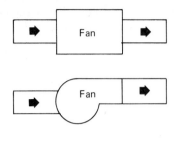

Fan or Blower

A *fan* or a *blower* is used to force the air through an air-type system. The kind of fan best suited to heating applications is the *centrifugal* fan. All fans must be carefully sized to match the ductwork and other equipment in the system [4]. To minimize noise, fans should be mounted on resilient bases (e.g., rubber) and connected to the duct system with canvas or other flexible material.

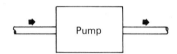

Pump

A *pump* is used to circulate the water or other heat-transfer fluid throughout a liquid-type system. For heating-system applications, centrifugal pumps are best. Pumps should always be sized to meet the heat and flow requirements of a system [2, 4]. For example, a drain-down-type system will require a pump with a large *head* (water-lifting capacity) since it will occasionally have to refill the entire collector loop—perhaps from the basement to the roof. All metal pump surfaces that are in contact with water should be of stainless steel or brass, to resist corrosion.

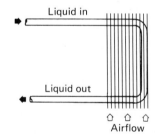

Liquid-to-Air Heat Exchanger

A *liquid-to-air heat exchanger* transfers heat energy from flowing liquid to flowing air, or vice versa. It consists of liquid-filled tubes connected to many closely spaced, flat metal plates, or *fins*. Note that the fins are only on the *air side* of the heat exchanger (the inside of the tubes is the *liquid side* of the heat exchanger). This is because air is not a very effective heat-transfer fluid, so more surface area is required. (Air has a lower heat capacity per unit volume than liquids, and does not conduct heat as well.) The exchanger is typically constructed of aluminum fins press-fitted to copper tubing. A car radiator is a liquid-to-air heat exchanger.

Liquid-to-Liquid Heat Exchanger

A *liquid-to-liquid heat exchanger* is used to transfer heat from one liquid to another without mixing the liquids. For example, it might be used to heat domestic hot

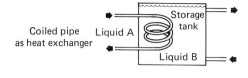

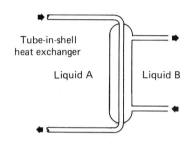

water without the water actually passing through the solar collectors. A metal surface separates the hotter liquid from the cooler one; heat naturally flows from the hotter to the cooler. In a solar system the separating surface is often a coil of pipe. One liquid (liquid A) flows inside the pipe and the other (liquid B) flows around the outside of it. Heat exchangers may be purchased in many varieties. A common one consists of many metal tubes in a shell (symbolized in the drawing by a single tube, for clarity). Fins are not generally needed on liquid-to-liquid heat exchangers since the two fluids are likely to have similar heat-transfer characteristics. *Tube-in-shell* heat exchangers should be sized according to the manufacturer's recommendations, and the *approach temperature* should be 5°C(10° F) (meaning that the cooler incoming fluid [liquid B] should heat to within 5° C of the hotter incoming fluid [liquid A] before it exits from the heat exchanger). Homemade coiled-pipe heat exchangers that are placed in tanks are more difficult to size since no simple guidelines are available. The pipe should be long enough to bring the liquid passing through to within a few degrees of the tank temperature.

Hot-Water Preheat System (Air Type)

The *hot-water preheat system* is used to heat domestic hot water in an air-type solar heating system. Cold supply water is passed through a *preheat tank* on its way to the regular hot-water tank. Hot air from the collector array passes over a liquid-to-air heat exchanger located in an air duct. Water from the preheat tank is circulated through this heat exchanger with a small pump, becoming warm. It would not be as effective to leave the preheat tank out and just pass cold supply water through the heat exchanger since, when hot water was not being used, the supply water would just sit inside the heat exchanger, get hot, and stop collecting heat. The preheat tank is generally sized to provide 1.5–2 times the estimated daily hot-water consumption. If the hot-water consumption is 75 l(20 gal) per person per day (which is typical consumption), then the tank capacity should be 110–150 l(30–40 gal) per person.

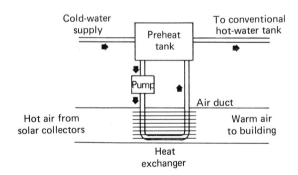

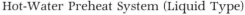
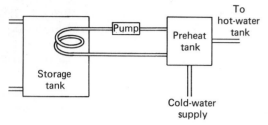

Hot-Water Preheat System (Liquid Type)

The hot-water preheat system for a liquid-type system is essentially the same as for an air-type system, except that the heat source used is the system's heat storage tank and the heat exchanger is a liquid-to-liquid type; for example, a coil submersed in the tank. Many building codes are strict regarding good separation of the water supply from a heating system. A double-walled heat exchanger may be required, for example. Also, hot water cannot usually be taken directly from the storage tank, as there are special requirements for potable-water tanks.

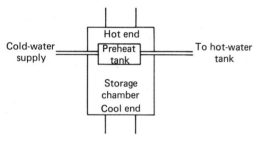

Embedded Preheat System

It is possible to preheat domestic hot water by embedding a preheat tank in a rock-pebble storage chamber or by submersing a preheat tank in a liquid heat-storage tank. The advantage of this is that no pump or heat exchanger is needed (the metal tank acts as its own heat exchanger). The disadvantages are that access and maintenance are difficult and, in the case of a water system, leaks would be difficult to detect.

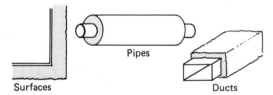

Insulation

Insulation is used throughout a solar heating or cooling system, essentially to keep the heat where it belongs. Particularly important are collector insulation, storage-unit insulation, and fluid-line insulation. Special insulation is available for pipes (foam rubber tubing) and ductwork (glass wool board) and should be liberally used, particularly on warmer lines such as the line leading in from the collector array. It is true that once well inside the building a heating pipe's loss is the building's gain; however, the purpose of a pipe or duct is to carry heat to where it is needed. For clarity, insulation is not shown in all the diagrams in this chapter. It would be present in a real system, however.

5.2.2 Fluid-Control Elements

The following components are used in solar systems to control the flow path and behavior of the heat-transfer fluid.

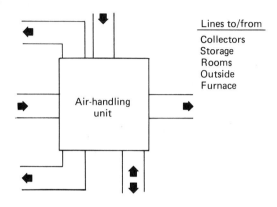

Air Handler, or Air-Control Unit

This device is sold by some solar-system manufacturers. The *air handler* consists of a fan, some motorized dampers, and often some system-control circuitry. Its purpose is to make an air-type system simpler to install by replacing all or most of the separate dampers and fans with one *black box* that does everything. Unfortunately, many air handlers on the market are not very practical due to (1) insufficient choice in fan sizes and speeds; (2) lack of flexibility (i.e., not enough modes possible); (3) poor efficiency (too many circuitous turns near the fan that reduce its performance); and (4) high cost.

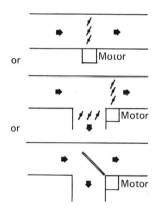

Motorized Damper

A *motorized damper* is used to change automatically the airflow path in an air-type system by shutting off airflow or switching it. For example, to switch from collectors-to-rooms mode to collectors-to-storage mode, some ducts must be closed and others must be opened. Motorized dampers are switched open and shut by the system-control circuit. The two basic types are single- and multiple-louvered dampers. In general and especially for large ducts, multiple-louvered dampers work better and can seal off the airflow more effectively when closed. For switching air from one path to another, two dampers can often be connected to the same motor. Dampers in ducts leading to or from the outside air should be weatherstripped and insulated. Motorized dampers may be purchased or made to order by heating and mechanical contractors. One or more is almost always needed in an active system, but the fewer the better.

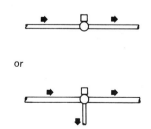

Solenoid Valve, or Motorized Valve

A *solenoid*, or motorized valve is used to control automatically the flow of liquid in a liquid-type solar system. The three common types of solenoid valves are

1. Normally open type. A current shuts off the liquid flow. When the current is shut off, the valve springs open.

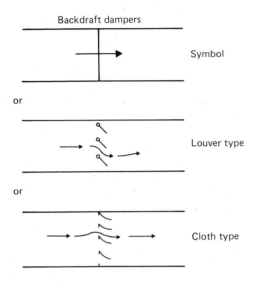

2. Normally closed type. A current opens the valve and lets liquid flow. When the current is shut off, the valve springs shut.
3. Two-way type. A current switches the liquid flow from one line to another. For durability, solenoid valves should be made of brass, stainless steel, or high-temperature plastic. Inexpensive PVC valves are available, but their durability is questionable for high-temperature lines.

Backdraft Damper, or One-Way Damper

A *backdraft damper* is a gravity- or spring-operated damper that allows air to flow in one direction only. It opens whenever the air pressure is higher "above" it than "below" it. This type of damper is inexpensive and relatively trouble free, and it does not rely on electricity for its operation. For these reasons, backdraft dampers rather than motorized dampers should be used whenever possible. (Motorized dampers give more control and are necessary in some cases.)

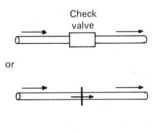

Check Valve, or One-Way Valve

A *check valve* allows a liquid to flow through a pipe in one direction only. Like backdraft dampers, check valves simplify a system because they are not operated electrically. There are several types of check valves, some of which depend partly on gravity for their operation; these should be installed the right way up.

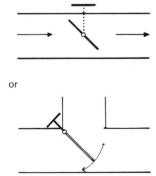

Manual Damper

A *manual damper* is operated by hand. Its two principal uses are (1) to balance the airflow through the various ducts in a system and (2) to make possible seasonal manual adjustments to the system (e.g., to open the collectors to outside air in summer to prevent overheating). Manual dampers are practical when only infrequent switching is needed. For more frequent switching, motorized dampers should be installed.

Manual Valve

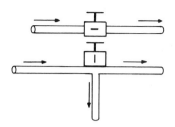

The *manual valve* is familiar and widely available. It is used in liquid systems to balance fluid flow throughout the system or to shut off or switch fluid-flow lines on a seasonal basis.

Turning Vanes

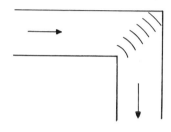

Turning vanes are used in air systems to minimize the turbulence and air resistance that occur in air ducts whenever they turn corners. The turning vane is simply a series of curved slats or louvers that direct airflow smoothly around a corner. A square corner with turning vanes is likely to exhibit less resistance than a rounded corner that does not have them. For solar energy to pay off, energy usage by fans, pumps, and so on should be minimized. Turning vanes and other devices that reduce resistance to airflow are strongly recommended—particularly where high-volume, high-velocity airflow is required.

Expansion Tank

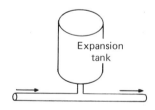

An *expansion tank* is necessary in many liquid-type systems as a place for expanding liquid to go as the system heats up. Without an expansion tank, pressure sufficient to cause extensive damage could easily build up in the system. Normally the tank is filled with air. When the liquid expands and partially fills the tank, it compresses the air. One problem is that over time air dissolves in the liquid and the tank becomes "waterlogged." To avoid this, some expansion tanks have a diaphragm of rubber or other material that separates the air from the liquid. Manufacturers of expansion tanks can provide guidelines for selecting a tank of proper size. Typical tanks hold 5–50 gal. Systems that contain antifreeze require larger tanks; for example, a system containing 50% antifreeze will require a tank 50–80% larger than a system filled with water. Expansion tanks are required in every loop of a liquid system, though one or two can sometimes be eliminated if the heat-storage tank is deliberately underfilled (the extra space serves the same purpose as an expansion tank).

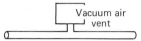

Vacuum Breaker, or Vacuum Air Vent

A *vacuum breaker* is used in a draindown-type freeze-protection system to let air into the system as the water drains out. Water cannot drain down out of the collector loop unless air is let into the top of the loop. A vacuum breaker, placed at the highest point in the system, will open and let air into the system as the falling water creates a vacuum in the pipes. It is important to select vacuum breakers carefully and to locate them indoors if possible since many have a tendency to freeze solid in the shut position in cold weather, thereby threatening the collectors with freezing.

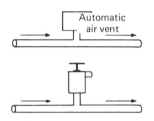

Air Vent, Manual or Automatic

An *air vent* is used in many liquid-type systems to let air out of the system as the system is being filled with liquid. Automatic air vents shut automatically in the presence of water (by means of a float valve or other device). Manual air vents are opened and closed by hand while the system is being purged of air. Air vents are generally installed at the highest point in the system.

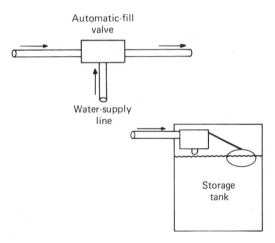

Automatic-Fill System

Automatic fill is used to keep a water-type system full automatically. It is not recommended for closed-loop water/antifreeze systems because if a leak occurred somewhere, the automatic-fill system would dilute the antifreeze, rendering it useless. The automatic-fill system may be prohibited by some building codes because of the remote possibility that its water may contaminate the water supply. The simplest automatic-fill installation is an automatic-fill valve that opens whenever system pressure falls below the valve setting. Or, in an open-loop system, the heat-storage tank may have a float-type valve that opens whenever the water level drops below a certain point (as in a toilet tank).

5.2.3 System-Control Elements

The following components are used in solar heating systems to control their electromechanical components, such as fans and dampers.

ELEMENTS OF SOLAR SYSTEMS

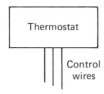

Thermostat

A *thermostat* is a temperature-controlled switch that can be used to turn a piece of electrical equipment on or off in response to temperature. For example, a thermostat can turn an air conditioner on when room temperature rises above 75° F, or turn a furnace on when room temperature falls below 65° F. Thermostats are generally not electronic; they are simple electromechanical devices that are readily available and fairly inexpensive. The thermostats commonly used for heating and cooling systems are designed to switch only low voltages; for example, a 24-V control signal. The higher-voltage fans, pumps, and so on are then switched on and off with relays.

Differential Thermostat

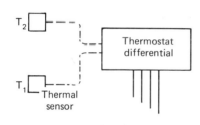

A *differential thermostat* is a switch that can turn electrical equipment on or off when a difference in temperature becomes larger than a set value. Two temperature sensors are put in different places. The thermostat reads the difference between the temperatures registered by the two sensors. When this difference becomes great enough (or small enough), the thermostat turns something on or off. A differential thermostat is typically used to compare the temperatures of the collectors and the heat-storage unit. No matter what the temperature of the storage unit—whether 50° F or 120° F—it is always desirable to put more solar heat in if the collectors are a few degrees hotter. (When the collectors are cooler than the storage medium, the fan or pump should never be run.) A differential thermostat can tell when one area is hotter or cooler than another. A conventional thermostat can only respond to the temperature of one area. Differential thermostats are electronic devices and are available from manufacturers of solar controls. They generally have thermoresistor-type sensors and are available in different power ratings (based on how much power they can safely switch).

Thermoresistor or Thermistor Sensor

A *thermoresistor* sensor is used with a differential thermostat as the temperature-sensing element. It is an

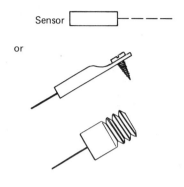

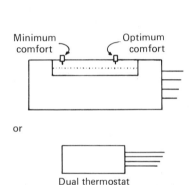

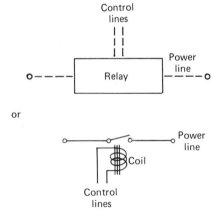

electrical resistor whose resistance varies as the temperature varies. The changes in resistance can be electronically detected. Other types of temperature sensors, *thermocouples* and *thermopiles,* are seldom used in heating systems. In general, temperature sensors are not interchangeable—only the sensors the control system was designed for should be used. Thermoresistors are available in many types of cases. Some can be screwed directly to absorber plates, some may be immersed in storage tanks, and some may be used outdoors.

Two-Stage or Dual-Temperature Thermostat

A *two-stage thermostat* is often used inside the conditioned space. One element is set at the optimum-comfort level and the other element is set at the minimum-comfort level (a little lower). When the rooms are warmer than the optimum-comfort level, all heat shuts off. When room temperature is between minimum- and optimum-comfort levels, only solar heat or storage heat is used. If these heat sources are not adequate and room temperature drops below the minimum-comfort setting, the backup heating system comes on to supplement solar heat. Two-stage thermostats are standard items, widely available. Generally, only the optimum-comfort temperature (stage 1) is adjustable. The minimum-comfort temperature (stage 2) automatically follows a few degrees below the setting for stage 1.

Relay

A *relay* is an automatic electrical switch. When it receives current through the control lines (coil) the switch closes, which can turn a large piece of equipment on or off. With a relay, a low-voltage, low-power control signal can turn on a high-voltage, high-power device such as a pump motor, fan, or electric heater. The large device can be turned on from a distance; only two small wires need lead from the device to the relay. Relays can also be used to build *logic circuits,* which are needed in a solar system to make the control decisions. Relays are not the only electrically powered switches—transistors, vacuum tubes, and silicon-controlled rectifiers can all be used as well. But relays are still widely used for switch-

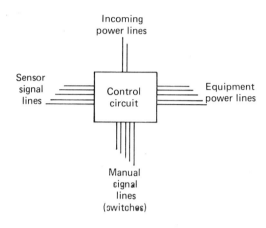

ing large devices on and off, even in circuits where the "decision making" is done by solid-state components such as microprocessors. For nonengineers, relays are relatively simple devices to understand and use.

Control Circuit

The *control circuit* "makes the decisions" about which fan to turn on when, which valve to open, and so forth; that is, the control circuit "decides" which mode the system should be in at any given time. The "decisions" the control circuit makes are based on (1) the temperature information it receives from the thermostats, limit switches, and so on and (2) commands the system operator gives by flipping switches. Decisions may also be based on other factors (e.g., sometimes the system is designed to pause for a short time before changing modes, to reduce excessive cycling of the system back and forth between modes). In a simple system the control circuit may be nothing more than a few relays. For solar heating and cooling systems, control circuits are available from manufacturers. Some can even be custom programmed and some include microprocessors. Proper system control is essential. The designer of a heating or cooling system must understand the control circuits inside and out. Faulty logic has resulted in systems that lose heat, instead of gaining it. It may also result in safety hazards.

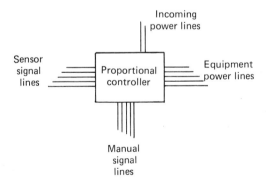

Proportional Controller

A *proportional controller* is a control system that can vary the speed of a pump or fan in order to increase the efficiency of solar collection on cloudy or hazy days. An ordinary control system can only turn motors on and off. If the pump or fan is on at full speed during low-intensity sunlight, it will quickly cool the collectors down and shut off. Greater heat collection is possible if the motor is operated at a speed proportional to the temperature difference between the collectors and the storage medium. The hotter the collectors are in relation to the storage medium, the more energy is available and the faster the fan or pump can be run. This is the reasoning behind proportional control. The actual gains that can be achieved are fairly small, however. One manufacturer of proportional controllers estimates that the

improvement will only be in the range of 6–10%, and even then only during periods of low-intensity sunlight. Pumps and fans must be selected with care since the speed control only works for certain types of motors.

Transformer

Most solar control systems require low voltages for thermostats, relays, and solid-state circuits (typically 5–30 V). A *step-down transformer* is used to reduce the 120-V-line voltage to the desired level.

5.2.4 Protective Components

The following components are used in solar heating systems for safety or to protect other components from damage or excessive wear.

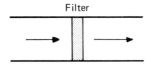

Air Filter

An *air filter* is needed in an air-type system to keep dust and suspended particles from collecting in the storage chamber and in the collectors, as well as to keep the building clean. Most furnaces used for backup heating contain air filters already, but because of the many possible air paths in a solar system, one or more additional filters are installed in other strategic locations. In most cases, high-quality furnace-type filters are adequate. Filters should not create a lot of resistance to airflow.

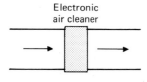

Electronic Air Cleaner

An *electronic air cleaner* may be used in place of an air filter. The air cleaner more effectively removes dust and particulate matter from the air and presents less air resistance than most filters. It works by charging the dust particles with electricity and then attracting them to electrodes, which must be cleaned periodically. Electronic air cleaners have a maximum operating temperature (typically 51° C, or 125° F) that should not be exceeded.

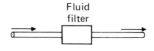

Fluid Filter

It is surprising to some that even in a sealed, liquid-type system a considerable amount of particulate matter develops in the fluid as time passes. This is due in part to erosion and corrosion of the metal surfaces. *Fluid filters* are sometimes included in liquid systems as a precautionary measure.

Limit Switch

A *limit switch* is a thermostat used for safety reasons. A high-temperature limit switch is used to detect overheating in the collector array. A low-temperature limit switch is used to detect freezing conditions in a drain-down-type liquid system.

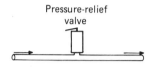

Pressure-Relief Valve

A *pressure-relief valve* is a valve used in liquid systems that opens whenever the pressure in the system is greater than some safe maximum. It guards against destructive pressures that could build up if the system overheated.

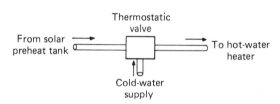

Thermostatic Mixing Valve, or Tempering Valve

The *thermostatic mixing valve* automatically mixes solar heated water with cold water so that the output does not exceed a certain temperature. It is used in systems where solar heated water could sometimes be dangerously hot for domestic use. The valve is normally installed between the hot-water preheat tank and the hot-water heater. (A similar effect can be had if a thermostat is installed in the preheat tank that turns off the water-preheat pump at some fixed temperature.)

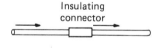

Insulating Connector, or Dielectric Union

The *insulating connector* is used to connect two pipes of electrochemically dissimilar metals, such as copper and iron, to prevent galvanic corrosion (corrosion caused by electrical contact between different metals).

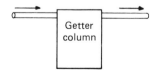

Ion-Getter Column

An *ion-getter column* is used to slow corrosion in some liquid-type systems. Galvanic corrosion can occur in systems having dissimilar metals, such as copper and aluminum, even if the metals are not in direct contact. This corrosion occurs because of the movement of metal ions in water (ordinary tap water and even softened water contain metal ions). An ion-getter column contains a *sacrificial* metal that will pick up ions from the water and corrode itself, thus slowing corrosion elsewhere. (An equally good or better approach is to use only distilled or deionized water in the system.)

5.2.5 Other Machinery Available

The following pieces of equipment are sometimes incorporated into solar heating systems. Some are quite common; others are relatively new and difficult to find or still rather expensive.

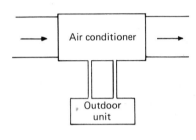

Air Conditioner

An *air conditioner* is a conventional air-cooling device that is powered by an electric motor and has freon as its working fluid. It removes heat from indoor air and exhausts it to the outside environment. A *central air conditioner* delivers cool air throughout a building by forcing it through the duct system. Conventional central air conditioning is commonly used in small- to medium-scale solar systems due to the expense of solar-powered equipment.

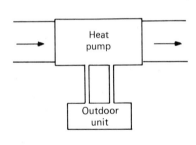

Heat Pump

A *heat pump* is used in some solar systems to provide backup heat and summer cooling. An air-to-air heat pump is essentially identical to an air conditioner except that it can be reversed to extract heat from cold outside air and "pump" the heat indoors. There are also liquid-to-air and liquid-to-liquid heat pumps available, though they are much less common. These may be used in liquid-type systems. In general, any heat pump collects low- to moderate-temperature heat from some source, raises the temperature, and puts the heat somewhere else. Heat pumps are expensive and it is usually not economical to combine them with solar systems (see section 5.3.2).

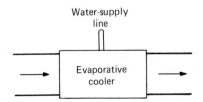

Evaporative Cooler

An *evaporative cooler* is a device used in dry climates for cooling building air. Dry air is blown through a wet element, whose water evaporates as the air passes through it. The evaporating water absorbs its heat of vaporization from the air, thereby cooling the air.

Evaporative coolers have been used in combination with rock-type heat-storage chambers to provide summer cooling for buildings with air-type solar systems. During the night relatively cool outside air is passed through the evaporative cooler and into the rock chamber, cooling the chamber. During the following day warm building air is cooled as it passes through the chamber.

Absorption-Type Solar Air Conditioner

The *absorption-type solar air conditioner* is a cooling device that can be powered by solar heat. Unlike an ordinary air conditioner, it has no mechanical compressor. Water may be its working fluid. Compression is achieved by a chemical *absorbent*, typically lithium bromide (see Chapter 9). The absorption-type solar air conditioner requires fairly high-temperature water (e.g., 95° C or 200° F) for efficient operation, so concentrating collectors or other high-temperature collectors must be used. Thus far this air conditioner has been used economically only in large-scale systems.

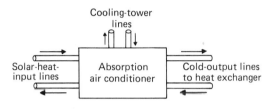

Rankine-Cycle Solar Air Conditioner

The *Rankine-cycle solar air conditioner* is like an ordinary air conditioner except that the compressor, instead of being run by electricity, is run by a solar-powered turbine (See Chapter 9). This air conditioner may be economical in large-scale systems, but thus far it has proved to be too costly for widespread use in smaller systems.

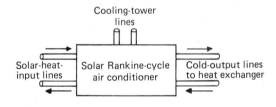

5.3 SYSTEM DESIGN

Systematic procedures can be used to define the modes necessary for an active solar system. Once the modes are determined, a simple equipment layout can be sketched. Next, sensors and controls can be specified; they will "make the decisions" and switch the appropriate equipment on at the appropriate times. At this point, a plan of the entire system can be drawn up. In other words, there are four essential phases in the design of solar heating systems:

1. Specification, or deciding what basic kinds of equipment the system must have (solar collectors, air conditioners, storage units, etc.), and determining exactly what modes will be needed (direct solar heating, solar heat storage, etc.)

2. Equipment layout, or determining a simple way of connecting the basic equipment that will allow for each of the desired modes and yet have a minimum of expensive subordinate equipment (fans, pumps, dampers, check valves, etc.)
3. Controls design, or deciding what exact set of conditions will trigger each mode on and off, determining what sensing devices (thermostats, switches, etc.) are needed, and specifying a control circuit that will perform the decision making
4. Physical design, or sizing the equipment to fit the specific building and location and drawing plans or blueprints

This chapter is primarily concerned with steps 1, 2, and 3. The sizing of solar components, such as collector arrays and storage units, is covered in Chapters 4 and 6. The sizing of ducts, pipes, fans, and so forth are standard engineering problems covered in ASHRAE manuals and the other works mentioned in this chapter's References and Bibliography.

5.3.1 Specification

Specifying an active solar heating/cooling system means deciding what the system must do and then choosing the major pieces of equipment needed. Even at the beginning it is likely that certain choices have already been made; for example, that the system is active and does not rely totally on natural heat transfer. It may have been decided that a certain brand of air-type collector will be used, perhaps because it is locally available and carries a good guarantee. Or, maybe it has been decided that solar energy will provide the heat and hot water, but that a conventional air conditioner will do the cooling. The first step in specifying a system is to clearly pin down what the owner (or future owner) has in mind, and what his or her values and priorities are. What is the building to be used for? Is the building for the owner's use or for resale? Why does the owner want solar heating or cooling? How much money is the owner willing to tie up?

It cannot be overemphasized that there is a wide variety of possible reasons for installing a solar system. Saving money is just one of them. Electrical wiring is rarely chosen on the basis of how cheaply it can be installed. The plumbing system of a building is never chosen on the basis of how soon it will pay for itself (it cannot pay for itself, except through convenience and satisfaction). Similarly, solar heat is just a feature of a building. Whether or not it will pay its way with energy savings is very important to some people but less important to others. Some owners may be solely interested in becoming independent from the gas company, at all costs. Others may wish to project a certain image

SYSTEM DESIGN   175

to their neighbors or the community. Economics is very important, but problems arise if an analysis is based solely on economic grounds since people are motivated by many other factors as well.

After analysis of the values of the system's owner, the system's modes must be chosen and the equipment selected. As we've seen, several modes of operation are needed for any active system. Specifying modes is not an easy task. The designer must imagine all of the possible conditions a system must respond to (and not omit any modes by mistake). The designer must also be familiar with the equipment that is available and with the standard ways of doing things. Specifying the modes of operation for a system involves

1. Clarifying the values and needs of the building owner, including how the building is intended to be used and to what extent economy is a consideration
2. Listing all of the possible sets of temperature (and other) conditions that can occur at different times (e.g., collectors hot, building warm enough, storage chamber not yet "full")
3. Sketching one or two very general equipment layouts for each set of conditions—layouts that provide for adequate response to those conditions but do not go beyond the values and needs of the building owner

So many different sets of conditions are usually possible, that the number of modes and the amount of equipment needed can easily get out of hand. It would be wonderful if a system could respond perfectly to every possible situation, but in order to do so it would have to be so complex and costly that it would be impractical. The mode-specification phase of system design is where a lot of "fat" can be cut out. Are the owners sure that they want solar-powered cooling when a standard electric air conditioner might cost one-tenth as much and eliminate a mode or two? Is an automatic collectors-to-environment mode, complete with fan, really needed to cool the collectors in the summer, or would a simple manual vent to allow thermosiphoning suffice? Is using a heat pump for backup really going to save any money? Proper specification of a system to begin with can save a lot of trouble later on.

The three examples that follow are for air-type systems. In each case, we start by thinking up a set of conditions. Then, looking at these conditions, we decide where heat should come from and call it the *heat source*. Next, we decide where heat should go and call it the *heat destination*. The name of the mode is based on the source and destination. Finally, we decide what basic pieces of equipment are needed to execute this mode and sketch some different ways this equipment might be arranged. Remember that every mode is a circular path, even a mode in which air is vented out to the environment (such as in example 3). Wherever air is withdrawn, other air must replace it. When sketching possible equipment layouts, it is wise to try to keep each piece of equipment in the same relative position each time; for example, collectors on the left and the rooms on the right. This will make the system-layout stage easier.

Example 1

Conditions:	It is cold outside.
	Building needs heat.
	Sun is shining.
	Collectors are warm.
Heat source:	Solar collectors.
Heat destination:	Rooms.
Mode name:	Collectors-to-rooms mode.
Equipment needed:	Solar collectors, fan, ductwork.
Possible flow paths:	

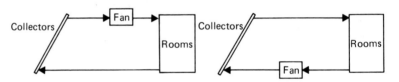

Example 2

Conditions:	It is cold outside.
	Building needs heat.
	Sun is not out.
	Collectors are cool.
	Storage chamber is warm.
Heat source:	Heat-storage chamber.
Heat destination:	Rooms.
Mode name:	Storage-to-rooms mode.
Equipment needed:	Storage chamber, fan, ductwork.
Possible flow paths:	

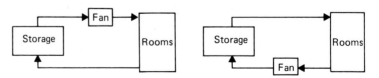

SYSTEM DESIGN 177

Example 3
Conditions: It is hot outside (summer).
Sun is out.
Building needs no heat.
Collectors hot and need venting.
Domestic water could be preheated.

Heat source: Solar collectors.

Heat destination: Domestic hot-water tank and outside environment.

Mode name: Collectors-to-preheater/environment mode.

Equipment needed: Solar collectors, fan, ductwork, hot-water preheat unit.

Possible flow paths:

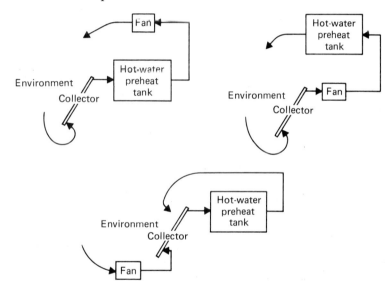

5.3.2 Equipment Layout

Once the system's modes have been worked out and the basic equipment needs are known, it is time to come up with a layout for the simplest possible system that has all of these modes and pieces of equipment. The layout does not specify where everything is to be placed in the building, just how all of the equipment can be connected together with least expense and redundancy. Any one system can be built in several ways, yet only one design will best combine simplicity, economy, performance, and reliability to match the needs of the building's owners.

It is not necessary to come up with a totally original system design every time. In fact, almost any new design will be similar to something that has already been done. Many manufacturers and solar-system professionals have published system layouts that anyone is free to use. To make things even easier, several manufacturers sell ready-made air-handling units for air-type systems—devices that contain fans, automatic dampers, and electronic controls, and are claimed to do all or most of the system control and air switching. Similar devices are available for liquid systems.

Unfortunately, these canned designs and packaged systems can easily create as many problems as they solve. Many of the published system layouts are not complete. Important equipment such as expansion tanks and air filters are often left out (presumably on the assumption that anyone actually building the system will know enough to include them, which is not true). Also, the published designs are based on one set of system modes that may or may not apply to every case. A ready-made air-handling unit will allow for certain modes, but to incorporate additional modes expensive modifications may be necessary.

By far the greatest danger with using ready-made system layouts is complacency. The tendency is to accept a published design without questioning whether it will do all that it should when it is supposed to. Whenever a builder uses a ready-made design without carefully following each mode through, it is likely that some element will be left out or that a control function will be hooked up wrong—and system errors tend to be very expensive errors. It is imperative that the designer fully understand any design to be used, even a design that "somebody else has already thought out." The designer must specify completely which modes will be needed and then check to see that the proposed design will allow these modes. A separate sketch of the layout should be made for each mode, with arrows showing where the fluid circulates and notes explaining which valves or dampers are open, and so on. Standard system designs can be helpful, but are no substitute for careful planning.

The alternative, and it can be an instructive one, is to design an equipment layout from scratch. We can do this in a fairly systematic way by taking the set of modes we desire the system to have, consolidating and joining them, and then replacing fluid-flow arrows with ducts, pipes, and hardware.

Suppose we wish to design an air-type system that can have only a small number of solar collectors because of limited space on the building. The air ducts leading to the collectors will be fairly small, yet we wish to be able to supply solar heat to the entire building even if backup heat will be necessary at times. The backup furnace will require fairly large air ducts. To provide for the simultaneous delivery of solar heat and backup heat, we will have to mix solar heated air with furnace air.

SYSTEM DESIGN

Figure 5.9(a) shows the set of six modes we desire the system to have. The thin arrows indicate ducts for solar-heated air; the thick arrows indicate the larger ducts for furnace-heated and/or solar-heated air.

The first step in laying out this system is to try to consolidate or combine any two modes that are similar to each other, or that can be overlapped. Figure 5.9(b) shows how this may be done. Mode 1 can be accomplished with the same equipment shown in mode 2, so we combine modes 1 and 2 into a single sketch. Modes 4 and 6 can be accomplished with the equipment in mode 5, so we combine modes 4, 5, and 6. As we combine modes this way, we can sometimes make the final system design simpler. Two modes can sometimes be consolidated in a way that is not immediately obvious but that eliminates an unnecessary duct.

After the modes have been consolidated somewhat, we put all of them together in a single sketch. To make the sketch shown in Fig. 5.10(a), we first draw each piece of equipment used in the mode sketches, then draw in all the arrows from the mode sketches. What we end up with is a *system-flow map*, which shows all of the different places air will have to flow in the final system.

Once we know where air will have to flow, we can sketch a layout of a duct system such as the one shown in Fig. 5.10(b). At this stage, the sketch does not have to be in true scale; it is only a schematic drawing. The duct system sketched should be as simple as possible and still allow air to flow in each of the paths shown on the system-flow map. The way to verify that a duct system works is simply to follow out each mode with a pencil or finger, checking to see that there are no conflicts.

In the layout in Fig. 5.10(b), notice that where the small duct meets the large duct, they come together at a "Y" angle. The layout was drawn this way for a reason. Without the "Y," if the furnace fan was much larger than the solar fan and both fans were on at the same time (as in mode 2), the larger fan in the furnace might overpower the smaller fan and stall it, or at least reduce the solar airflow. But if the air from both fans comes together in a "Y," as indicated by the diagram, this problem is reduced. For example, as furnace air passes through the large arm of the "Y" on its way back to the building, it can create a suction in the small arm and assist the smaller fan.

Any system in which more than one fan can operate at the same time must be carefully laid out and balanced, or strange airflow problems are likely to occur. At this stage of design, the duct layout is only preliminary—final layout and duct sizing come later—but it is nevertheless important to develop an awareness of how flowing air behaves and of potential problems that must be avoided. Such knowledge is available only through practical experience with duct-system engineering and sizing procedures such as those found in SMACNA [1] and ASHRAE [2] manuals.

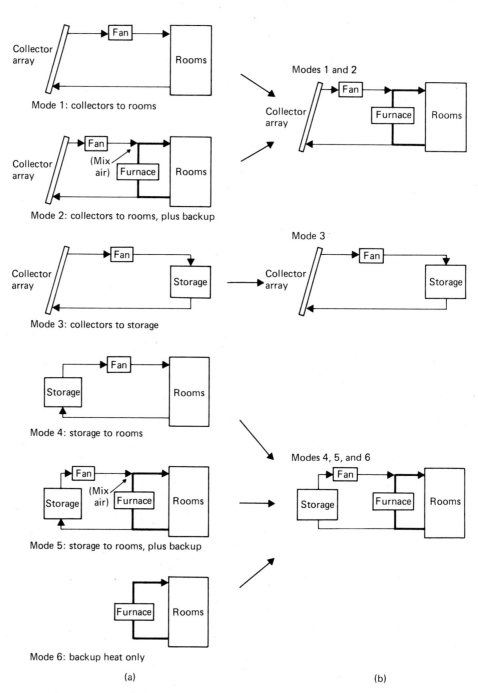

Fig. 5.9 Consolidation of similar modes: (**a**) desired modes for system, (**b**) similar modes consolidated.

SYSTEM DESIGN 181

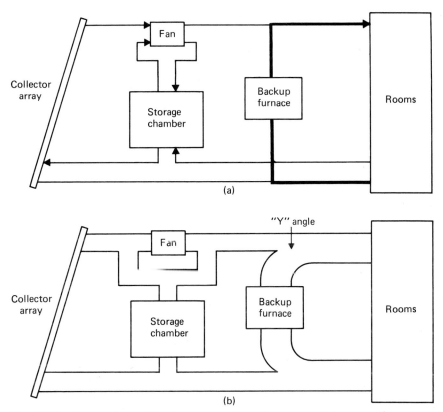

Fig. 5.10 Duct system laid out using a system-flow map: (a) system-flow map, (b) one possible duct layout.

Now we have drawn a duct layout, but the system will not function as drawn. In each mode the air must take a different path through the system. We need to put dampers in strategic places to direct and switch the airflow. We need dampers so that we can switch the system from one mode to another. (Valves perform this function for liquid systems.)

There are three basic classes of dampers. A backdraft damper is chosen whenever possible. A backdraft damper is like a one-way air valve. Air can flow through it in one direction, but not in the reverse direction. These dampers are preferred because they are simple, inexpensive, and do not have to be operated by hand or by a motor. A motorized damper must be used where air must be prevented from flowing in both directions. Sometimes a backdraft damper is not enough or does not allow enough control of the airflow. Motorized dampers are expensive and must be controlled by electric signals, like fans. Sometimes a duct will have to be opened or closed so seldom that it may be acceptable to do the job by hand. Then a manual damper may be used if it is not too inconvenient. In small systems and in areas where essentially no heating is needed in the summer, a system may be designed to be manually set for summer operation. Manual dampers can eliminate the expense of several motorized dampers, relays, and associated equipment.

For example, perhaps some late spring day an alarm will sound, indicating that the building is warm, that the storage medium is full of heat, and that the temperature of the collectors is rising toward the overheating point. As the alarm sounds, the system might also send newly collected heat into the building—for safety and as an additional hint to set the system for summer operation. This is done by opening and closing some manual dampers and flipping some switches.

Deciding on the type and placement of dampers is largely a matter of trial and error. Each mode must be traced through the system so we can determine which ducts need to be shut off during that mode. As one mode is traced, it might be decided that a backdraft damper is needed in one duct. Later, as another mode is traced, it might be found that the backdraft damper is not sufficient and that a motorized damper is needed for more control. If two motorized dampers are required at the same intersection, sometimes a double motorized damper with a single motor can be used instead (Fig. 5.11). This drawing shows two motorized, two-way dampers and three backdraft dampers (perhaps the cloth type). If the system pictured here looks familiar, it is because it is very similar to the basic air-type system shown in Fig. 5.4.

It should always be verified that the system layout will work, even if a published system layout is chosen. A separate sketch of the system should be made for each mode, indicating where air (or liquid) flows, which dampers are open, and which equipment is on. The sketches in Fig. 5.12(a–f) verify that our example system will function.

Fig. 5.11 Example of a completed layout for an air-type system.

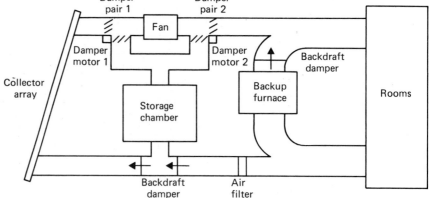

SYSTEM DESIGN

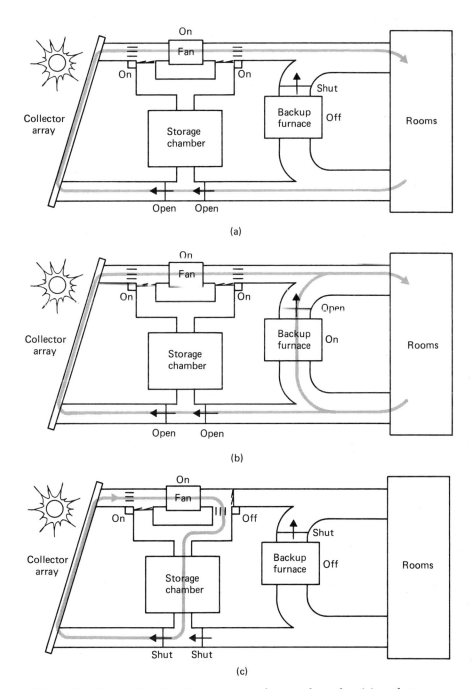

Fig. 5.12 Verification that the system works in each mode: (**a**) mode 1, collectors to rooms; (**b**) mode 2, collectors to rooms (with backup heat); (**c**) mode 3, collectors to storage; (continued)

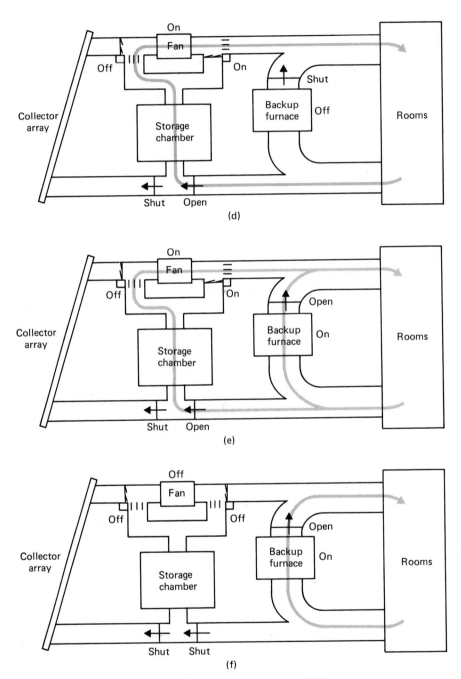

Fig. 5.12 (continued) (**d**) mode 4, storage to rooms; (**e**) mode 5, storage to rooms (with backup heat); (**f**) mode 6, backup heat only.

SYSTEM DESIGN

In anything as complicated as solar-system design, it is essential to be as systematic and organized as possible. What we have done in these few pages is to design a relatively simple, trouble-free layout for an air-type solar heating system. If we had attacked the problem on a totally random basis we might have eventually ended up with a system that worked, but it would probably have had 50% more ductwork and twice as many motorized dampers, fans, and so forth as our example system. Essentially the same techniques can be used for the design of liquid systems, heat-pump systems, and solar cooling systems. The sections that follow contain system diagrams that show some of the many ways various types of solar heating and cooling systems can be connected.

Backdraft-Damper-Controlled Air System (With No Motorized Dampers)

The air-type system shown here can operate in four modes:

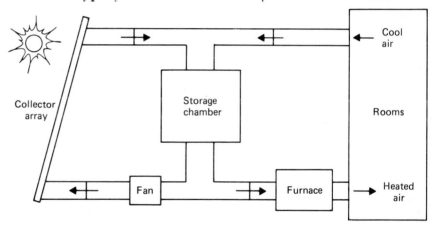

1. Storing solar heat (collector fan on)
2. Storing and using solar heat (both fans on)
3. Using solar heat (furnace fan only on)
4. Backup heating (furnace and heat on)

The system's advantage is that:

1. It is very simple, with no motorized dampers.

Its disadvantages are that:

1. No direct solar heating. Heat is always stored before it can be used.
2. It is difficult to size fans properly. Airflow to rooms is reduced when collector fan comes on, and vice versa.

Standard, Single-Fan Air System
with Air-Handling Unit

The air-type system shown here can operate in five modes:

1. Collectors to rooms, with preheat
2. Collectors to storage, with preheat
3. Storage to rooms, with preheat
4. Collectors to preheat (summer setting)
5. Backup heating

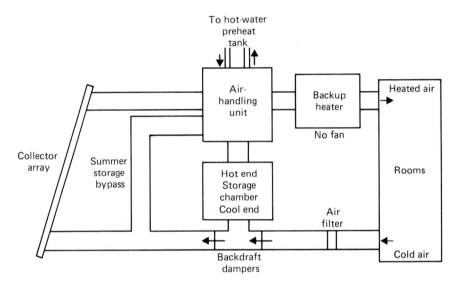

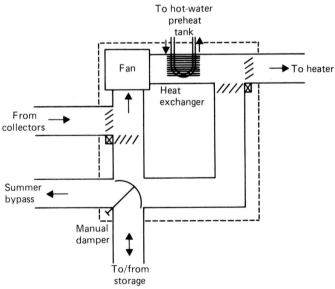

SYSTEM DESIGN 187

The system's advantages are that:

1. It only requires one fan.
2. Only two damper motors are needed (even though there are actually four automatic dampers).

Its disadvantages are that:

1. No outside venting of collectors is possible in the summer.
2. The backup heater is usually a furnace with separate fan (there is little advantage to having only one fan).

Air System With Water Preheat, Conventional
Air Conditioning, and Summer Collector Venting

This air-type system can operate in six modes:

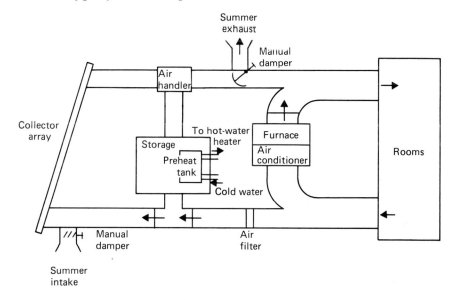

1. Collectors to rooms
2. Storage to rooms
3. Collectors to storage and preheat tank
4. Backup heating
5. Air conditioning
6. Collector overheat venting

Its advantages are that:

1. The embedded preheat system is simple and inexpensive.

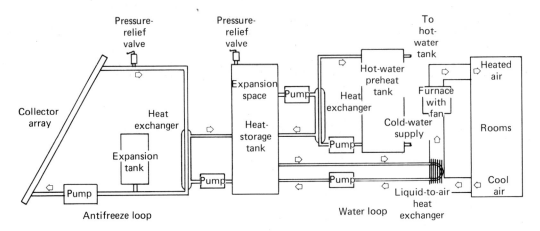

2. Air conditioning is possible, summer or winter.
3. The collectors are efficiently ventilated when overheated.

This system's disadvantage is that:

1. It requires manual changeover in spring and fall.

Standard, Closed-Loop Liquid System

This liquid system's advantage is that:

1. Collectors are protected from freezing.

Its disadvantages are that:

1. It's very complex, requiring five pumps and three heat exchangers.

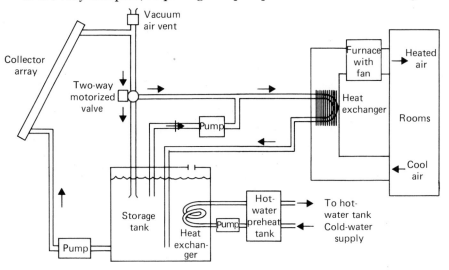

SYSTEM DESIGN

2. The heat exchanger in the collector loop reduces performance.
3. The system requires more maintenance than the draindown types.
4. There is no direct collectors-to-rooms mode. All solar heat generated must be stored before being used.

A Practical, Draindown-Type
Liquid System

This liquid system's advantages are that:

1. It has a direct solar heating mode (collectors to rooms), unlike many liquid systems.
2. The collector loop drains during freezing conditions or power outages, protecting collectors fom damage due to freezing.
3. The system operates near atmospheric pressure. It requires no expansion tanks, and so on.

Its disadvantage is that:

1. It requires a larger main pump than closed-loop systems; this pump uses more electricity.

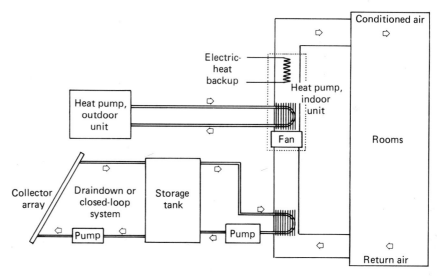

Standard, Parallel-Type
Solar/Heat-Pump System

When a standard, parallel-type solar/heat-pump system is used, the following must be kept in mind:

1. Combining solar systems with heat pumps does not always result in economic savings, due to the high costs of both solar systems and heat pumps. Combined systems can be economical, however, particularly in buildings that require summer cooling.
2. In general, the parallel configuration gives the best overall performance. It is, for example, better than the series configuration, in which the heat pump uses the storage tank instead of the outside air as the heat source.
3. In a parallel system, single-glazed collectors may be used. In general, the use of double-glazed collectors will result in only a slight improvement in performance.
4. Intuition is not a good guide in predicting how various heat-pump systems will perform. Unfortunately, no simple method of predicting performance yet exists [5].

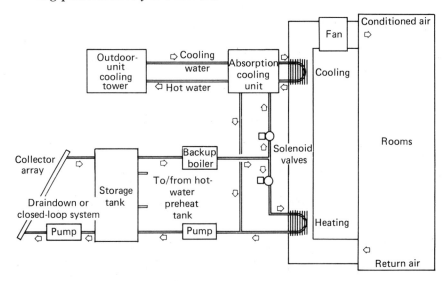

A Basic Solar Heating and
Absorption-Cooling System

The following points about absorption cooling[1] should be kept in mind:

1. Solar absorption cooling is not generally economical at this time, except possibly for large solar installations.

---

1. For an explanation of how absorption cooling works, see Chapter 9.

SYSTEM DESIGN

2. Absorption cooling requires high temperatures (> 80° C or 180° F) in order to be efficient. For this reason, flat-plate collectors do not work as well as evacuated-tube collectors or concentrating collectors.

3. Having a backup boiler in the load loop (as shown) reduces system efficiency, since some heat from the boiler goes toward warming the storage tank (thereby forcing collectors to operate at a higher temperature and with lower efficiency).

4. Many solar cooling systems have one or more additional storage tanks to store coolness. This is done so that the absorption cooler can operate continuously for long periods without cycling on and off. Absorption coolers perform best when operated in this manner.

5.3.3 Controls Design

Once the equipment needs are known and a system diagram exists, it is possible to specify an electronic control system that will automatically adjust the system to meet changing conditions. The control system "decides" which mode the system should be in at any given time, and then switches on the appropriate equipment. As shown in Fig. 5.13, a basic control system has three parts: sensors, a control circuit, and electronic switches. The sensors (such as thermostats) detect the various temperatures and other conditions. The control circuit, or logic circuit, "makes the decisions" and turns on the appropriate equipment, making use of electronic switches (such as relays). To specify a control system means to:

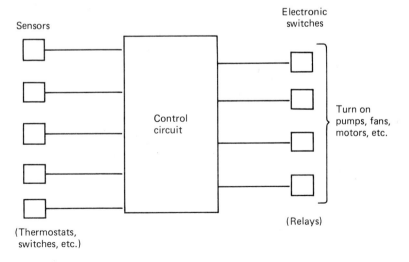

Fig. 5.13 The three basic parts of a control system.

1. Determine what sensors are needed to recognize each mode.
2. Decide which equipment should be actuated during each mode.
3. Summarize all of this information in table form.

Once these things are determined, the control circuit can be designed and built or the job can be given to a competent electrical engineer.

Electrical design and wiring are jobs for professionals. Basic control circuits are not extremely difficult to design, but the designer needs specialized experience to know what safety features to include in a circuit, what kinds of wiring practices are required by the National Electrical Code, and so forth. Unless the solar designer is very experienced in electrical engineering, his or her function should be to specify for an engineer what the system should do (the task of specifying control circuits is difficult enough as it is).

The first step in specifying a control system is to list each system mode and then, for each mode, decide what sensors can be used to recognize that the system should be in that mode. For example, at this point it is no longer enough to say, "The system should be in the collectors-to-rooms mode when the building needs heat and the collectors are hot." We must choose one sensor to indicate when the building needs heat and another to indicate when the collectors are hot. This requires that we know something about the different types of sensors that are available and how they are commonly used.

Example Specify a set of sensors that can be used to recognize that an air-type system should be in the storage-to-rooms mode.

Solution
1. List all of the important conditions, as they occur to you:
   a) It is cold outside.
   b) The building needs heat.
   c) No heat is available from the collectors.
   d) Heat is available in the storage chamber.
2. Choose a sensor to recognize each of the important conditions:
   a) This condition doesn't really matter. All that matters is whether or not the building needs heat inside.
   b) The building needs heat. Place a regular heating thermostat in a central location in the building. The building needs heat if the temperature drops below the set point on the thermostat.
   c) No heat is available from the collectors. Mount a small, probe-type thermostat near the top of the absorber plate in one of the

solar collectors. Choose a thermostat that shuts off whenever the temperature drops below 27° C (75° F). Whenever it shuts off, we know that the collectors are too cool to supply any significant heat.

d) Heat is available in the storage chamber. Place another small, probe-type thermostat in the hot end of the storage chamber. Choose one that turns on whenever the temperature in the storage chamber rises above 27° C (75° F), signaling that stored heat is available.

Example Specify a set of sensors that can be used to recognize that the draindown-type liquid system shown in section 5.3.2 should be in the collectors-to-storage mode (storing heat).

Solution
1. List all of the important conditions, as they occur to you:
   a) The building needs no heat at the moment.
   b) The collectors are warmer than the storage tank.
   c) The collectors are not too close to the freezing point.
2. Choose a sensor to recognize each of the important conditions:
   a) The building needs no heat at the moment. Install a standard heating thermostat in a central location in the building.
   b) The collectors are warmer than the storage tank. Use a differential thermostat, with one probe in the collectors and the other in the storage tank. Set it to turn on whenever the collectors are 3° C (5° F) hotter than the storage medium, and shut off when the collectors drop to within 0.6° C (1° F) of the storage-medium temperature. This way, it only comes on when there is a fair amount of sun, and it does not shut off until the collectors really start to cool down (it won't cycle on and off too much).
   c) The collectors are not too close to the freezing point. To prevent freezing, fasten a probe-type thermostat to one of the absorber plates. Set it to stay off until the collectors drop down to about 4° C (40° F). If it comes on, the system should shut off the collectors-to-storage mode and go into a draindown mode.

Thermostats and differential thermostats are discussed in section 5.2.3. Standard heating thermostats deserve some special attention here. Most conventional (nonsolar) heating and cooling systems are too simple to require a separate-control logic circuit. Both the sensing and decision-making functions are taken care of by a single box—the thermostat—which operates at 24 volts AC (VAC). For example, when the temperature in the building drops below the set point, the thermostat switches on, which sends off a 24-VAC signal to a relay, which turns the furnace on (Fig. 5.14).

Fig. 5.14 Conventional-heating-system thermostat: (**a**) basic operation; (**b**) heat anticipator, a common feature.

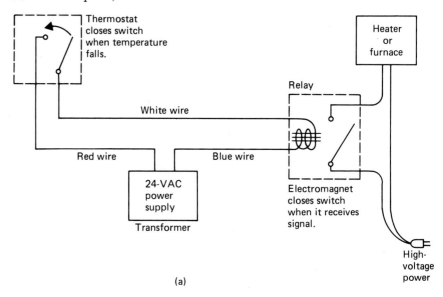

(a)

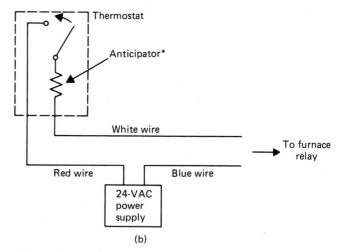

(b)

*The anticipator is a tiny heater that heats up the thermostat to speed up the heating cycle. Heat from the furnace takes time to arrive at the thermostat. The anticipator helps the thermostat to "anticipate" heat's arrival.

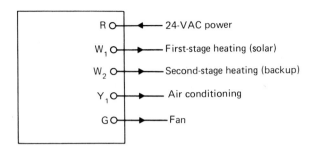

Fig. 5.15 A typical two-stage thermostat.

Room thermostats are available in many different models and can trigger the functioning of air conditioning, fan switches, and so forth. One type of thermostat that is widely used in solar heating systems is the two-stage thermostat discussed previously. A two-stage thermostat is really two thermostats in one package. The first stage is set by the user to the desired temperature level. The second stage follows along automatically a few degrees below the first stage. The first stage controls solar heat. When it is on, the building is heated by solar energy. But if there is not enough solar heat available the temperature continues to fall, the second-stage temperature is reached, and backup heat comes on. Figure 5.15 shows the terminals found on a common two-stage thermostat. Table 5.1 lists the standard color codes for wiring conventional thermostats.

A simple manual switch can also be thought of as a sensor, since by flipping the switch a person is giving the control circuit some information it needs to know. One often-used manual switch is a summer/winter switch. Many systems are designed to respond differently at different times of the year.

Table 5.1 **Standard Meanings of Terminals and Wires**

Letter	Wire Color	Meaning
R	Red	24-VAC power
W	White	Heating relay or valve
Y	Yellow	Cooling relay
G	Green	Fan relay
O	Orange	Cooling damper
B	Brown	Heating damper
X		Clogged filter switch
P		Heat-pump contactor coil
	Blue	Common ground for transformer

Table 5.2 **The Status of Each Piece of Equipment During Each Mode**

	Motor 1	Motor 2	Fan	Furnace
Mode 1	On	On	On	Off
Mode 2	On	On	On	On
Mode 3	On	Off	On	Off
Mode 4	Off	On	On	Off
Mode 5	Off	On	On	On
Mode 6	Off	Off	Off	On
Mode 7	Off	Off	Off	Off

The second step in specifying a control system is to list precisely which pieces of equipment should be activated during each system mode. Usually, for simplicity, systems are designed so that the equipment is either on or off. The fan is either running or it is not. A damper is either powered (in the open position) or it is off (a spring returns it to the closed position). One exception to this is *proportional control*, in which the speed of a fan or pump is varied to match the amount of sunlight available. We can specify what equipment is activated for each mode by making a table (Table 5.2).

Example For the following air system, specify what equipment is on and what is off for each mode (Fig. 5.16):

Mode 1: collectors to rooms

Mode 2: collectors to rooms, with backup

Mode 3: collectors to storage

Fig. 5.16 A complete basic air-type system.

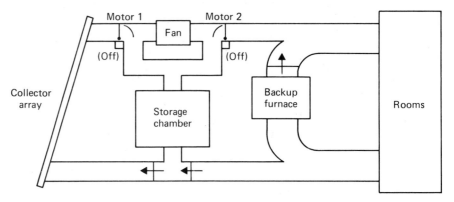

SYSTEM DESIGN

Mode 4: storage to rooms
Mode 5: storage to rooms, with backup
Mode 6: backup heat only
Mode 7: standby

Solution Let us define the "up" position for each damper, shown in Fig. 5.16 to be the "off" position. When all power is off, the storage chamber is closed off and no hot air can leak out. We can make a table that shows the status of each device during each mode (Table 5.2).

The final step in specifying a control system is to put all the information that has been developed into one table that can be referred to easily. Table 5.3 is a table for the system shown in Fig. 5.16. First, the modes were listed. Second, sensors were decided upon for recognizing each mode. Third, it was decided which equipment must be on for each mode. Fourth, all of the information was tabulated. With this table, a competent electrical engineer should readily be able to design a control logic circuit that will meet these specifications.

Table 5.3 appears to be complete and it nearly is, but an electrical engineer is likely to find some information missing or not quite accurate. The first time a control circuit is specified, it is very difficult to imagine

Table 5.3 **Basic Control-Circuit Specification* for System in Fig. 5.16**

	Sensors Needed	State of Sensors	Equipment States			
			Motor 1	Motor 2	Fan	Furnace
Mode 1	W_1, T_c, W_2	W_1 on, T_c on, W_2 off	On	On	On	Off
Mode 2	W_2, T_c	W_2 on, T_c on	On	On	On	On
Mode 3	W_1, T_d	W_1 off, T_d on	On	Off	On	Off
Mode 4	W_1, T_c, T_s, W_2	W_1 on, T_c off, T_s on, W_2 off	Off	On	On	Off
Mode 5	W_2, T_c, T_s	W_2 on, T_c on, T_s on	Off	On	On	On
Mode 6	W_2, T_c, T_s, T_d	W_2 on, T_c off, T_s off, T_d off	Off	Off	Off	On
Mode 7	W_1, T_d	W_1 off, T_d off	Off	Off	Off	Off

Definition

W_1 = first-stage heating thermostat; on when solar heat required

W_2 = second-stage heating thermostat; on when backup heat required

T_c = collector thermostat; on when collector temperature > 75°F (57° C)

T_s = storage thermostat; on when storage temperature > 75°F (57° C)

T_d = differential thermostat; on when collector temperature > storage temperature + 5°F, off when collector temperature < storage temperature + 1°F

*Not a complete specification, but a good starting point for the circuit designer.

every possible combination of sensor states that can occur. But the engineer will find and consider every possible state while designing the circuit. Any errors are likely to be caught in this phase of design. Errors are not always caught, however, and it is important to remember that a system having poor controls or bad logic wastes solar energy that it could otherwise be using.

The control circuit, or logic circuit, is the "brain" of any system. Most active control systems are simple enough so that the control circuit can be built from a few relays. Larger systems may require microprocessors or other solid-state circuitry, but basic systems often do not benefit from the use of solid-state electronics.

A relay is an electrically powered switch. If the coil is supplied with current, the switch will close (Fig. 5.17[a]). Most relays are *double-throw*

Fig. 5.17 Various types of relays: (**a**) a normally open switch, (**b**) a normally closed switch, (**c**) a three-switch (three-pole) relay.

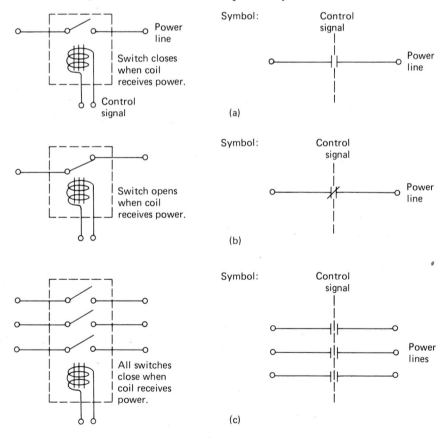

SYSTEM DESIGN

types, meaning that they can also be wired so that the switch opens when the coil is supplied with current (Fig. 5.17[b]). Logic circuits often have relays with more than one switch or *pole* (Fig. 5.17[c]).

A simple relay control circuit is shown in Fig. 5.18. This circuit could be used to control the basic air-type system shown in Fig. 5.16. It matches the control circuit that was specified in Table 5.3. This relay control circuit consists of only four relays. It is best understood if we observe that:

1. Each piece of equipment can receive power only if the right combination of switches is closed.
2. The sensors close the switches (i.e., the sensors provide the control signals that energize the relay coils).

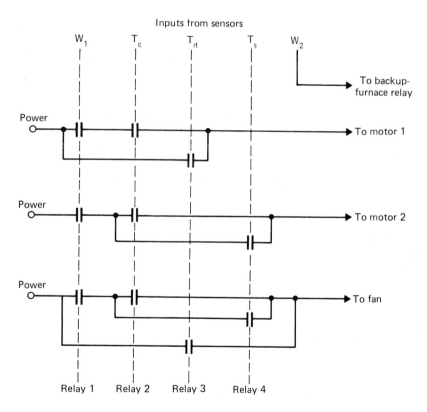

Fig. 5.18 A simple relay-type control system.

W_1 = first-stage (solar) heating
W_2 = second-stage (backup) heating
T_c = collector thermostat
T_s = storage thermostat
T_d = differential thermostat

Therefore, it takes the proper combination of sensor inputs to turn on any piece of equipment. For example, during the collectors-to-rooms mode we want motor 1 to be on so that the collector damper is open. When W_1 is closed (building needs heat) and T_c is closed (collectors are warm), the system should be in the collectors-to-rooms mode. Figure 5.18 shows that when these two switches are closed motor 1 does receive power.

The relay-type control circuit in Fig. 5.18 was designed using the principles of *logic design.* This is the field of the electrical engineer, and is beyond the scope of this book. For more information, refer to the Bibliography at the end of the chapter.

Once a solar heating system has been fully specified and laid out, and once the control circuitry has been thought out, it is time to begin the final phase of design. Collector arrays, fans, pumps, ducts, pipes, and heat-storage units must all be sized to fit the building, and detailed system plans must be drawn to scale. Once this phase is complete, the system design is complete.

5.4 A WORD ABOUT OBSOLESCENCE

In a field such as solar heating, where new advances are a daily occurrence, how is it possible to know that a system installed today will not be obsolete in 10 years, 5 years, or even 2 years? It is not possible. Many areas are under study and we cannot guarantee that there will be no major breakthroughs. Many important discoveries are likely to be made within 10–20 years. Today's exotic materials and methods will be made practical.

The system or component that becomes obsolete does not necessarily lose its original value, however. An automobile may become "obsolete" within five years, yet it will certainly remain in running order if well maintained. A computer may have lost 95% of its resale value in seven years, yet if it can still perform the job it was originally set up to do and if the owner bought a large enough system to begin with, it should retain its original value for that owner. The important point is not whether a thing will become outdated and outmoded, but whether it was properly chosen to begin with and continues to fulfill its original purpose.

Consider heat-storage technology. Today, the two most economical solar-heat storage units are (1) a tank of water and (2) a bin of rocks. Both units are fairly large and heavy, yet both work well and will continue to work well for many years. It is likely that methods for storing solar heat in reversible chemical reactions will soon be discovered, which will dramatically reduce the size and cost of heat-storage units and possibly even reduce the complexity of solar-heating systems. Does this mean that buildings that now have rock or water storage will be forced to scrap these systems and buy new ones in five years? Of course not—not if the old system was properly designed to begin with.

If part of the original specification of a system is that it must retain all of its resale value for the next 20 years, then no system now in existence can fill the bill. But if the goal is to design a solar heating system that works and will continue to work, then there are a large number of choices available. It is wise to anticipate the future but not to wait for it. After all, the best system 3 years from now may be outdated in 3 more years. When thinking of the future, the questions to ask are

1. Will the system do everything it should do?
2. Will it be adequate to handle current needs and all anticipated future needs?
3. Will it pay for itself over its lifetime (at least in owner satisfaction)?
4. Does it work well? Well enough so that there is a margin of physical safety for occupants?
5. Will it continue to work well, without extensive maintenance?
6. Can it be fixed when parts are no longer available?

If these questions can be answered in the affirmative, obsolescence should pose no problem.

REFERENCES

1. SMACNA (1977). *HVAC Duct System Design.* Tyson's Corner, Vienna, Va.: SMACNA.
2. ASHRAE (1977). *1977 Handbook of Fundamentals.* New York: ASHRAE.
3. NESCA (1977). *Equipment Selection and System Design Procedures.* Washington, D.C.: NESCA, Manual Q.
4. ASHRAE (1976). *1976 Systems Handbook.* New York: ASHRAE.
5. Freeman, T., J. Mitchell, and T. Audit (1979). Performance of combined solar-heat pump systems. *Solar Energy* 22(2):125.

BIBLIOGRAPHY

Generally Useful Works

ASHRAE (1974). *1974 Applications Handbook.* New York: ASHRAE.

Brumbaugh, J. (1976). *Heating, Ventilating, and Air Conditioning Library,* vol. 1–3. Indianapolis: Theodore Audel & Co.

Duffie, J. A., and W. A. Beckman (1974). *Solar Energy Thermal Processes.* New York: John Wiley.

Honeywell Corporation (1977). *Solar Energy Systems and Controls Guide.* Minneapolis: Honeywell Corporation.

Jordan, R. C., and B. Y. H. Liu (ed) (1977). *Applications of Solar Energy for Heating and Cooling of Buildings.* New York: ASHRAE.

SMACNA (1978). *Fundamentals of Solar Heating* Washington, D.C.: U.S. Government Printing Office, stock no. 061-000-00043-7.

Solar Energy Applications Laboratory, Colorado State University (1977). *Solar Heating and Cooling of Residential Buildings: Sizing, Installation, and Operations of Systems.* Washington, D.C.: U.S. Government Printing Office, stock no. 003-011-00085-2.

Solar Energy Applications Laboratory, Colorado State University (1977). *Solar Heating and Cooling of Residential Buildings: Design of Systems.* Washington, D.C.: U.S. Government Printing Office, stock no. 003-011-00084-4.

System Components, Solar Equipment

Anderson, B. (ed.) (1976). *The Solar Age Catalog.* Port Jervis, N.Y.: Solar Vision, Inc.

ASHRAE (1976). *1976 Systems Handbook.* New York: ASHRAE.

ASHRAE (1977). *1977 Handbook of Fundamentals.* New York: ASHRAE

NESCA (1977). *Equipment Selection and System Design Procedures.* Washington, D.C.; NESCA, Manual Q.

SMACNA (1977). *HVAC Duct System Design.* Tyson's Corner, Vienna, Va.: SMACNA.

Solar Age Magazine (1979). *Solar Products Specifications Guide;* Church Hill, Harrisville, N.H.: Solar-Vision, Inc.

Solar Energy Applications Laboratory, Colorado State University (1977). *Solar Heating and Cooling of Residential Buildings: Design of Systems.* Washington, D.C.: U.S. Government Printing Office, stock no. 003-011-00084-4.

Solar Controls, Logic Design

Anderson, B. (ed.) (1976). *The Solar Age Catalog.* Port Jervis, N.Y.: Solar Vision, Inc.

Blakeslee, T. (1975). *Digital Design with Standard MSI and LSI.* New York: John Wiley.

Honeywell Corporation (1977). *Solar Energy Systems and Controls Guide.* Minneapolis: Honeywell Corporation.

Honeywell Corporation (1980). *Tradeline Catalog for the Electrical Industry.* Minneapolis: Honeywell Corporation.

Kohavi, Z. (1970). *Switching and Finite Automata Theory.* New York: McGraw-Hill.

Solar Age Magazine (1979). *Solar Products Specifications Guide.* Church Hill, Harrisville, N.H.: Solar-Vision, Inc.

Solar Energy Applications Laboratory, Colorado State University (1977). *Solar Heating and Cooling of Residential Buildings: Design of Systems* Washington, D.C.: U.S. Government Printing Office, stock no. 003-011-00084-4.

PROBLEMS

5.1 Explain why an air system with rock storage performs best when heat is always put in and taken out of the same end of the chamber (e.g., solar heated air is blown in the top and hot air is later removed from the top). How does this consideration affect liquid systems?

5.2 Explain some factors that must be considered when we connect a small air-type solar heating system to the duct system of a large conventional system heated by a furnace.

5.3 Explain some of the advantages and disadvantages of (1) hot-water preheat systems that have a heat exchanger and (2) those that have a preheat tank embedded in the storage unit.

5.4 Pressure drop in an air-duct system occurs when air must make sudden or circuitous turns or pass an obstruction. The greater the pressure drop in a duct system, the larger the fan must be and the more power it will consume. Give several examples of locations in an air-type solar heating system where pressure drop might occur, and describe some possible measures for reducing pressure drop in these areas.

5.5 The vacuum breakers used in draindown-type liquid systems can freeze up in cold weather, preventing the system from draining. Devise a draindown-type liquid system that uses no vacuum breakers but that still drains during the night and drains for safety during power outages.

5.6 Water is a more compact heat-storage medium than rock pebbles.

a) For an air system, specify the modes in which a large water tank might be efficiently used for heat storage.

b) Draw a layout for this system.

c) Make a sketch of the system for each mode to verify that the chosen layout will work.

5.7 Draw a layout for an air system with rock storage that has the following modes:

Collectors to rooms

Collectors to rooms, with backup

Collectors to storage

Storage to rooms

Storage to rooms, with backup

Summer collector venting, natural convection

Summer-night storage of coolness from outside air

Air conditioning, assisted by coolness storage

Backup heating, summer and winter

5.8 Specify and lay out a closed-loop, liquid-type solar heating system that has a storage unit but allows direct solar heating through a liquid-to-air heat exchanger in an air-duct system.

5.9 The air system with water preheat, conventional air conditioning, and summer collector venting described in section 5.3.2 has a mode for forced ventilation of the collectors in the summer, when they might become overheated. Specify all of the sensors necessary for recognition that the system should be in this mode.

5.10 For the system in problem 5.6, specify a complete control system. Use the types of sensors discussed in this chapter.

5.11 Specify a control system that might be used for the simple, backdraft-damper-controlled system with no motorized dampers, described in section 5.3.2.

5.12 For the draindown-type liquid system shown in section 5.3.2, make a list of all potentially useful modes. Specify what equipment is on and what is off for each mode.

5.13 For the system in problem 5.7, make a fairly complete specification for a control system. Use the types of sensors discussed in this chapter.

5.14 Specify and lay out a single air-type solar heating system that can be used to heat two separate conditioned spaces (each with its own backup furnace).

6 SIZING ACTIVE SYSTEMS

The proper size for a solar heating system is determined by evaluating the heating and the economic performance of the system. The heating performance of a solar heating system depends on the physical makeup of the system, its location, and the heat load imposed on it by the building.

Economic performance depends on the system's heating performance, its initial cost, and its operating costs. Total costs include additional mortgage costs, property tax, insurance and maintenance costs, and the cost of backup fuel and operation, minus tax credits for installing a solar system and minus tax savings from deductible items. Inflation of some of these costs over time must also be considered.

A properly sized system will pay for itself with interest over its lifetime. Typically such a system will provide 30–80% of a building's yearly heating and hot-water needs.

6.1 WHY IS SIZING DIFFICULT?

Sizing a solar heating system is quite an intricate task because of the many different variables that must be considered. Some of the most accurate models that simulate solar systems to determine performance (e.g., TRANSYS [1]) require hourly solar radiation data and are so complex to use that they are suitable only for sophisticated computers. This complexity and the special equipment required preclude the use of these sizing methods by most people.

Efforts to simplify sizing procedures without sacrificing accuracy have led to the development of methods that require monthly daily average radiation measurements rather than hourly data. One of the most notable of these methods is the f-Chart method developed by Beckman, Klein and Duffie of the University of Wisconsin Solar Energy Laboratory [2, 3, 4, 5]. In the f-Chart method a series of calculations is used to determine monthly f values, the fraction of each month's heating load provided by solar energy. Though this method requires only monthly data, because of the large number of calculations involved it, too, is ideally suited for use with a computer. For those of us who do not have easy and inexpensive access to a computer, these calculations are difficult, tedious, and time consuming.

Several even simpler methods have been developed [6, 7, 8, 9, 10]. Most of these methods do not require as much input information as the f-Chart method and, since fewer calculations are necessary, they are less time consuming to perform, but they are also tedious.

All of us would like to have the optimum solar heating system; one that would be perfectly sized for our needs and would actually earn a "profit" over its lifetime by saving us money. Fortunately the last-mentioned methods for determining solar performance can be used for sizing this optimum collector area and then the system.

To find the optimum area for a solar collector we have to consider (1) economic factors and (2) solar system performance. In most simplified methods these two considerations are combined and virtually inseparable. Why and how they are connected?

The optimum area for a solar collector is the area that delivers the maximum return per dollar invested over the life of the system. This type of economic analysis is often referred to as *life-cycle costing, net-benefits analysis* or *savings analysis*. These analyses are just different ways of looking at the same situation.

With life-cycle costing, the total costs associated with purchasing and using a solar heating system with a conventional backup heater are compared with the total costs of having a conventional heater alone. The system with the lowest life-cycle cost is a better investment. With net-benefits analysis, only the cost differences between the two types of systems are compared. If the savings associated with the solar system are greater, it is a better investment than the conventional system by itself. If the life-cycle costs for the solar system with conventional backup are subtracted from the life-cycle costs for the conventional system, the result is the life-cycle savings, or the same result obtained from the net-benefits analysis.

What is really involved in these cost calculations? With a net-benefits calculation, the cost of the conventional system alone is subtracted from the purchase price of the complete solar heating system, including backup (or the extra mortgage costs), extra taxes, extra insurance and maintenance costs, and so on. The higher cost of the solar heating system would be reduced by extra tax savings (due to federal and state solar incentive laws and/or the deductible interest portion of the extra mortgage payment) and the savings in energy costs (based on the simulated solar performance).

Many costs are annual and may increase each year due to inflation. Life-cycle costing incorporates a special way of comparing present with future costs and their relative values. The *present value* of these costs and the equivalent (in today's dollars) of future costs are compared. The process of bringing a future cost back to its present value is called *discounting*.

To discount a future cost a *discount rate*, or rate for the best alternative investment, must be defined. Generally, if the system is purchased with a mortgage, the discount rate should equal the mortgage interest rate.

Example The present value, P, of a future amount F, in year N is defined as

$$P = F\left(\frac{1}{(1+d)^N}\right)$$

where d is the discount rate. What is the present value of a $500 cost in year four if the discount rate is 12%?

Solution $P = 500 \,\dfrac{1}{1.12^4} = \$318.$

If $318 were invested today at the discount rate, $500 would be accumulated in four years. This problem can also be solved as

$P = F \times$ discounting factor.

The discounting factor would be defined as $1/(1+d)^N$, or 0.636 in this case.

To discount future costs that are inflating over time, a special type of function called a *cumulative inflating-discounting function* is used. To use this function, only the cost in the first year and the inflation and discount rates need be known to calculate the sum of the present values of the cost for *N* years. An *inflating-discounting factor* is determined and is multiplied by the first-year cost. The product is the present value of the series of inflated costs. The cumulative inflating-discounting function is used in the sizing process presented in this chapter and is defined in section 6.3.2.

Example The first-year cost of insurance and maintenance is $140, which inflates at 8% per year. What is the present value for that expense for five years if the discount rate is 12%? The inflating-discounting factor for five years at 8% inflation with a discount rate of 12% is 4.157.

Solution $140 × 4.157 = $582. Even though a total of $821, or $140 + 140(1.08) + 140(1.08)^2 + 140(1.08)^3 + 140(1.08)^4 has been spent, the total of the present value of these costs is only $582.

Future savings due to the presence of the solar heating system are treated in the same manner as costs.

Example Your new solar heating system (mortaged at 12%) should provide 62% of your yearly heating requirements. If your heating bill would have been $1200 without the system, what are your first-year energy-cost savings? If the cost of energy increases at 13% per year, what is the present value of these savings over the next 20 years? The inflating-discounting factor for 20 years, with 13% inflation and a 12% discount rate is 19.456.

Solution If the solar system provides 62% of your heat, you should save $744, or 0.62 × 1200 on your heating bill in the first year. If you use the inflating-discounting factor, you will find that the present value of all your energy cost savings over 20 years is 14,475, or 744 × 19.456.

Now we can see why economics and the solar performance are inseparably linked for the calculation of optimum collector area. A bigger solar system will cost more to purchase and maintain, but it will save more heating fuel than a smaller system. The optimum area can be determined by repeating performance calculations and then economic calculations for many system sizes (iterating) to find the area size that leads to the highest savings. This procedure is very time consuming, not to mention boring.

Fortunately such tedious calculations are not necessary. The sizing method presented in this chapter is based on that developed by John C. Ward of the Solar Energy Applications Laboratory at Colorado State University, Fort Collins [10]. In Ward's method the performance and economic calculations for the system are combined into a series of calculations that result in the optimum area. Even though this is a simplified method, it involves many calculations. All calculations are explained step by step and, taken individually, require only simple mathematical skills.

Ward's method was developed to give results comparable to those of the f-Chart method. In one study involving 108 collector sizings in 108 different cities [9], the f-Chart result was compared with the Ward result. The deviations between the optimum specified areas derived as a result of the two methods were less than 20% in about 90% of the cases. Deviations in total costs over the system life were less than 3%. Since total costs do not vary much if the areas compared are near optimum, there was a much wider variation in optimum areas than in total costs. No tax credit was included in these comparisons, as it is in this chapter. The inclusion of tax credit should not change the accuracy of the method, however. The Ward method is sufficiently accurate for sizing small solar heating systems. (See Appendix B for the mathematical derivation of the economic portion of this method.)

6.2 HEAT LOAD

6.2.1 What Is Heat Load?

A building's total *heat load* is the amount of energy it takes to heat the building and provide hot water. The heat load for the building's space is based on heat loss for that space. The hot-water heat load depends on the number of persons living in the building and how much hot water they use. The total heat load is the sum of the space heat load and the water heat load.

The performance of a solar heating system is usually represented by the fraction of the total heat load supplied by solar energy annually. The recommended size for a solar heating system depends on the heat load. Therefore the total yearly heat load for a building must be known before the optimum collector area can be determined. With the sizing method presented in this chapter, both the total January and total yearly heat loads are used in the calculations.

The building's heat load, which depends on building structure and local weather conditions, can be determined approximately by using the worksheets in Fig. 6.1 (pp. 209–212). The hot-water load and total load can also be determined approximately by using these worksheets. Since

Fig. 6.1 Approximate-heat-load calculation sheet: (**a**) metric units; (**b**) English units. (Adapted from American Society of Heating, Refrigerating and Air-Conditioning Engineers [1979]. *Cooling and Heating Load Calculation Manual.* New York: American Society of Heating, Refrigerating and Air-Conditioning Engineers, Inc.; and Air Conditioning Contractors of America [1975]. *Load Calculation for Residential Winter and Summer Air Conditioning.* 4th ed. Washington, D.C.: Air Conditioning Contractors of America.)

HEAT LOAD

[1]
Building _____
Location _____
Number of occupants _____

Date _____
January modified DD* _____
Total modified DD* _____

[2]	Windows (including sliding glass doors)	Type	N ×	ft ×	ft =	area	× U =	U × A	Approximate U values and notes	
									Type	U
									Single	1.10
									Double ¼" space	0.62
									½" space	0.49
									Triple ¼" space	0.39
									½" space	0.31
									Storm /¼" space	0.50
				Total area						

[3]	Doors	ft ×	ft =	area	× U =	U × A	U values	Solid wood, no storm	Storm door Wood	Storm door Metal
							1"	0.64	0.30	0.39
							1.25"	0.55	0.28	0.34
							1.5"	0.49	0.27	0.33
							2"	0.43	0.24	0.27
			Total area							

[4]	Above-grade walls (excluding window and door areas)	ft ×	ft =	area	× U =	U × A	Notes
							1. $U \approx \dfrac{1}{(R+3)}$ where $R = R$ value of insulation used.
							2. Exclude exterior walls, ceilings, and floors of unheated rooms. Include house walls, ceilings, and floors next to these rooms. Effective U values for these bordering partitions appear in note 3.
	Ceilings						3. If the adjacent unheated space is — then — U equals
							Unheated basement $\tfrac{1}{3}U$
							Garage, crawl space, attic U
	Floors (not including slab)						

[5]	Slab (or shallow below-grade walls 3 ft or less)	Perimeter measurement ft	× U =	effective U × A	Perimeter insulation	U for plain slab	U for slab with embedded ducts
					2" rigid	0.52	0.93
					1" rigid	0.62	1.14
					None	0.81	1.9

Fig. 6.1 (a) metric units (continued).

210 SIZING ACTIVE SYSTEMS

6	Below-grade basement walls (extending at least 1 m below grade)	m × m = area × U =	effective U × A	Approximate U values and notes	
				Wall insulation	U value
				None	0.35
				R - 0.88	0.24
				R - 2.3	0.14

7	Basement floor (at least 1 m below grade)	m × m = area × U =	effective U × A	Basement floor U = 0.18
		0.18		
		0.18		
		0.18		
		0.18		

8	Infiltration	Volume of house m³ ×	0.17 =	effective U × A	1. Assume ½ air change/hr
			0.17		
			0.17		
			0.17		

9	U × A column total			
		(1.15)		1. If ducts run through unheated space
		× 0.0864	sec/day MJ/J	
		Heat loss MJ/DD 10⁶ J = 1 MJ		

10	Daily hot water	Number of occupants ×	15.4 MJ/person day =	MJ/day	1. Assumes 75 l HW/person/day
			15.4		2. Assumes $T_{water} - T_{main}$ = 49° C

11	January load	Daily HW × 31 =	Water heat load	+	heat loss ×	January DD =	heat load	=	total load (L) (water + heat) MJ/Jan
		31							
	Yearly load	Daily HW × 365 =	Water heat load	+	heat loss ×	total DD =	heat load	=	MJ/yr
		365							

Note: This calculation method is designed only to give an estimate of heat loads and should not be used as the basis for sizing or selecting conventional heating equipment.
*Degree days: see Appendix C.

Fig. 6.1 (a) metric units (continued)

HEAT LOAD

1 Building _____ Date _____
Location _____ January modified DD* _____
Number of occupants _____ Total modified DD* _____

2 Windows (including sliding glass doors)

Type	N	×	m	×	m	=	area	×	U	=	U × A
			Total area								

Approximate U values and notes

Type		U value
Single		6.25
Double	0.6 cm space	3.5
	1.3 cm space	2.8
Triple	0.6 cm space	2.2
	1.3 cm space	1.8
Storm	3.2 cm space	2.8

3 Doors

	m	×	m	=	area	×	U	=	U × A
			Total area						

U values	Solid wood, no storm	Storm door Wood	Metal
2.5 cm	3.6	1.7	2.2
3.2 cm	3.1	1.6	1.9
3.8 cm	2.8	1.5	1.9
5.1 cm	2.4	1.4	1.6

4 Above-grade walls (excluding window and door areas)

	m	×	m	=	area	×	U	=	U × A

1. $U \approx \dfrac{1}{(R + 0.53)}$ where $R = R$ value of insulation used.
2. Exclude exterior walls, ceilings, and floors of unheated rooms. Include house walls, ceilings, and floors next to these rooms. Effective U values for these bordering partitions appear in note 3.
3. If the adjacent unheated space is – then – U equals
   Unheated basement $\tfrac{1}{3}U$
   Garage, crawl space, attic U

Ceilings

Floors (not including slab)

5 Slab (or shallow below-grade walls 1 m or less)

Perimeter measurement, m	×	U	=	effective U × A	Perimeter insulation	U for plain slab	U for slab with embedded ducts
					5 cm rigid	0.90	1.7
					2.5 cm rigid	1.1	1.97
					None	1.4	3.3

Fig. 6.1 (a) English units (continued)

									Approximate U values and notes	
6	Below-grade basement walls (extending at least 3 ft below grade)	ft.	×	ft.	= area	× U	=	effective $U \times A$	Wall insulation	U value
									None	0.061
									R-5	0.043
									R-13	0.025

7	Basement floor (at least 3 ft below grade)	ft.	×	ft.	= area	× U	=	effective $U \times A$	Basement floor $U = 0.032$
						0.032			
						0.032			
						0.032			
						0.032			

8	Infiltration	Volume of house ft³	×	0.009	=	effective $U \times A$	1. Assume ½ air change/hr
				0.009			
				0.009			
				0.009			

9	$U \times A$ column total	(1.15)	1. If ducts run through unheated space
		× 0.024	hr/day
	Heat loss (Btu/DD) × 10³		10³

10	Daily hot water	Number of occupants	×	14.6 10³ Btu/person day	=	10³ Btu/day	1. Assumes 20 gal HW/person/day
				14.6			2. Assumes $T_{water} - T_{main} = 88°$ F

													Total load (L) (water + heat)
11	January load	Daily HW	×	31	=	water heat load	+	heat loss	×	January DD	=	heat load	=
				31									(Btu/Jan) × 10³
	Yearly load	Daily HW	×	365	=	water heat load	+	heat loss	×	total DD	=	heat load	=
				365									(Btu/yr) × 10³

Note: This calculation method is designed only to give an estimate of heat loads and should not be used as the basis for sizing or selecting conventional heating equipment.

*Degree days: see Appendix C.

Fig. 6.1 (**b**) English units

individual living habits and solar heat gains to the building are not included, however, the result of these calculations may be off by as much as a factor of three.

For existing buildings the space and water heat loads can usually be very accurately estimated from old heating bills. At least one year of gas, oil, or electric bills should be used in the calculation. Often energy companies keep records on hand for several years, in which case consumption can be averaged for a typical year. With this method, which is described in section 6.2.3, the water heat load may have to be calculated using Fig. 6.1.

6.2.2 Calculating Heat Load

January and yearly heat loads can be determined using either of the charts in Fig. 6.1. The total heat load consists of the space heat load, which depends on building construction and outdoor temperature, and the water heat load, which is assumed constant over the year and depends on the number of occupants.

The amount of heat lost through a building depends on its construction—windows lose more heat than insulated walls per unit area. As we stated in Chapter 1, the rates at which different materials lose (or transmit) heat depend on each material's resistance to heat flow, or its R value (m² °C/W or hr ft² °F/Btu). R values are commonly used to rate insulation; a high R value means a low heat-loss rate. Table 6.1 shows English-unit R values of several common insulation materials.

Table 6.1 R **Values (in English Units) for Selected Insulation Materials**

R value of insulation	Batts or Blankets		Loose Fill		
	Fiberglass	Rock wool	Fiberglass	Rock wool	Cellulosic fibers*
7	2¼–2¾	2	3–3½	2¼–2¾	2–2¼
11	3¼–4	3	4¾–5¼	3½–4	3–3½
13	3⅝	3½	5¹/₆–6	4¼–4¾	3½–4¼
19	6–6½	5¼	8¼–9	6¼–7	5–6
22	7	6	9½–10¼	7¼–8¼	6–7
30	9½–10½	8¼	13–14	10–11	8–9½
38	12–13	10½	16½–18	12½–14	10¼–12

Note: Thicknesses of insulation materials measured in inches.
*Milled paper or wood pulp.

An English-unit R value of 5 (hr ft^2 °F/Btu) means that it will take five hours for 1 ft^2 of the insulation to lose 1 Btu of energy if the difference between the indoor and outdoor temperatures is 1° F. If the temperature difference is 5° F, it will take only one hour for 1 ft^2 of insulation to lose 1 Btu. The equivalent metric R value of 0.88 (m^2 °C/W) means that it will take 0.88 sec for 1 m^2 of the insulation to lose 1 J of energy (1 W = 1 J/sec) if the temperature difference is 1° C.

R values are additive. For example, two layers of R-11 insulation have an R value of 22. The U value is the inverse of the R value and is measured in W/m^2 °C or Btu/hr ft^2° F.

To determine the heat loss of a whole building, each type of construction is considered separately. The U value for each construction type is multiplied by its area, giving a $U \times A$ product. The sum of these products is multiplied by other factors to yield the heat loss in megajoules (MJ) per modified *degree day* (DD) (or Btu per modified DD × 10^3).

A degree day is a unit based on time and the difference between indoor and outdoor temperatures. Degree days are used to estimate heat loads and fuel consumption. The indoor temperature used for calculating degree days is arbitrarily set at 18.3° C (65° F). It is assumed that appliances, lights, and people provide sufficient heat to raise the indoor temperature to a more comfortable level. A degree day is a one-degree difference between the indoor temperature of 18.3° C (65° F) and outdoor temperatures that exists for one day. If the average outdoor temperature is 0.3° C, 18 – 0.3, or 18 metric heating degree days are measured for that day. One English-unit degree day would be measured if the temperature were 64° F for one day, and 45 degree days would be measured if the temperature were 20° F for one day.

The January and yearly space heat loads are calculated by multiplying the MJ/DD (or [Btu/DD] × 10^3) by the January and yearly modified degree days for the site. The degree days listed in Appendix C include a heat-loss/degree-day factor that relates measured degree days to energy consumption. These are called *modified degree days* here.

The following paragraphs describe heat-loss calculations for different construction types and are meant to supplement the brief instructions given in Fig. 6.1. Note that Fig. 6.1(a) is in metric units and Fig. 6.1(b) is its equivalent in English units. The method presented here estimates the actual heat used and is therefore not suitable for sizing conventional heating equipment or any heating equipment that must be sized for severe winter conditions.

Design Conditions

Design conditions are set forth in block 1 of Fig. 6.1(a) and (b). The January and yearly modified degree days for many cities are listed in Appen-

HEAT LOAD

dix C. These degree-day listings include a factor that relates degree days based on temperature to actual heat load [13]. If your town is not listed, use data from a nearby location with similar weather.

Example What are the January and yearly degree days for Cleveland, Ohio? Consulting Appendix C, we see that in Cleveland there are 491 metric degree days in January (491 °C DD/Jan) and 2558 metric degree days in the year (2558 °C DD/year).

Windows

Heat loss through windows and sliding glass doors (block 2, Fig. 6.1[a] and [b]) depends on the type of window (single, double [insulated], triple, or window with storm window) as well as its area. U values for these different types of windows are given in the right-hand column of block 2.

Example Calculate the $U \times A$ value for two 1.5 m × 2 m single windows with storm windows.

Solution The calculation is performed as follows:

Type	N	× m	× m =	area	× U =	U × A
Window with storm	2	1.5	2	6	2.8	16.8

and the $U \times A$ value is 16.8.

Doors

Heat loss through a door (block 3, Fig. 6.1[a] and [b]) depends on the dimensions of the door as well as the presence or absence of a storm door. U values for doors are given in block 3.

Example Calculate the $U \times A$ value for a 2.5-cm-thick door measuring 1 m × 2 m with no storm door.

Solution The $U \times A$ calculation is performed as follows:

m	× m =	area	× U =	U × A
1	2	2	3.6	7.2

and the $U \times A$ value is 7.2.

Above-Grade Walls, Ceilings, Floors

For simplicity the U value for walls, ceilings, and floors (block 4, Fig. 6.1[a] and [b]) can be assumed to be $1/(R + a)$, where R = the R value of the insulation used and a = 0.53 in metric units and 3 in English units. The a accounts for siding, paper, sheetrock, studs, and so on. This approximation is fairly accurate for all types of construction, including brick veneer and concrete block.

Only exterior-area measurements (net dimensions of window and door areas) are needed. Determination of the heat loads of individual rooms is unnecessary.

Heat loss is not calculated for unheated areas such as the garage, crawl space, or attic. Heat lost from heated rooms to these spaces must be considered, however. Therefore the exterior measurements of garages, crawl spaces, and attics are not considered, but interior walls, ceilings, and floors bordering these spaces must be included (Fig. 6.2).

If the unheated space is a basement, the U value of the floor above the basement is approximately one-third of the $1/(R + a)$ value. For all other partitions between heated and unheated spaces, the full $1/(R + a)$ U value should be used.

Example Calculate the $U \times A$ value for a wall measuring 2.5 m × 16 m that is insulated to an R value of 3.47 m² °C/W. Assume that the window and door from the previous examples are part of this wall.

Fig. 6.2 Building surfaces used for heat-loss calculations.

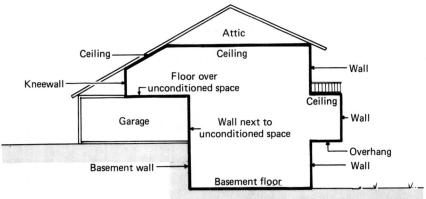

The heavy line shows which surfaces must be considered for heat-loss calculations in this chapter.

HEAT LOAD

Solution As we see from block 4, the U value for the insulated wall portion is equal to $1/(R + 0.53)$, or 0.25. Only the net wall area (excluding window and door areas) is used in this calculation. The $U \times A$ calculation is performed as follows:

m	X	m	=	area	X	U =	$U \times A$
2.5		16		40			
				-6			
				-2			
				32		1/4	8

and the $U \times A$ value for this wall is 8.0.

Slab

The heat loss from a concrete-slab foundation (block 5, Fig. 6.1[a] and [b]) is proportional to the perimeter measurement of the slab. U values for plain slabs and slabs incorporating embedded heating ducts are given in the right-hand column of block 5. Below-grade basement walls that do not extend more than 1 m (3 ft) below ground level are treated as a slab.

Example Calculate the effective $U \times A$ value for heat loss through the plain slab floor of a house measuring 11 m × 15 m. The perimeter of the slab is insulated with 5 cm of rigid insulation.

Solution The effective U value for the plain slab is 0.90. The $U \times A$ calculation is performed as follows:

Perimeter measurement m	X	U =	effective $U \times A$
$2(11+15) = 52$		0.90	46.8

and the effective $U \times A$ value is 46.8.

Basement Walls and Floor

Above-grade basement walls are treated as regular walls described previously. Basement walls that extend 1 m (3 ft) or less below grade are treated as slabs.

Heat loss for basement walls actually varies with depth and ground temperature. An average value is assumed for both the basement walls and floor (blocks 6 and 7, Fig. 6.1[a] and [b]). An allowance for the difference between the outdoor air temperature and ground temperature is included in the approximate U values listed in blocks 6 and 7.

Example Calculate the effective $U \times A$ value for a below-grade basement wall measuring 2 m × 9 m. The wall is not insulated.

Solution The effective U value for the wall is 0.35. The effective $U \times A$ calculation is performed as follows:

m	×	m	=	area	×	U	=	effective U × A
2		9		18		0.35		6.3

and the effective $U \times A$ value is 6.3.

Infiltration

Cold, fresh air enters into the building through cracks around windows and doors (block 8, Fig. 6.1[a] and [b]). The cold incoming air forces some heated air to escape through these cracks. The cold air is treated as a heat loss because it must be warmed to room temperature. One-half of a complete air change per hour throughout the year for the entire building can be considered average for new construction.

Example Calculate the effective $U \times A$ value for infiltration into a house measuring 10 m × 20 m × 2.5 m.

Solution The effective $U \times A$ calculation is performed as follows:

Volume of house m^3	×	0.17	=	effective U × A
10 × 20 × 2.5		0.17		85

and the effective $U \times A$ value is 85.

$U \times A$ Total and Heat Loss

After the $U \times A$ values for each part of the building have been calculated, they are totaled to determine the overall $U \times A$ value for the building (block 9, Fig. 6.1[a] and [b]). This total is then multiplied by 1.15 if heating ducts for the building run through an unheated space such as the crawl space or attic. There is no adjustment to the figure if the ducts run through heated space because heat lost to the interior of the building from the ducts is not lost but serves to heat the space.

The $U \times A$ total is also multiplied by 0.0864 (sec/day)(MJ/J) to convert the total to MJ/DD (or by 0.024 [hr/day]/10^3 to convert to [Btu/DD] × 10^3). This final figure represents the building's heat loss.

Example The $U \times A$ total is 350. Calculate the heat loss if the ducts run through the crawl space.

Solution The calculation is performed as follows:

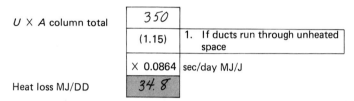

and the heat loss is 34.8 MJ/DD.

Hot-Water Load

The average use of hot water per person per day is near 75 l (20 gal) (block 10, Fig. 6.1[a] and [b]). The hot-water heat load of the building is then

$$HWL = (\text{number of persons}) \times \frac{75\ l}{\text{person day}} \times 1\frac{\text{kg}}{l} \times \frac{0.00419\ \text{MJ}}{\text{kg °C}} \times (T_w - T_m),$$

where

HWL = hot-water heat load,

number of persons = number of persons living in building,

1 kg/l = density of water (8.34 lb/gal),

$\dfrac{0.00419\ \text{MJ}}{\text{kg °C}}$ = heat capacity of water (0.001 [Btu/lb °F] × 10^3),

T_w = usable-hot-water temperature,

T_m = water main's temperature.

If a temperature difference of 49° C between the hot-water main's temperature and the usable-hot-water temperature is assumed (60–11°, for example), this equation becomes

HWL = (number of persons) × 15.4 MJ/day (or 14.6 [Btu/day] × 10^3).

Example There are three people in your family. What is the daily hot-water load?

Solution The calculation is performed as follows:

Daily hot water	Number of occupants ×	15.4 MJ/person day = MJ/day		1. Assumes 75l HW/person/day
	3	15.4	46.2	2. Assumes $T_{water} - T_{main} = 49°$ C

and the hot-water load is 46.2 M J/day.

Total Heat Loads

The total heat load, L, is the sum of the space and water heat loads (block 11, Fig. 6.1[a] and [b]). The total January and yearly loads are calculated as indicated and entered in block 11. These values are used in the calculation of the optimum solar collector area described in section 6.3.3.

Example Your daily hot-water use is 46.2 M J/day and your heat loss is 28 M J/DD. There are 377 modified degree days for your city in January. What is your total January load?

Solution The calculation is performed as follows:

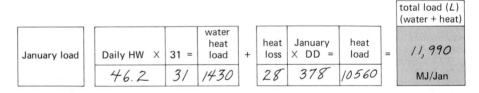

and the total January load, L_{jan}, is 11,990 M J.

Practice Calculation of Heat
Load for a House

Example Calculate the total January and yearly heat loads for the house shown in Fig. 6.3. (You may make a copy of Fig. 6.1 to use in working this example.)

The house is located in Yakima, Washington. It has a full basement: 1.6 m of the basement walls are below grade and 0.8 m are above grade. The above-grade walls are insulated to a metric R value of 3.4 and the below-grade walls to $R = 0.88$. The ceiling is insulated to $R = 5.5$. The ceil-

HEAT LOAD 221

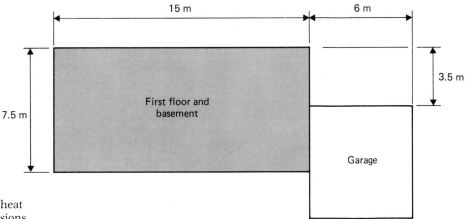

Fig. 6.3 Calculation of heat load for a house: dimensions.

ing height is 2.4 m throughout. All ducts run within the heated portions of the building. A family of four lives in the house.

The windows and doors are as follows:

	Windows	
Type	Number	Measurements
Double (1.3 cm)	4	1.2 × 2 m
Double (1.3 cm)	2	0.6 × 1.2 m
Double (1.3 cm)	2	1.2 × 2.4 m
Single	1	0.4 × 1.2 m
Single with storm	6	0.8 × 0.8 m

	Doors	
Type	Number	Measurements
3.8 cm	2	0.9 × 2 m
Sliding glass, double (1.3 cm)	2	2.1 × 2 m

Solution All calculations are shown in Fig. 6.4, which is the same worksheet as Fig. 6.1. The calculations are self-explanatory. Note that the sliding glass door is treated as a window and the placement of the unheated garage next to the house has no effect on the simplified heat-load calculations used here. The total January load is 12,050 MJ, and the total yearly load is 74,920 MJ.

[1]

Building: Example, section 6.2.2
Location: Yakima, WA
Number of occupants: 4
Date: _____
January modified DD*: 483
Total modified DD*: 2497

[2] **Windows (including sliding glass doors)**

Type	N	×	m	×	m	=	area	×	U	=	U × A
S	1		0.4		1.2		0.48		6.25		3.0
D	4		1.2		2		2.8		2.8		51.5
D	2		.6		1.2		1.44				
D	2		1.2		2.4		5.76				
D/SGD	2		2.1		2.0		8.4				
with storm	6		0.8		0.8		3.84		2.8		10.8
				Total area			22.7				

Approximate U values and notes:

Type	U value
Single	(6.25)
Double 0.6 cm space	3.5
1.3 cm space	(2.8)
Triple 0.6 cm space	2.2
1.3 cm space	1.8
Storm 3.2 cm space	(2.8)

[3] **Doors**

	m	×	m	=	area	×	U	=	U × A
2	0.9		2.0		3.6		2.8		10.1
			Total area		3.6				

U values	Solid wood, no storm	Storm door Wood	Metal
2.5 cm	3.6	1.7	2.2
3.2 cm	3.1	1.6	1.9
3.8 cm	(2.8)	1.5	1.9
5.1 cm	2.4	1.4	1.6

[4] **Above-grade walls (excluding window and door areas)**

m	×	m	=	area	×	U	=	U × A
15		3.2						
7.5		3.2						
15		3.2						
7.5		3.2		144				
				−22.7				
				−3.6				
				117.7		1/3.93		29.9

Ceilings

| 15 | 7.5 | 112.5 | 1/6.03 | 18.7 |

1. $U \approx \dfrac{1}{(R + 0.53)}$ where R = R value of insulation used.
2. Exclude exterior walls, ceilings, and floors of unheated rooms. Include house walls, ceilings and floors next to these rooms. Effective U values for these bordering partitions appear in note 3.
3. If the adjacent unheated space is — then — U equals

Unheated basement	$\frac{1}{3}U$
Garage, crawl space, attic	U

Floors (not including slab)

[5] **Slab (or shallow below-grade walls 1 m or less)**

Perimeter measurement, m	×	U	=	effective U × A

Approximate U values and notes:

Perimeter insulation	U for plain slab	U for slab with embedded ducts
5 cm rigid	0.90	1.7
2.5 cm rigid	1.1	1.97
None	1.4	3.3

Fig. 6.4 Calculation of heat load for a house: worksheet. (Adapted from American Society of Heating, Refrigerating and Air-Conditioning Engineers. [1979]. *Cooling and Heating Load Calculation Manual.* New York: American Society of Heating, Refrigerating and Air-Conditioning Engineers, Inc.; and

6	Below-grade basement walls (extending at least 1 m below grade)	m	×	m	=	area	×	U	=	effective U × A	Approximate U values and notes	
											Wall insulation	U value
		15		1.6		28.1		0.24		6.7	None	0.35
		7.5		1.6							R - 0.88	(0.24)
		15		1.6							R - 2.3	0.14
		7.5		1.6								

7	Basement floor (at least 1 m below grade)	m	×	m	=	area	×	U	=	effective U × A	Basement floor U = 0.18
		15		7.5		112.5		0.18		20.2	
								0.18			
								0.18			
								0.18			

8	Infiltration	Volume of house m^3	×	0.17	=	effective U × A	1. Assume $\frac{1}{2}$ air change/hr
		$2 \times 2.4 \times 15 \times 7.5$		0.17		91.8	
				0.17			

9		U × A column total				242.7		
						(1.15)		1. If ducts run through unheated space
						× 0.0864		sec/day MJ/J
			Heat loss MJ/DD 10^6 J = 1 MJ			21.0		

10	Daily hot water	Number of occupants	×	15.4 MJ/person day	=	MJ/day	1. Assumes 75 l HW/person/day
		4		15.4		61.6	2. Assumes $T_{water} - T_{main} = 49°$ C

11	January load	Daily HW	×	31	=	Water heat load	+	heat loss	×	January DD	=	heat load	=	total load (L) (water + heat)
		61.6		31		1910		21.0		483		10140		12,050 MJ/Jan
	Yearly load	Daily HW	×	365	=	Water heat load	+	heat loss	×	total DD	=	heat load	=	74,920 MJ/yr
		61.6		365		22480		21.0		2497		52440		

Note: This calculation method is designed only to give an estimate of heat loads and should not be used as the basis for sizing or selecting conventional heating equipment.

*Degree days: see Appendix C.

Air Conditioning Contractors of America [1975]. *Load Calculation for Residential Winter and Summer Air Conditioning.* 4th ed. Washington, D.C.: Air Conditioning Contractors of America.)

6.2.3 Simple Heat-Load Calculation

If we have an oil or gas furnace and intend to put a solar heating system on our existing building, calculating heat load can be very simple. Annual heating costs from past years can be averaged to determine costs for a typical year, and January and yearly heat loads can be calculated without the more detailed heat-loss calculation described in section 6.2.2.

Gas Furnace and Hot-Water Heater

If we have a gas furnace and water heater, monthly gas bills can be totaled and the average January and yearly gas consumption and heat loads can be determined. If we also have a gas stove and/or a gas air conditioner, it might be simpler to calculate space heat and water heat loads as described in section 6.2 than to separate heating from cooling and air conditioning costs. The total heat loads can be calculated using the following equation:

$$HL = C \times HV \times 1/n,$$

where

HL = total January or yearly heat load for heat and hot water in MJ or 10^3 Btu,

C = average January or yearly consumption of gas for heat and hot water (m^3 or thm),[1]

HV = heating value of gas from the local gas company (average 37 MJ/m^3 or 100 [Btu/thm] × 10^3),

n = running efficiency of furnace (average 0.5–0.6, even for rated efficiencies of 0.7–0.8).

Gas Furnace, Other Water Heater

If we have a gas furnace but another type of water heater, we may use the equation just given to calculate January and yearly space heat loads by substituting for C our average January or yearly consumption of gas for heating. We must calculate water heat loads and total loads using the method described for blocks 9 and 10 into Fig. 6.1. In the economic-sizing method presented in this chapter it is assumed that the furnace and hot-water heater use the same fuel.

---

[1] One therm (thm) equals 100,000 Btu.

HEAT LOAD

Oil Furnace and Hot-Water Heater

If we have an oil furnace and water heater, yearly oil bills can be used to estimate January and yearly heat loads. The total yearly heat load is

$$HL = C \times HV \times 1/n,$$

where

HL = yearly heat load for heat and hot water in MJ or Btu $\times 10^3$,

C = average yearly consumption of oil for heat and hot water (liters or gallons),

HV = heating value of oil (35 MJ/l or 140 [Btu/gal] $\times 10^3$),

n = running efficiency of furnace (average 0.5–0.6 even though rated efficiency may be 0.7–0.8).

To calculate our January heating load we must first calculate January and yearly hot water heat loads by the method used for blocks 9 and 10 in Fig. 6.1. Then the total January load can be estimated as

$$HL_{jan} = (HL_{yr} - HW_{yr}) \times DD_{jan}/DD_{yr} + HW_{jan},$$

where

HL_{jan} = total January heat load (MJ or 10^3 Btu),

HL_{yr} = total yearly heat load (MJ or Btu $\times 10^3$),

HW_{yr} = yearly hot-water load from Fig. 6.1 (MJ or Btu $\times 10^3$),

DD_{jan} = modified January degree days from Appendix C,

DD_{yr} = Modified yearly degree days from Appendix C,

HW_{jan} = January hot-water load from Fig. 6.1 (MJ or Btu $\times 10^3$).

Oil Furnace, Other Water Heater

If we have an oil furnace but another type of water heater, we may use the equation given for total yearly load to calculate total yearly space heat load by substituting for C our average yearly consumption of oil for heating. The January space heating load is

$$HL_{jan} = HL_{yr} \times DD_{jan}/DD_{yr},$$

where

HL_{jan} = January space heat load (MJ or Btu $\times 10^3$),

HL_{yr} = yearly space heat load (MJ or Btu $\times 10^3$),

DD_{jan} = modified January degree days from Appendix C,

DD_{yr} = modified yearly degree days from Appendix C.

We must calculate our January and yearly water heat and total loads using the method described for blocks 9 and 10 in Fig. 6.1. In the economic-sizing method presented in this chapter it is assumed that the furnace and hot-water heater use the same fuel.

Example What is your average January heat load if you normally use 300 m³ of gas for January in your very efficient furnace?

Solution
$HL = C \times HV \times 1/n$
$= 300 \text{ m}^3 \times 37 \text{ MJ/m}^3 \times 1/0.6$
$= 18{,}500 \text{ MJ/Jan.}$

6.3 ECONOMIC-HEATING-PERFORMANCE SIZING

6.3.1 Introduction

A solar heating system should save money over its lifetime. To get the best economic performance from a solar heating system, all costs associated with its operation must be considered. As a rule, these costs vary with collector area. Economic-heating-performance sizing minimizes these costs by determining the collector area that will give the best heating performance for the dollar.

Sometimes, even with an optimum collector area, the costs associated with operating a solar heating system are greater than those associated with the operation of a conventional heating system. A check for this occurrence is included in the sizing analysis. The system that costs less to operate should be chosen.

Economic-heating-performance sizing provides an estimate of the yearly fraction of the total heat load provided by solar energy for a solar heating system at a specific location in just a few relatively simple steps. The derivation of the method is shown in Appendix B. All calculations are presented in a simple, logical order. The area-optimization procedure is divided into two main parts: (1) the calculation of economic factors (Table 6.2 and Fig. 6.5) and (2) the calculation of the optimum collector area and the fraction of the yearly heating load provided by that area (Fig. 6.6, p. 229). The graphs in Figs. 6.7, 6.8, and 6.9 (pp. 230, 231) provide additional data. A check to make sure that the solar heating system is more economical than a conventional heating system is also shown in Fig. 6.6. (The January and yearly total heat loads from Fig. 6.1 must be known.)

To simplify the analysis, several assumptions have been made:

1. The operating costs of fans, pumps, and controls are considered negligible.

Table 6.2 **Cost and Inflation Factors**

Factor	Amount
C_a area-dependent cost (\$/m² or \$/ft²)	____ C_a
C_b area-independent cost (\$)	____ C_b
d discount rate (interest rate of best alternative investment)	____ d
e energy inflation rate	____ e
i general inflation rate	____ i
J mortgage rate	____ J
K down-payment fraction (if purchased with cash)	____ K
M mortage term (years)	____ M
N analysis term (years)	____ N
P property-tax rate	____ P
R income-tax rate (state + federal − state × federal)	____ R
U insurance and maintenance fraction	____ U
Z depreciation lifetime (commercial only; years)	____ Z

	F (Year, Row, Column)	Inflating-Discounting Factor*
AA[†]	$F(\underline{},\ \underline{},\ \underline{})$ $\quad\ N\quad\ e\quad\ d$	____ AA
BB	$F(\underline{},\ \underline{},\ \underline{})$ $\quad\ N\quad\ i\quad\ d$	____ BB
CC	$F(\underline{},\ 0,\ \underline{})$ $\quad\ M\ \ \text{zero}\ \ d$	____ CC
DD	$F(\underline{},\ 0,\ \underline{})$ $\quad\ M\ \ \text{zero}\ \ J$	____ DD
EE	$F(\underline{},\ \underline{},\ \underline{})$ $\quad\ M\quad\ J\quad\ d$	____ EE
FF	$F(\underline{},\ \underline{},\ 0)$ $\quad\ M\quad\ J\quad\ \text{zero}$	____ FF
GG	$F(\underline{},\ 0,\ \underline{})$ $\quad\ Z\ \ \text{zero}\ \ d$	____ GG (commercial only)

*Inflating-discounting factors are listed in the tables in Appendix D.
[†]If we are purchasing the solar heating system with cash, only AA, BB, and GG (if applicable) need be determined.

228

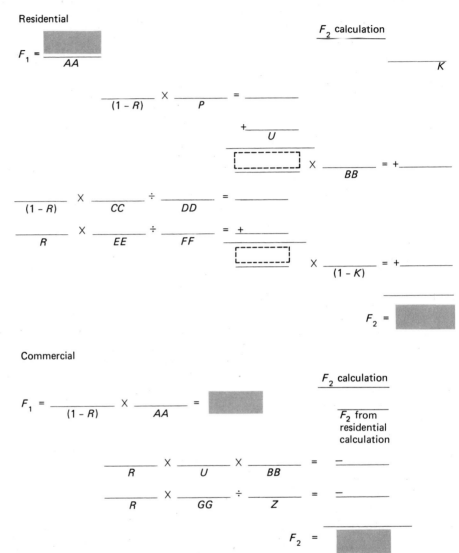

Variables are defined in Table 6.3.

Fig. 6.5 Calculation of economic factors F_1 and F_2.

2. The same fuel is used for space heating and water heating (i.e., if we have an oil furnace, we also have an oil water heater).
3. The entire federal tax credit for the installation of a solar heating system is claimed the first year.
4. The mortgage term is shorter or equal to the analysis term.

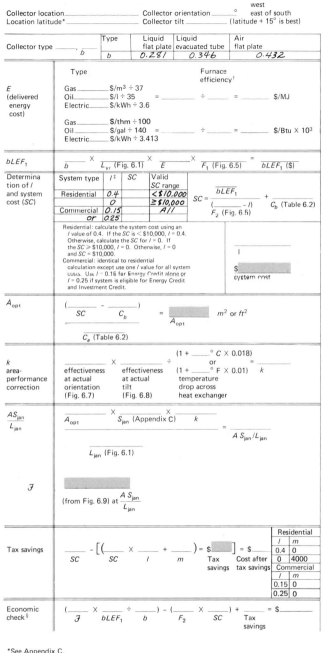

Fig. 6.6 Determination of A_{OPT}, yearly solar fraction, and total-system cost.

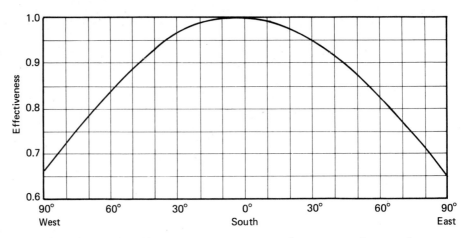

Fig. 6.7 Yearly collector effectiveness at actual orientation. (From Solar Energy Applications Laboratory [14]; reprinted courtesy of the Economic Development Administration, U.S. Department Of Commerce.)

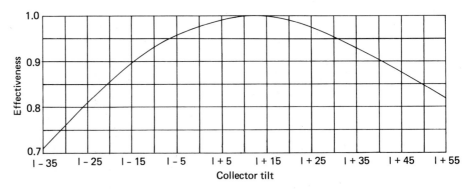

I = degrees north latitude.

Fig. 6.8 Yearly collector effectiveness at actual tilt. (From Solar Energy Applications Laboratory [14]; reprinted courtesy of the Economic Development Administration, U.S. Department of Commerce.)

5. The salvage value of the system at the end of the analysis term is assumed to be zero (the actual salvage value for scrap materials is equal to the cost of removing the system from the building).

6. The depreciation lifetime is shorter than or equal to the analysis term and straight-line depreciation only is used.[2]

---

2. Item six applies to commercial systems only.

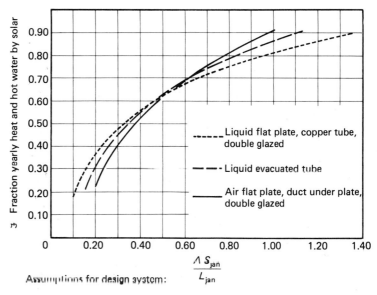

Fig. 6.9 Determination of yearly solar fraction. (From Ward [10].)

Assumptions for design system:
1. Collector oriented due south.
2. Collector tilt = latitude + 15°.
3. No heat exchanger between Collector and storage.

6.3.2 Calculation of Economic Factors

The following sections describe how to calculate the economic factors F_1 and F_2 using Table 6.2 and Fig. 6.5. The cost and inflation variables are specified in the table and F_1 and F_2 are calculated in the figure.

C_a and C_b: System Costs

The purchase cost of a solar heating system can be broken down into collector-area-dependent costs, C_a, and independent costs, C_b. The total-system cost is given by $(AC_a + C_b)$, where A is the collector area. Certainly the cost of the collector and its installation are area dependent, and so is a portion of the storage and controls cost. Some fixed costs for engineering, construction, and control equipment do not depend on the collector area, however.

Area-dependent costs generally range from $160 to $375 per square meter of collector area ($15–35 per ft²) and are even higher for some commercial systems. Fixed costs of about $500 to $2000 can be considered typical.

d, e, and *i:* Discount and
Inflation Rates

In this economic analysis we assume that the solar heating system is an investment earning money at the discount rate, *d*. The discount rate is usually the rate we would earn on our best alternative investment, or a rate equal to the mortgage rate. The discount rate is usually 1–2% above the general inflation rate for individuals and 3–4% above the inflation rate for businesses.

Inflation of operating costs must also be considered. Separate inflation rates—the energy inflation rate, *e*, for backup oil, gas, or electricity, and the general inflation rate, *i*, for property taxes, insurance, and maintenance—must be specified. The energy inflation rate is expected to be higher than the general rate.

If we live in an area with very low electricity rates, we might expect a significantly higher energy inflation rate for electricity than would the rest of the country. Similarly, if we live in an area with very high gas rates, we might expect a lower than average energy inflation rate for gas.

J, K, M and *N:* Mortgage Costs and
Analysis Term

The mortgage interest rate, *J*, down-payment fraction, *K* (if we must put 10% down, $K = 0.10$), mortgage term, *M*, and analysis term, *N*, are all listed in Table 6.2.. If the purchase of the solar system does not involve a mortgage, *J* and *M* are 0 and *K* is 1. The analysis term, *N*, must be as long as or longer than the mortgage term, *M*, and is usually set equal to the expected system life (generally 20–30 years).

P and *R:* Tax Rates

The property-tax rate, *P*, as a fraction of total property value (which may be different from assessed value) applicable to the solar heating system must be specified. Many states now exempt solar heating systems from property tax, in which case $P = 0$. (See the summary of state solar legislation at the end of Chapter 10.)

The income-tax rate, *R*, must be known to allow for calculations of tax savings due to payment of interest on the mortgage (all operating costs plus depreciation are deductible expenses for businesses). The income-tax rate can be calculated as follows:

R = state rate + federal rate − (state rate × federal rate).

For example, if the federal rate is 30% and the state rate is 10%, the overall rate, *R*, is

$R = 0.10 + 0.30 - (0.1 \times 0.30) = 0.37$.

ECONOMIC-HEATING-PERFORMANCE SIZING

U: Insurance and Maintenance Costs

Insurance and maintenance costs are combined as a fraction, U, of the total-system cost. Values ranging from 0.01 for some air systems to 0.05 for some liquid systems are common.

Z: Depreciation Lifetime

The depreciation lifetime, Z, is used for commercial systems only. In this analysis, straight-line depreciation only is used and the salvage value of the system is assumed to be zero. The depreciation lifetime must be shorter than or equal to the analysis term, N.

Inflating-Discounting Factors

Items AA through GG in Table 6.2 are inflating-discounting factors determined by the equations that follow. Note that other variables besides N, i, and d may also be used in these equations.

$$F(N,i,d) = \frac{1}{d-i}\left[1 - \left(\frac{1+i}{1+d}\right)^N\right] \text{ if } i \neq d,$$

or

$$F(N,i,d) = \frac{N}{1+i}$$

if $i = d$, where

> N = the analysis term N (M and Z are also used),
> i = the general inflation rate i (e, J, and 0 are also used),
> d = the discount rate d (J and 0 are also used).

Values for these functions can be found in Appendix D for years 5, 10, 15, 20, 25, and 30. AA–GG are listed in Appendix D as F (year), (inflation), (discount), and can be looked up in the tables as F (year), (row), (column). For example, AA is F (N), (e), (d). If N is 25, e is 0.12 and d is 0.09, then AA or F 25 (year), 0.12 (row), 0.09 (column) is 32.382. If we will be purchasing the solar heating system with cash, only AA, BB, and GG (if applicable) need be determined.

Economic Factors F_1 and F_2

The economic factors F_1 and F_2 are calculated in Fig. 6.5 using the variables from Table 6.2. F_1 accounts for the inflation of energy costs and F_2 accounts for the mortgage, taxes, and upkeep of the system (with inflation). These factors take into account the time value of our money by including the discount rate.

6.3.3 Determination of Actual Yearly Solar Fraction and Total-System Cost

Figure 6.6, the determination of actual yearly solar fraction and total-system cost, is nearly self-explanatory. As before, computations are completed as indicated in the figure.

b: Collector-Dependent Constant

b is a constant that is dependent only on collector type. Typical values for *b* for different collectors can be found in Fig. 6.6.

E: Delivered-Energy Cost

E is the cost of energy for backup fuel delivered by the furnace or hot-water heater. In the first blank, we enter the price presently paid for fuel for our type of backup furnace and hot-water heater. (In this analysis we assume that the furnace and hot-water heater use the same type of fuel.) Conversion of the price of the fuel to its MJ or Btu $\times 10^3$ value is accomplished by dividing by the indicated conversion factor, as shown. The delivered-energy cost is this value divided by our furnace's running efficiency (usually 0.5–0.6 for gas and oil, 1 for electric resistance, and 1.4 for electric heat pumps). For example, if we have electric resistance backup and pay $0.04/kWh, our delivered energy cost is

$$\left(\frac{0.04}{3.6}\right) \div 1 = 0.01 \ \$/MJ.$$

$bLEF_1$

$bLEF_1$ is the product of the collector constant, b, the yearly load, L (from Fig. 6.1), the delivered energy cost, E, and the economic factor, F_1 (from Fig. 6.5).

Determination of *l* and System Cost

The system cost is calculated as indicated in Fig. 6.6, using the values listed. For residential solar heating systems *l* can equal 0.4 or 0, depending on the system cost. The values are related to the federal income tax credit for installing a solar heating system (40% of the first $10,000 of the system's cost, with a maximum credit of $4,000).

For a residential system, the system's cost is first calculated using $l = 0.4$ and then checked to see if the resulting system cost is less than $10,000. If it is, $l = 0.4$ and we may proceed to the calculation of A_{opt}. If, however, the system's cost is greater than $10,000, the cost should be calculated using $l = 0$. If the resulting system cost is again greater than $10,000, $l = 0$ and we may proceed to the next step. Otherwise the correct system cost is $10,000, with $l = 0$.

For commercial buildings, the system's cost is determined by similar calculations using a single l value. Most active solar systems on commercial buildings are eligible for both the commercial-systems Energy Credit of 15% and the Investment Tax Credit of 10%, for a total credit of 25%. A major exclusion is the apartment house, which is eligible for the Energy Credit alone. Eligibility should be carefully checked with the Internal Revenue Service. The appropriate l value of 0.15 or 0.25 is used in calculating the cost of the system.

Alternate Determination of l and System Cost

The income-tax benefits from states are not considered in this collector-sizing method. If we wish to include state tax benefits, we may be able to develop the necessary formulas. The system's cost is calculated in the same manner as indicated in Fig. 6.6, but the values for l will be different because they will contain provisions for our state tax as well as federal tax benefits. See Appendix B for more details on how the tax-credit formula is derived and used.

A_{opt}: Optimized Collector Area

A_{opt} is calculated as $(SC - C_b)/C_a$, using the system cost, SC, determined above, and C_a and C_b from Table 6.2. A_{opt} represents the collector area that will give the best economic heating performance over the lifetime of the system.

k: Area-Performance Correction Factor

In general, the best overall heating and economic performance by a solar heating system occurs when the system is facing due south (in the Northern Hemisphere) and tilted toward the equator at an angle equal to degrees north latitude + 15°. If the collectors will not be oriented due south or will be tilted at an angle other than latitude + 15°, corrections for these deviations must be made in the performance-prediction equation. Also, if the system will have a heat exchanger between the collectors and the storage unit, a correction for this must be included. The collector area A_{opt} as calculated in the preceding section represents the collector area that will give the best economic performance for all system configurations. If the system is ideal, with perfect orientation and tilt and no heat exchanger, the best economic performance will be realized. If the system is not ideal, the collector area A_{opt} is still the best area but the economic performance will not equal that of an ideal system.

The performance penalty associated with a nonideal system is k, the area-performance correction factor. For an ideal system, $k = 1$ and for nonideal systems $k = < 1$. k relates the area A_{opt} to an equivalent but lower A_{ideal}, which is then used for calculating system performance. k is simply the product of

1. The yearly collector effectiveness at our orientation
2. The yearly collector effectiveness at our tilt
3. The heat-exchanger correction factor

If the collectors do not face due south, the collector effectiveness can be determined from the graph in Fig. 6.7. For example, if the collectors are oriented 20° west of south, the yearly effectiveness at that orientation is 0.98.

If the collectors are not tilted at an angle of collector latitude + 15°, the collector effectiveness can be determined from the graph in Fig. 6.8. For example, if the collectors are tilted at latitude + 35°, the yearly effectiveness at that tilt is 0.93.

If we have a liquid collector system that has a heat exchanger between the collectors and the storage unit, the correction factor is equal to

$$\frac{1}{(1 + Z\,dT)},$$

where

dT = temperature drop across the exchanger,

z = 0.018 for °C and 0.01 for °F.

If we do not know the temperature drop across the exchanger, an average value of 5.6° C (10° F) can be assumed. For example, if the temperature drop is 7° C, the heat-exchanger correction factor is 0.888. Air-type collector systems do not have a heat exchanger between collectors and the storage unit, so the heat-exchanger correction factor does not apply to them. Though there is a penalty involved with the inclusion of a heat exchanger between collectors and the storage unit, if we plan to have liquid collectors and live in an area where freezing temperatures are encountered we must have either a draindown system or an antifreeze loop with a heat exchanger (and the accompanying penalty). There are advantages and disadvantages associated with each type of system and all factors (including the penalty) should be weighed before a system type is chosen.

All corrections are cumulative. If our system is oriented at 20° west of south, is tilted at latitude + 35°, and includes a heat exchanger with a

7° C temperature drop, k, the area-performance correction factor is calculated as follows:

$k = 0.98 \times 0.93 \times 0.888$
$k = 0.809$

AS_{jan}/L_{jan}

AS_{jan}/L_{jan} is calculated using the area, A_{opt}, the January solar radiation on a horizontal surface, S_{jan} (Appendix C), and the area-performance correction factor, k.

$\mathcal{J}$: Fraction of Yearly Total Heat
Load Supplied by Solar Energy

$\mathcal{J}$, the fraction of the total yearly heat load provided by solar energy, is found on the graph in Fig. 6.9 by use of the AS_{jan}/L_{jan} value. $\mathcal{J}$ is read off of the curve for the appropriate collector type and represents the yearly heating performance for the system design.

All the curves on Fig. 6.9 cross near a solar fraction of 0.65. At low AS_{jan}/L_{jan} values, since the air-type-collector curve is the lowest, it might seem that air-type collectors are not as desirable as the other types until a solar fraction of 0.65 can be obtained. Comparison of the collectors and their overall performances is not that simple, however. Since the collectors are different and the costs associated with them are different, the calculated AS_{jan}/L_{jan} and $\mathcal{J}$ values will also be different.

Tax Savings

Federal-income-tax savings and system cost after tax savings are calculated as indicated in Fig. 6.6 using the system cost and the l value already determined and the m value from the tax-savings block in Fig. 6.6 corresponding to the type of system and the l value used.

If we include state income tax benefits in our calculation of the system cost and determination of l, the m values listed in Fig. 6.6 will not correspond to our values. We must determine m values for our tax situation.

Economic Check

Whether or not our solar heating system is economical is the last calculation necessary in this sizing procedure. $\mathcal{J}$, $bLEF$, b, F_2, the system cost, and the amount of tax savings are used in this calculation, which is performed as indicated in Fig. 6.6.

If the result of the economic-check equation is positive, the solar heating system is a good investment. If the result is negative, the system is not as economical as a conventional heating system and is not recommended. (The numerical result represents the amount we would receive today if all our future payments and savings due to the presence of the solar heating system were totaled and discounted to today's dollars.)

In this sizing method we use fixed values for all economic factors (inflation, discount rates, etc.). The economic feasibility of the system also depends on predicted performance, which is based on average weather conditions. It is also assumed that corrections to the collector area do not change the performance level of the system. If all the specified conditions are met, the performance of the solar heating system should be as indicated. In reality, however, things seldom remain constant. Any deviations from the specified values could result in a different level of performance. A slight change in an economic factor (a 1% increase in the insurance and maintenance fraction, for example) may make the difference in the economic-check equation between an economical and an uneconomical system. The result of the economic-check equation should therefore be taken only as a rough estimate of the system's total economic performance.

6.3.4 Sizing Example

You own the house in Yakima, Washington used in section 6.2.3 for the practice calculation of heat load. The total January load for your house, L_{jan}, is 12,050 MJ and the total yearly load, L_{yr}, is 74,920 MJ.

You plan to use air-type flat-plate collectors and the anticipated system costs are \$175/m² of collection surface plus \$1000. You plan to finance the system with a 25-year loan at 8% interest and 10% down payment and you expect the system to last 25 years. You expect energy inflation to be 11% and general inflation to be 7%.

Solar heating systems are exempt from property tax in Washington state for the next seven years, but you expect that the legislature will reenact the property-tax exemption when it expires. The state has no other solar incentive programs.

Your income tax rate is 30%. The insurance and maintenance fraction is expected to be 0.5% per year. Electric resistance backup heat will be used and the hypothetical price of electricity is \$0.027/kWh.

The actual collector orientation will be 10° west of south at a tilt of latitude + 25°.

Using the above information, determine

1. The system cost and collector area
2. The yearly fraction of heat that will be provided by solar energy
3. Whether the system is economical

All calculations are shown in Table 6.3 and Figs. 6.10 and 6.11, which correspond to Table 6.2 and Figs. 6.5 and 6.6:

ECONOMIC-HEATING-PERFORMANCE SIZING

Table 6.3 Cost and Inflation Factors

Factor	Amount
C_a area-dependent cost (\$/m² or \$/ft²)	175 C_a
C_b area-independent cost (\$)	1000 C_b
d discount rate (interest rate of best alternative investment)	0.08 d
e energy inflation rate	0.11 e
i general inflation rate	0.07 i
J mortgage rate	0.08 J
K downpayment fraction (if purchased with cash)	0.10 K
M mortage term (years)	25 M
N analysis term (years)	25 N
P property tax rate	— P
R income-tax rate (state + federal − state × federal)	0.30 R
U insurance and maintenance fraction	0.005 U
Z depreciation lifetime (commercial only; years)	— Z

F (Year, Row, Column)	Inflating-Discounting Factor*
AA† $F(\frac{25}{N}, \frac{0.11}{e}, \frac{.08}{d})$	32.791 AA
BB $F(\frac{25}{N}, \frac{0.07}{i}, \frac{0.08}{d})$	20.750 BB
CC $F(\frac{25}{M}, \frac{0}{\text{zero}}, \frac{0.08}{d})$	10.675 CC
DD $F(\frac{25}{M}, \frac{0}{\text{zero}}, \frac{0.08}{J})$	10.675 DD
EE $F(\frac{25}{M}, \frac{0.08}{J}, \frac{0.08}{d})$	23.148 EE
FF $F(\frac{25}{M}, \frac{0.08}{J}, \frac{0}{\text{zero}})$	73.106 FF
GG $F(\frac{}{Z}, \frac{0}{\text{zero}}, \frac{}{d})$	_____ GG (commercial only)

*Inflating-discounting factors are listed in the tables in Appendix D.
†If you are purchasing the solar heating system with cash, only AA, BB, and GG (if applicable) need be determined.

Residential

$$F_1 = \frac{\boxed{32.791}}{AA}$$

F_2 calculation

$$\frac{0.70}{(1-R)} \times \frac{}{P} = \frac{0}{}$$

$$\frac{0.10}{K}$$

$$+\frac{0.005}{U}$$

$$\boxed{0.005} \times \frac{20.750}{BB} = +\frac{0.104}{}$$

$$\frac{0.70}{(1-R)} \times \frac{10.675}{CC} \div \frac{10.675}{DD} = \frac{0.70}{}$$

$$\frac{0.30}{R} \times \frac{23.148}{EE} \div \frac{73.106}{FF} = +\frac{0.095}{}$$

$$\boxed{0.795} \times \frac{0.90}{(1-K)} + \frac{0.715}{}$$

$$F_2 = \boxed{0.919}$$

Commercial

$$F_1 = \frac{}{(1-R)} \times \frac{}{AA} = \boxed{}$$

F_2 calculation

F_2 from residential calculation

$$\frac{}{R} \times \frac{}{U} \times \frac{}{BB} = \frac{}{}$$

$$\frac{}{R} \times \frac{}{GG} \div \frac{}{Z} = \frac{}{}$$

$$F_2 = \boxed{}$$

Variables are defined in Table 6.3.

Fig. 6.10 Calculation of economic factors F_1 and F_2.

ECONOMIC-HEATING-PERFORMANCE SIZING   241

Collector location __Yakima, WA__   Collector orientation __10°__ (west) east of south
Location latitude* __46.6°__   Collector tilt __Lat +25__ (latitude + 15° is best)

Collector type __air__ 0.432/b	Type b	Liquid flat plate 0.281	Liquid evacuated tube 0.346	Air flat plate 0.432

E (delivered energy cost):

Type		Furnace efficiency†	
Gas ____ $/m³ ÷ 37			
Oil ____ $/l ÷ 35	= 0.0075	÷ 1.0	= 0.0075 $/MJ
Electric .027 $/kWh ÷ 3.6			
Gas ____ $/thm ÷ 100			
Oil ____ $/gal ÷ 140	= ____	÷ ____	= ____ $/Btu × 10³
Electric ____ $/kWh ÷ 3.413			

$bLEF_1$:
$$0.432 \times \frac{74{,}920}{L_{yr}\,(\text{Fig. 6.1})} \times \frac{0.0075}{E} \times \frac{32.791}{F_1\,(\text{Fig. 6.5})} = \frac{7960}{bLEF_1\,(\$)}$$

Determination of I and system cost (SC):

System type	$I^{\ddagger}$	SC	Valid SC range
Residential	0.4	16,337	<$10,000
	0	9662	≥$10,000
Commercial	0.15 or 0.25		ALL

$$SC = \frac{7960\,(bLEF_1)}{0.919 - I\,(F_2,\text{Fig. 6.5})} + \frac{1000}{C_b\,(\text{Table 6.2})}$$

Residential: calculate the system cost using an I value of 0.4. If the SC is < $10,000, I = 0.4. Otherwise, calculate the SC for I = 0. If the SC ≥ $10,000, I = 0. Otherwise, I = 0 and SC = $10,000.
Commercial: identical to residential calculation except use one I value for all system costs. Use I = 0.15 for Energy Credit alone or I = 0.25 if system is eligible for Energy Credit and Investment Credit.

I = 0
system cost = $10,000

A_{opt}:
$$\frac{(10{,}000 - 1000)/SC \quad C_b}{175 \; C_a\,(\text{Table 6.2})} = \underline{51.4}\;A_{opt}\; (m^2) \text{ or } ft^2$$

k area-performance correction:
$$0.99 \times 0.98 \div (1 + \_\_°C \times 0.018) \text{ or } (1 + \_\_°F \times 0.01) = 0.97\;k$$
(effectiveness at actual orientation (Fig. 6.7); effectiveness at actual tilt (Fig. 6.8); temperature drop across heat exchanger)

AS_{jan}/L_{jan}:
$$\frac{51.4\,A_{opt} \times 128.4\,S_{jan}(\text{App. C}) \times 0.97\,k}{12{,}050\,L_{jan}(\text{Fig. 6.1})} = 0.531\;AS_{jan}/L_{jan}$$

$\mathcal{F}$ = 0.64 (from Fig. 6.9) at AS_{jan}/L_{jan}

Tax savings:
$$10{,}000 - \left[(10{,}000 \times 0 + 4000)\right] = \$4000 \;\text{Tax savings} \;= \$6000\;\text{Cost after tax savings}$$
(SC; SC; I; m)

Residential	
I	m
0.4	0
0	4000
Commercial	
I	m
0.15	0
0.25	0

Economic check§:
$$\left(\frac{0.64}{\mathcal{F}} \times \frac{7960}{bLEF_1} \div \frac{0.432}{b}\right) - \left(\frac{0.919}{F_2} \times \frac{10{,}000}{SC}\right) + \frac{4000}{\text{Tax savings}} = \$6603$$

*See Appendix C.
† Usually 0.5–0.6 for gas and oil, 1.0 for electric, and 1.4 for heat pump.
‡ Federal tax credit only.
§ If the result of the economic check is negative the proposed solar heating system is not as economical as a conventional heating system.

Fig. 6.11 Determination of A_{opt}, yearly solar fraction, and total-system cost.

1. The 51.4 m² system costs $10,000 before the $4000 federal income-tax credit.
2. Approximately 64% of the year's heat and hot water will be provided by solar energy.
3. The system is economical.

6.4 SUMMARY Sizing of a solar heating system can be done relatively quickly and accurately using the method described in this chapter. The actual size of a solar heating system depends on the physical makeup of the system as well as all the costs associated with it. Table 6.4 shows sizing recommendations based on collector area (A_{opt}) for both air and liquid systems.

Table 6.4 **Summary of Sizing Recommendations for Air-and Liquid-Type Solar Heating Systems**

		Air-Type Systems	
Collectors	Orientation	Due south	Due south
	Slope	Latitude + 15°	Latitude + 15°
	Airflow rate	5–15 l/s m² collector	1–3 cfm/ft² collector
	Pressure drop	50–200 Pa	0.2–0.8 in water
Storage unit	Volume of pebbles	0.15–0.35 m³/m² collector	0.5–1.15 ft³/ft² collector
	Pebble size	0.019–0.038 m concrete aggregate	0.75–1.5 in concrete aggregate
	Bed length in airflow direction	1.2–2.4 m	4–8 ft
	Pressure drop	25–75 Pa	0.1–0.3 in water
Water preheat tank	Tank size	1.5–2 times conventional water heater	
		Liquid-Type Systems	
Collectors	Orientation	Due south	Due south
	Slope	Latitude +15°	Latitude +15°
	Flow rate	0.015 l/s m² collector	0.02 gpm/ft² collector
	Pressure drop	3.4–6.9 kPa/collector	0.5–10 psi/collector
Storage unit	Volume of water	50–100 l/m² collector	1.25–2 gal/ft² collector
Water preheat tank	Tank size	1.5–2 times conventional water heater	

Note: Pa = pascal; kPa = kilo pascal; cfm = cubic feet per minute; gpm = gallons per minute; psi = pounds per square inch.

REFERENCES

1. Solar Energy Laboratory (1976). *TRNSYS, A Transient Simulation Program.* Madison, Wis.: University of Wisconsin Solar Energy Laboratory, report 38.
2. Klein, S. A., W. A. Beckman, and J. A. Duffie (1976). A design procedure for solar heating systems. *Solar Energy,* 18 (2):113.
3. Klein, S. A., W. A. Beckman, and J. A. Duffie (1976). Design procedure for air heating systems. *Sharing the Sun* 4:271.
4. Beckman, W. A., J. A. Duffie, and S. A. Klein (1977). Simulation of solar heating systems. In R. C. Jordan and B. Y. H. Liu (eds.), *Applications of Solar Energy for Heating and Cooling of Buildings.* New York: ASHRAE.
5. Beckman, W. A., S. A Klein, and J. A. Duffie (1977). *Solar Heating Design by the f-Chart Method.* New York: John Wiley.
6. Balcomb, J. D., and J. C. Hedstrom (1976). A simplified method for calculating required solar collector array size for space heating. *Sharing the Sun* 4:281.
7. Stickford, G. H., Jr. (1976). An averaging technique for predicting the performance of a solar energy collector system. *Sharing the Sun* 4:295.
8. Solar Energy Design Corporation of America (1976). *G-Chart.* Denver, Colo.: Solar Energy Design Corporation of America.
9. Barley, C. D., and C. B. Winn (1978). Optimal sizing of solar collectors by the method of relative areas. *Solar Energy* 21 (4):279.
10. Ward, J. C. (1976). Minimum cost sizing of solar heating systems. *Sharing the Sun* 4:336.
11. ASHRAE (1979). *Cooling and Heating Load Calculation Manual.* New York: ASHRAE.
12. Air Conditioning Contractors of America (1975). *Load Calculation for Residential Winter and Summer Air Conditioning.* 4th ed. Washington, D.C.: Air Conditioning Contractors of America.
13. ASHRAE (1977), *ASHRAE Handbook of Fundamentals.* New York: ASHRAE.
14. Solar Energy Applications Laboratory, Colorado State University. (1977). *Solar Heating and Cooling of Residential Buildings—Sizing, Installation and Operation of Systems.* Washington, D.C.: U.S. Government Printing Office, no. 003-011-00085-2.

BIBLIOGRAPHY

Heat-Load Calculation

Air Conditioning Contractors of America (1975). *Load Calculation for Residential Winter and Summer Air Conditioning, Manual J.* Washington, D.C.: Air Conditioning Contractors of America.

ASHRAE (1979). *Cooling and Heating Load Calculation Manual.* New York: ASHRAE.

Barley C. D., and C. B. Winn (1978). Optimal sizing of solar collectors by the method of relative areas. *Solar Energy* 21(4):279.

NAHB Research Foundation (1971). *Insulation Manual.* Rockville, Md.: NAHB.

SMACNA (1975). *Load Calculation Guide (1 and 2 Family Dwellings) for Heating and Air Conditioning.* Tyson's Corner, Vienna, Va.: SMACNA.

Solar Economics

Boer, K. W. (1976). Payback of solar systems. *Sharing the Sun* 9:1.

McGarity, A. (1976). *Solar Heating and Cooling—An Economic Assessment.* Washington, D.C.: U.S. Government Printing Office, no. 038-000-003000-3.

Ruegg, R. T. (1975). *Solar Heating and Cooling in Buildings: Methods of Economic Evaluation.* Washington, D.C.: U.S. Department of Commerce, NBSIR 75-712, COM 75-11070.

Sizing and Minimum—Cost Sizing

Balcomb, J. D., and J. C. Hedstrom (1976). A simplified method for calculating required solar collector array size for space heating. *Sharing the Sun* 4:281.

Barley, C. D., and C. B. Winn (1978). Optimal sizing of solar collectors by the method of relative areas. *Solar Energy* 21(4):279.

Beckman, W. A., J. A. Duffie, and S. A. Klein (1977). Simulation of solar heating systems. In R. C. Jordan and B. Y. H. Liu (eds.), *Applications of Solar Energy for Heating and Cooling of Buildings*. New York: ASHRAE.

Beckman, W. A., S. A. Klein, and J. A. Duffie (1977). *Solar Heating Design by the f-Chart Method*. New York: John Wiley.

Klein, S. A., W. A. Beckman, and J. A. Duffie (1976). A design procedure for air heating systems. *Sharing the Sun* 4:271.

Klein, S. A., W. A. Beckman, and J. A. Duffie (1976). A design procedure for solar heating systems. *Solar Energy* 18 (2):113.

Solar Energy Applications Laboratory, Colorado State University. *Solar Heating and Cooling of Residential Buildings—Design of Systems*. Washington, D.C.: U.S. Government Printing Office, no. 003-011-0084-4.

Solar Energy Design Corporation of America (1978). *G-Chart*. Denver, Colo.: Solar Energy Design Corporation of America.

Solar Energy Laboratory (1976). *TRNSYS, A transient Simulation Program*. Madison, Wis.: University of Wisconsin Solar Energy Laboratory, report 38.

Stickford, G. H., Jr. (1976). An averaging technique for predicting the performance of a solar energy collector system. *Sharing the Sun* 4:295.

Ward, D. S. (1977). The mathematics for sizing a solar heating system. *Solar Age Catalog*. Port Jervis, N.Y.: SolarVision, p. 124.

PROBLEMS

6.1 Describe in your own words the meaning of *optimum collector area*.

6.2 The savings attributable to your solar heating system will equal $2435 in year 20. What is the present value of that amount if the discount rate is 11%?

6.3 Your solar heating system provides 47% of your heat and hot water. Your heating bill this winter would have been $700 without a solar system. The cost of energy is increasing at 13% per year and your discount rate is 12%. Use the inflating-discounting functions from Appendix D to determine the present value of your heating-bill savings for the next 10, 15, and 20 years. What would the savings have been if the discount rate had been 10%? If the inflation rate had been only 11%?

6.4 Why does a discount rate equal to the borrower's mortgage rate make sense?

6.5 Use Fig. 6.1(a) to calculate the approximate heat-loss rate for a house that is a single-story slab on grade, measuring 12 m × 20 m. All walls are 2.4 m high and are insulated to a metric R value of 3.5. The ceiling is insulated to R-5.5 and the slab is surrounded by 5 cm of rigid insulation. There are eight double windows (1.3-cm air space) measuring 1.3 m × 2 m. There are four double windows (1.3-cm air space) measuring 0.4 m × 1 m. The two doors, each 5 cm thick × 2 m square, have metal storm doors. Ducts run through the attic. What is the heat loss in MJ/DD for this house? If three people live in the house, what is their daily hot-water use?

6.6 Use Fig. 6.1(b) to calculate the January and yearly heat loads for a house that is two stories tall, with each story measuring 20 ft × 30 ft. It has 8-ft-high ceilings. The upstairs ceiling is insulated to an English R value of 38. The walls are R-24 and the floor is R-19. The house has 6 double (1/4-in air space) windows measuring 18 in × 4 ft and 12 single windows with storm windows measuring 2 ft × 3 ft. There is a 6 ft 8 in × 7 ft insulated (1/4-in) sliding glass door in the back, and the 1.5-in-thick front door is 6 ft 8 in × 3 ft. Five people live in the house, which is located in Boise, Idaho.

6.7 How much energy will it take to heat your hot water if you use only 40 l of 55° C water per day? The water main's temperature is 9° C.

6.8 After examining past years' heating bills, you conclude that heating your house requires 700 gal of oil per year. Hot water for four people is also provided by the furnace. What are the January and yearly heat loads for your house if there are 1300° F heating degree days in your town for January and 6700 for the year? (The furnace is 50% efficient.)

6.9 Suggest a method that could be used to calculate January and yearly heat loads from old electric bills, even though charges for ordinary lights and appliances are included in these bills.

6.10 What is the optimum area for an air-type solar heating system on a commercial building whose statistics are as given in Table 6.3? The depreciation lifetime is 20 years. Item GG must be determined for this problem. The building's January heat load is 40,000 MJ and the yearly heat load is 200,000 MJ. Electricity costs $.06/kWh. The solar collectors are tilted at latitude + 15° and face due south. There is no extra heat exchanger. Is this system economical? Why are F_1 and F_2 different for a residence and a business?

6.11 A residence has a January heat load of 15,000 Btu × 10^3, and yearly load of 50,000 Btu × 10^3. A liquid-type system will be used and its cost is estimated at $30/ft² plus $1500. The system will carry a 90% mortgage at 12% interest for 20 years. Energy inflation is expected to be 14% while regular inflation is 10%. The property-tax rate is 0.03% and the income tax rate is 40%. Insurance and maintenance will cost approximately 2% of the purchase cost per year and will inflate over time. Backup heat is by gas at $.47/thm. The solar collectors will face 15° west of south and will be tilted at an angle of latitude + 10°. The heat exchanger between the collector and the storage unit has a temperature drop of 12° F. Calculate the optimum collector area and system performance for this Tulsa, Oklahoma residence. Is this system economical?

6.12 Show that the federal tax credit (as a fraction of system cost for a residence) of 40% of the first $10,000 of cost can be calculated from the equation

$$T = l + \frac{m}{AC_a + C_b}.$$

l and m are constants that can change with the system cost ($AC_a + C_b$). What are the values for l and m and their valid system-cost ranges?

6.13 Develop the l, m, and system-cost ranges for use in the formula for a tax credit shown in problem 6.12 for a combination of the residential federal tax credit and a flat 10% state tax credit.

6.14 Your air-type solar system has 20m² of collectors. What volume of rock pebbles should be used for solar heat storage?

7 PASSIVE SOLAR HEATING

7.1 WHAT IS PASSIVE SOLAR HEATING?

Passive solar heating is the capture, storage, and use of the sun's energy for heating without the use of fans or pumps to aid in heat circulation. The solar energy is collected and stored directly in or with the aid of the building itself. In truly passive systems convection, conduction, and radiation are the only means of heat circulation. Some passively heated buildings use elements of active solar heating systems such as fans to help circulate warm air. These are called *hybrid* systems.

The incorporation of passive solar heating features into a building should not be confused with simple energy-conserving construction. All structures should be well insulated and well constructed. Windbreaks, solar orientation, partial burying of structures, airlock entries, storm doors, insulated or storm windows, installation of fewer and smaller windows on the north side, and zoning (installation of separate thermostats for warmer and cooler areas of the building) are features that help cut the heating costs of every building. (See Chapter 8 for details on how to incorporate these energy-conserving features into your house.)

What does constitute a passive solar heating system? The keys to an effective passive heating system are glazing, mass, and insulation. The most common type of passive system is a *direct-gain* system. This type of passive heating involves large areas of south-facing glazing that admit sunlight directly into the living space, where it strikes massive structures and is stored as heat. *Indirect-gain* systems heat the space indirectly; the massive structures are placed between the sun and the living space.

The massive structures, collectively called the building's *thermal mass*, are often very thick floors or walls of concrete, adobe, or stone, or large containers of water. The solar radiation (sunlight) passes through the glazing and strikes the thermal mass, where the heat is converted to internal energy. This energy is stored for later use as an increase in temperature of the mass. (See Chapter 4 for a review of how heat is stored, causing a change in temperature.)

The thermal mass helps modulate temperature extremes within the building by absorbing extra heat during the day and releasing it at night (Fig. 7.1). Without the thermal mass, the building would quickly overheat during sunny winter periods (because of the large glazing area) and need to be vented. At night and during cloudy periods, heat stored in the thermal mass is slowly released to the building as the mass cools.

To ensure that the heat is released primarily to the building and not to the out-of-doors, insulation of the exterior of the thermal mass is desirable. If the building structure itself provides the mass (e.g., as concrete or adobe walls), the exterior of these walls must be insulated. If a slab or stone floor provide mass, at least the sides and preferably the sides and bottom should be insulated with rigid, rodent-proof and waterproof insulation. Insulation of the bottom of the slab isolates the interior of the building from the earth. Although the earth is warmer than the outside environment, it is still colder than the interior of the building and nonrecoverable heat can be lost to it. If a slab or massive wall continues outside the house to form a patio or windbreak, the inside and outside sections should be thermally isolated from one another by insulation to prevent heat loss. Also, tightly fitting nighttime insulation of the large glazed area is very important for reducing heat losses.

Thermal mass also helps modulate the indoor temperatures during the summer. The massive structures remain cooler than the air during the day and absorb extra heat. Summer shading of the mass and cross ventilation often are enough to ensure comfort without air conditioning (Fig. 7.2).

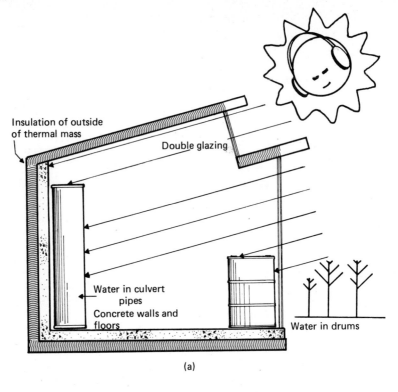

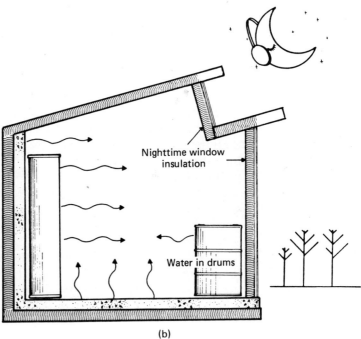

Fig. 7.1 Winter operation of passive systems: (**a**) thermal mass absorbs solar energy during the day, (**b**) thermal mass radiates heat to house during the night.

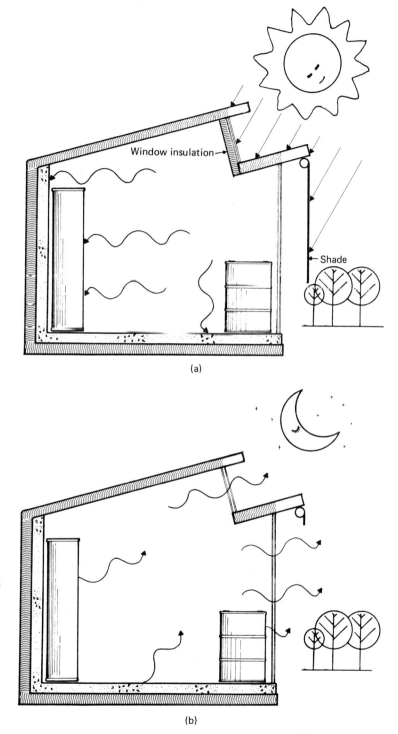

Fig. 7.2 Summer operation of passive systems: (**a**) covered glazing prevents solar energy collection and thermal mass absorbs excess heat from house during the day, (**b**) uncovered glazing allows thermal mass to radiate heat to the out-of-doors during the night.

7.2 THERMAL MASS FOR SIMPLE PASSIVE SYSTEMS

Concrete, adobe, stone, brick, and contained water are all used in passive solar heating systems to provide the thermal mass necessary for their efficient and comfortable operation. Water has a very high heat capacity (see Chapter 4) so it is well suited for use as a relatively compact heat-storage medium. Heat is circulated through water quickly by natural convection currents, so it is also quickly available to heat the room. To store a given amount of heat takes approximately one-half as much water as solid concrete, adobe, or stone (by volume). Some experimentation with phase-change materials for passive solar heat storage has also been done (see Chapter 4).

7.2.1 Massive Walls and Floors

Direct-gain massive walls and floors, often 20–45 cm (8–18 in) thick and insulated on the outside to prevent heat loss to the earth and outside air are effective as heat-collection and heat-storage units (Fig. 7.3). The walls and floors can be of concrete, adobe, stone, or brick. Often the floor may be covered with dark tile or slate to improve collection efficiency. Wall-to-wall carpeting on floors and paneling on walls are not recommended

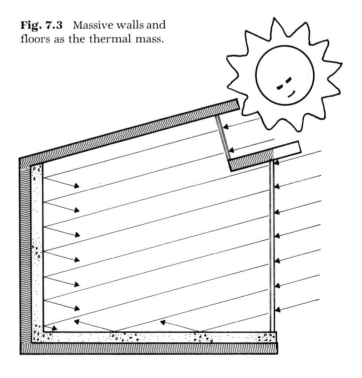

Fig. 7.3 Massive walls and floors as the thermal mass.

Fig. 7.4 Direct-gain passive solar residence located in Santa Fe, New Mexico. (Photographs courtesy of David Wright and Dennis Andrejko, SEAgroup, Nevada City CA.) (Continued)

in such systems as they interfere with the direct collection of solar energy by the mass and act as insulation between the mass and the room.

Figure 7.4 shows a direct-gain passively heated home with removable shading louvers. The multilevel house is constructed primarily of adobe, with interior division walls made of water vessels covered with adobe plaster. This solar home needs only 3/4 cord of wood to meet winter backup heating requirements.

7.2.2 Trombe Walls

The Trombe wall was developed during the 1960s at the Centre Nationale de la Recherche Scientifique (CNRS) in Odeillo, France and named after one of its developers, Dr. Felix Trombe. The Trombe wall is a very thick wall usually made of concrete (but also of adobe, stone, or brick) and located directly behind a continuous curtain wall

Fig. 7.4 (continued)

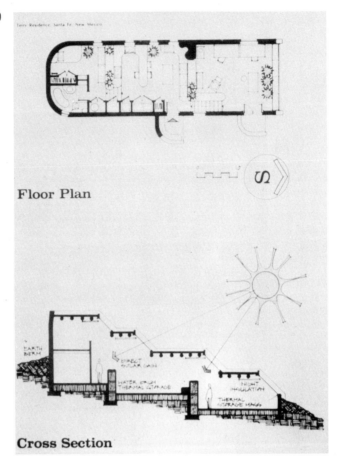

of double glazing (Fig. 7.5). The Trombe wall is actually an indirect-gain system because the solar radiation does not enter the living space. Solar energy does not have to pass completely through the mass to heat the space (because of the circulation vents) as it does in other indirect-gain systems.

The outside of the 20–45 cm (8–18 in) thick wall is usually slightly rough in texture and painted black or another dark color to allow efficient collection of solar energy. There are vents at the top and bottom of the wall, with dampers to control air circulation. During sunny periods the wall becomes warm and heats the air in the space between the wall and the windows. The warmed air rises through the vents at the top, drawing cool air in through the ducts at the bottom. The warm air circulates through and warms the house. This *thermosiphoning* action

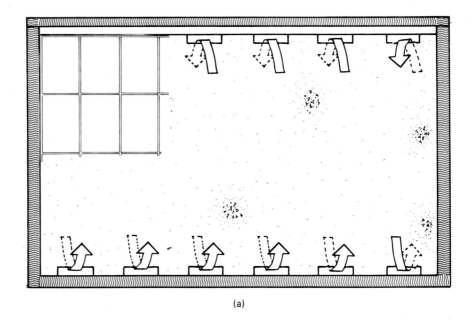
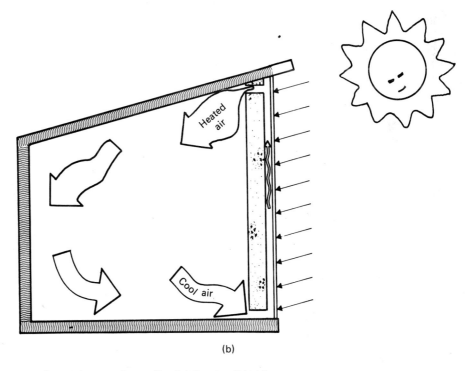

Fig. 7.5 Trombe wall: (**a**) front, (**b**) side.

continues until the exterior surface of the wall is approximately room temperature, about two to three hours after sunset. Since the movement of the thermosiphoned air is slow, the dimension of the air space between the glazing and the wall is important. A space at least 10 cm (4 in) deep is recommended for proper airflow.

At night, heat stored in the wall is slowly released into the room as the thermal mass cools. Most of the solar heat eventually reaching the rooms (about 70%) arrives via thermal radiation from the warmed mass. Heat is also released to the out-of-doors through the glazing if nighttime insulation is not used to isolate the Trombe wall from the outside. Removable blow-in insulation between the glazing layers or inflatable-retractable insulation between the Trombe wall and the glazing work very well.

At night during the heating season the vents must be blocked by manual or automatic dampers to prevent reverse thermosiphoning and cooling of the room. If the vents are not blocked, the warm room air cools as it comes through the vents and into contact with the cold glass. The air cools and falls, drawing more air through the cold collector, effectively cooling the room. An undampered Trombe wall can lose significant amounts of heat at night. During the summer this cooling action may be desirable.

Summer Cooling with a Trombe Wall

Windows placed at the top of the glazing wall aid in daytime summer cooling (Fig. 7.6). Heated air exits out the top of the collector wall, drawing air through the room for ventilation and cooling.

Trombe-Wall Variation

A variation of the Trombe wall is shown in Fig. 7.7 (p. 256). This is different from the standard Trombe wall in that it automatically prevents reverse thermosiphoning at night, during early morning hours, and on cloudy days. Cold air settles in the sunken portion of the collector and prevents reverse thermosiphoning. Summer operation is the same as for the regular Trombe wall.

7.2.3 Water Walls

Water for storing solar heat can be contained in vertical culvert pipes, ducting, fiberglass tubes, recycled large drums, or specially constructed wall-to-wall, ceiling-to-floor containers (Fig. 7.8, p. 257). These contain-

Fig. 7.6 Cooling with a Trombe wall.

ers can be placed directly behind south-facing windows for an indirect-gain system, or at the back of the room for a direct-gain system.

Inexpensive, used steel drums can be stacked horizontally or vertically. The floor should be specially reinforced to support the weight of the water. Drums placed directly behind windows are usually painted black on the side facing the windows to permit efficient energy collection and white on the inside for increased interior light. Water walls located directly behind south windows are more effective than Trombe walls of equal storage capacity [1].

Exterior and interior views of a *drumwall* are pictured in Figs. 7.9 and 7.10. Black-painted drums filled with water sit behind glass, creating an indirect-gain Drumwall passive solar heating system for the house shown in Fig. 7.9. The open insulated glazing cover serves as a reflector during winter days and is closed at night to reduce heat losses from the warm drums. The drums contain 80 kg of water per square meter of glass (120 lb/ft^2) and usually increase 4–5° C (7–9° F) in temperature during the day. The stored heat is released gradually by the

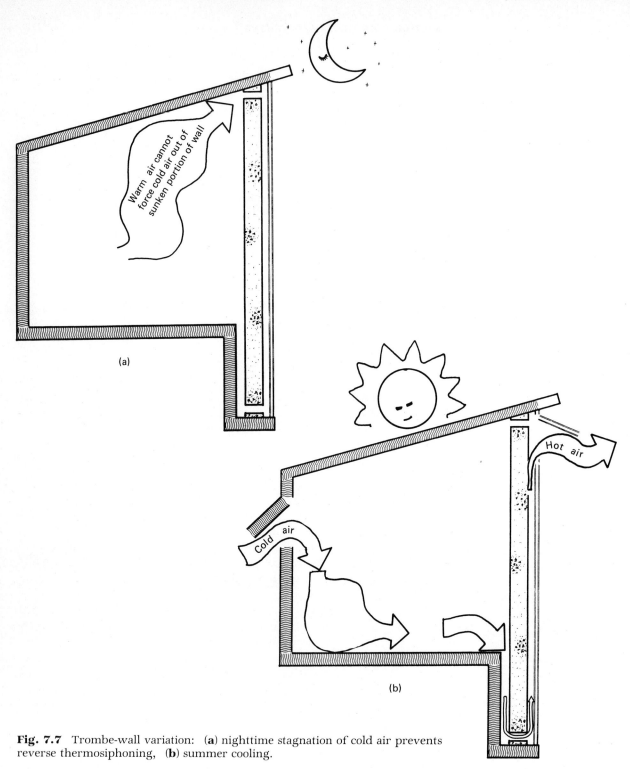

Fig. 7.7 Trombe-wall variation: **(a)** nighttime stagnation of cold air prevents reverse thermosiphoning, **(b)** summer cooling.

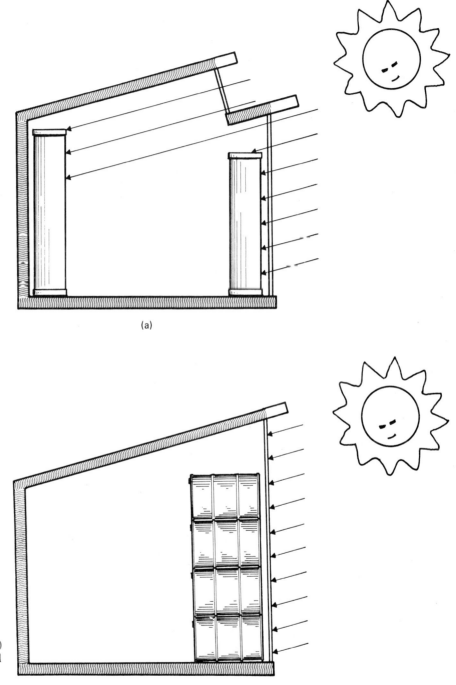

Fig. 7.8 Water walls: (**a**) culvert pipes, (**b**) recycled drums.

Fig. 7.9 House with an indirect-gain, Drumwall passive solar heating system. (Photograph courtesy of Zomeworks Corporation, Albuquerque, NM.)

Fig. 7.10 Interior view of a Drumwall. (Photograph courtesy of Zomeworks Corporation, Albuquerque, NM.)

drums at night, lessening the temperature swing inside the house. While the outside ends of the Drumwall drums should be painted a dark color to absorb sunlight efficiently, the inside may be painted almost any color (Fig. 7.10). In the summer the drums help cool the building because they absorb excess heat from the house during the day. Windows are opened at night, allowing this excess heat to escape.

The retrofitted solar greenhouse shown in Fig. 7.11 was added to a light-commercial-manufacturing factory in Deerfield, New Hampshire. The glazing and black-painted water-filled storage tubes are of Kalwall Sunlite ® fiberglass-reinforced polymer. The lightweight "venetian" insulating shutters were constructed by the owner to reduce nighttime heat losses.

The problem with using water for passive solar heating systems is that it must be contained. Evaporation, corrosion, and leaking are therefore potential problems. Anti-corrosive agents (such as potassium dichromate) and plastic liners for metal containers should permit them to last at least 15–30 years, however. High-quality fiberglass containers should also last 15–30 years. Evaporation from imperfectly sealed or open containers may be slowed by application of a thin layer of mineral oil over the water. Copper sulfate or chlorine can be added as an algicide.

Fig. 7.11 Passive solar greenhouse with water-filled storage tubes. (Photograph courtesy of Kalwall Corporation, Manchester, NH.)

7.2.4 Solid Walls

Indirect-gain solid masonry or water walls placed directly behind double south-facing glazing (Trombe walls without vents) work well in some climates (Fig. 7.12). All solar heat must pass through the solid wall. The performance of such walls is usually lower than that of Trombe walls with vents and dampers, but higher than that of Trombe walls with vents but without dampers [1]. Without vents, thermosiphoning cooling of the building in summer is not possible.

7.2.5 Combination of Materials

If the efficiency and compactness of water is desired without the (perhaps unsightly) appearance of drums or pipes, the side of a water vessel that faces the room can be disguised by other heat-storage materials such as concrete or adobe (Fig. 7.13). Water in containers or plastic bags can also be enclosed by concrete or stonework. Some combinations are more efficient than the masonry alone but less efficient than an equal amount of water.

7.2.6 Phase-Change Materials as Thermal Mass

For all passive solar heating systems, one key to maintaining comfort is a thermal mass that slowly stores solar heat and later releases it. Just as phase-change storage for active systems is more compact than water or rock storage (see Chapter 4), it can provide a more compact thermal mass for passive systems. Aside from problems stemming from the stability of the materials, this new approach to passive solar heat storage looks very promising. The advantage of the compact size of a phase-change thermal mass is greater with passive systems than with active systems because of the lower temperature range experienced, but the true advantage is constant temperature behavior at the melting point.

With a regular sensible-heat thermal-mass system, the exterior surface of the mass may go through extreme temperature changes that are reflected to a lesser degree in the temperature of the room. Since the temperature of the phase-change material remains essentially constant for long periods of time, a more constant air temperature and greater comfort for occupants is to be expected. Successful experiments involving impregnating the phase-change salts in concrete and incorporating them into ceiling tiles have been reported [2, 3]. Use of ceiling tiles with special window blinds that reflect sunlight toward the ceiling frees valuable floor and wall space for furniture.

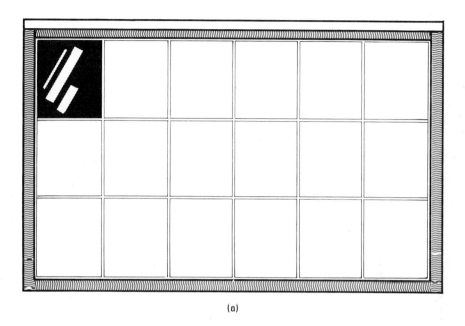

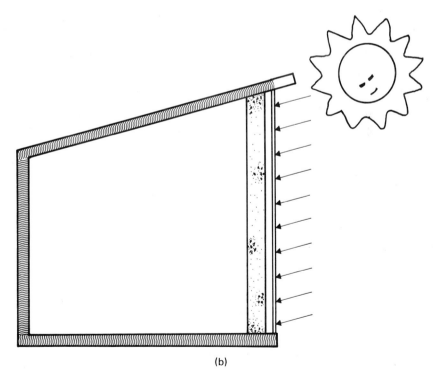

Fig. 7.12 Solid-wall thermal mass: (**a**) front view, (**b**) side view.

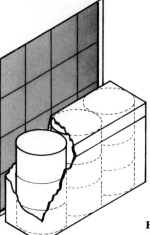

Fig. 7.13 Combination thermal mass: drums of water disguised by decor.

7.3 SIZING SIMPLE PASSIVE SYSTEMS

Sizing a passive system for a building is more complicated than simply supplying as much south-facing glass and thermal mass as possible. The amount of thermal mass and glazing area must be matched to the building's size, heating needs, and location.

A passive solar building is often elongated along its east–west axis to maximize the potential for solar gain along the long south side. The building is ideally rather shallow from front to back, often only 4.5–6m (15–20 ft) deep. This depth allows efficient solar penetration and/or passive heat distribution. The depth of the building can easily be increased if a *cooler zone,* which will not receive much solar heat, is added along the north side. Bedrooms, hallways, stairs, bathrooms and closets are good cooler-zone rooms.

Since the performance of passive solar systems is still a new area of study, only rough guidelines on how much glazing and mass are necessary are presented here. More specific sizing details can be obtained elsewhere [4].

The amount of direct-gain glazing needed depends on the climate. Generally the glazing area should equal about one-fifth to one-fourth the floor area of the space to be heated [4]. For warmer climates the glass area can be reduced and for cold climates the glass area can be increased. An indirect-gain system requires about twice as much glazing but approximately the same amount of thermal mass as a direct-gain system [4].

The amount of thermal mass needed relative to floor area is approximately 300–450 kg of masonry per square meter of floor area (60–90 lb/ft² floor) or 80 kg of water per square meter of floor area (16 lb/ft²) for either direct- or indirect-gain systems [4]. Massive walls and floors should be at least 10 cm (4 in) thick, and concrete or brick Trombe or solid walls should be 0.3–0.4 m (10–18 in) thick for best performance [4].

If a storage mass is located in the sun, the temperature swing of the room will be about one-half the temperature swing of the storage medium. If the storage mass is not located in the sun, the temperature swing of the room will be about twice the temperature swing of the storage medium. Doubling the storage mass decreases the temperature swing of the room by a factor of two. Therefore if thermal-mass storage cannot be located in the sun, four times as much mass is required to realize the same allowable temperature swing in the room. Table 7.1 sets forth guidelines for sizing passive systems.

Example Your one-story passively heated home will contain 20 m × 4.5 m of solar heated space. You would like to incorporate a Trombe wall into the design. How big should the wall be?

Solution The solar heated portion of the house is 90 m². Approximately 36–45 m² of indirect-gain glazing should be provided. If the maximum window height is 2.4 m, you need a Trombe wall 15–19 m long.

Table 7.1 **Sizing of Simple Passive Systems**

	Direct-Gain System	
Glazing	1/5–1/4 of floor area	
	Metric Units	*English Units*
Water	80 kg/m² floor	16 lb/ft² floor
Masonry	300–450 kg/m² floor	60–90 lb/ft² floor
	Minimum 10 cm thick	Minimum 4 in thick
	Indirect-Gain System	
Glazing	2/5–1/2 of floor area	
	Metric Units	*English Units*
Water	80 kg/m² floor	16 lb/ft² floor
Masonry	300–450 kg/m² floor	60–90 lb/ft² floor
	0.3–0.4 m thick	12–16 in thick

7.4 OTHER PASSIVE SOLAR HEATING SYSTEMS

Passive solar heating systems do not have to be simple in-house systems. Like the passive systems already described, other systems rely on thermal-mass storage and do not require fans or pumps for their operation, though some other special equipment may be necessary.

7.4.1 Roof Ponds

Roof ponds are large plastic bags of water set on top of a sturdy, level metal ceiling (Fig. 7.14). The water in the 15–30 cm (6–12 in) deep ponds supplies thermal mass for the solar heating and cooling of the building. A dark plastic liner is usually used under the bags to protect the ceiling from leaks and aid in solar energy collection. To protect the bags from deterioration caused by exposure to ultraviolet light, they are usually covered by a clear plastic sheet. Due to the greenhouse effect, this top sheet also helps trap the heat inside the bags. Tight-fitting, movable, rigid insulation (either a roll-away or tilt-up cover with a reflective underside) is used at night during winter and during the day in summer.

On sunny days during the winter the insulation is removed and solar radiation heats the water bags. Tilt-up insulation with a reflective underside can increase the amount of solar energy reaching the water, though precautions may be necessary to guard against destruction of the insulation by high winds. Heat is conducted through the ponds to the ceiling. The heat is then transferred to the rooms below by thermal radiation from the ceiling. At night the insulation is replaced and the water continues to heat the rooms as it slowly cools. Only the rooms directly below the ponds are heated, so buildings heated in this manner can be only one story tall unless provisions are made for heat transfer.

During the summer the process is reversed. The insulation remains over the pond during the day so the water can absorb extra heat from the rooms. At night the insulation is removed so the heat can be dissipated to the cooler night sky. The cooling performance is improved if a thin layer of water is added over the ponds at night. The stored heat in the ponds helps to evaporate the water, dissipating more heat energy than the ponds could alone.

Since collection of solar energy on horizontal surfaces during the winter is poor, especially in northern latitudes, and since freezing and snow loading are potential problems, roof ponds are most useful in warm climates and at 35° north latitude or less. In these southern climates summer cooling provided by the roof ponds may be more important than winter heating.

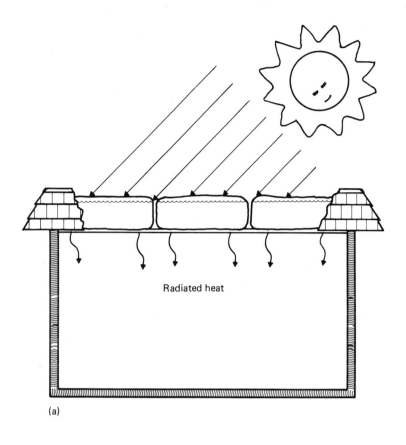

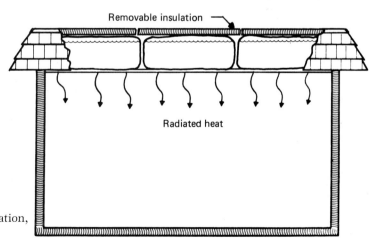

Fig. 7.14 Roof ponds: (**a**) daytime operation, (**b**) nighttime operation.

265

Roof ponds allow much architectural freedom because the only architectural restraints are the necessity for a level roof and adequate wall and ceiling strength to support the water. Roof ponds can be hidden easily by decorative parapets, making the solar heating system nearly invisible to passersby.

For winter heating in colder climates, roof ponds can be placed in the attic under tilted glazing (Fig. 7.15). The tilted glazing accepts more winter solar radiation than a horizontal pond. A reflective surface on the other half of the roof can help increase the solar energy received by the water bags. The attic must be reinforced to support the tremendous weight of the water. Insulation of the glazing at night and during cloudy periods to prevent heat loss is very important.

7.4.2 Thermosiphoning Air-Type Collectors

One type of thermosiphoning collector is the Trombe wall discussed in section 7.2. With a Trombe wall, as with all thermosiphoning air-type collectors, air is heated in the space between the glazing and the absorber (the thermal mass for a Trombe wall) and rises, drawing cool

Fig. 7.15 Attic roof pond for northern climates.

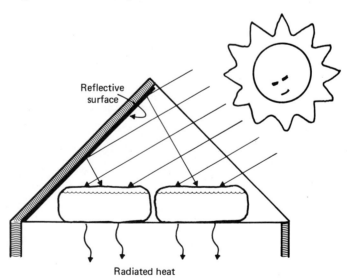

air into the collector to be heated. Most thermosiphoning or natural-convection systems resemble active airtype collectors, the primary difference being the lack of forced-air circulation. The position and construction of the thermosiphoning collectors determines how well the system will work.

Since air movement in thermosiphoning systems is slow, the sizing of the air spaces and ducts is crucial. Expert advice is needed for sizing anything besides a very simple thermosiphon system.

Storage can be included in the thermosiphoning loop (Fig. 7.16). As much storage as possible should be located above the collectors, and the house should be located above the storage unit so that natural convection can charge the storage material and also heat the house (from storage). Dampers between the collector and the storage unit and between the storage unit and the house are often necessary. During the summer tilted collectors must be covered, shaded, or vented to prevent overheating.

The passive thermosiphoning air-type collector system shown in Fig. 7.17 is similar to that illustrated in Fig. 7.16. The system contains two flow loops controlled by dampers—collectors to storage and storage to house. Domestic hot water is heated by the black plastic pipes in the middle of the collector. The water circulates by natural convection to a tank upstairs.

Fig. 7.16 Thermosiphoning collector with rock storage.

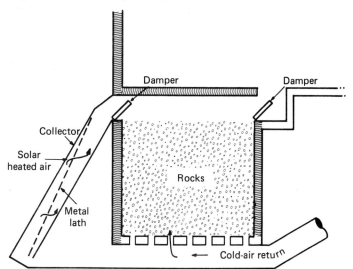

Fig. 7.17 Thermosiphoning-collector system. (Photograph courtesy of Zomeworks Corporation, Albuquerque, NM.)

7.4.3 U-Tube Thermosiphon

The simple U-tube wall heater works in a manner similar to the Trombe wall variation described in section 7.2. During the day the air heated inside the glazing rises, pulling cool room air into the collector to be heated (Fig. 7.18[a]). Excess heat is stored in the thermal mass of the heated rooms or can be directed through a storage area. At night cold air settles equally into both sides of the collector and stagnates, preventing the cooling of the room by reverse thermosiphoning (Fig. 7.18[b]).

7.4.4 Greenhouses

A solar greenhouse can provide inexpensive food and flowers during much of the year (Fig. 7.19, p. 270). An attached solar greenhouse can also provide some warmed and humidified air to the house during the winter and help with cooling in the summer.

Greenhouses can be big heat losers during cold weather. Lack of proper orientation to the sun, combined with large glass areas, make for

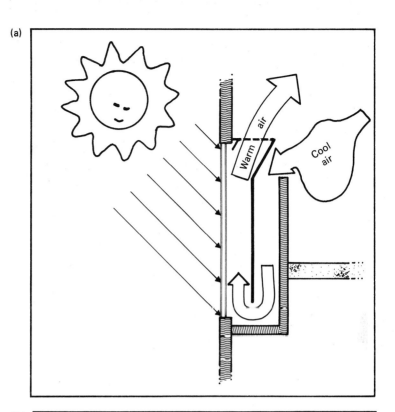

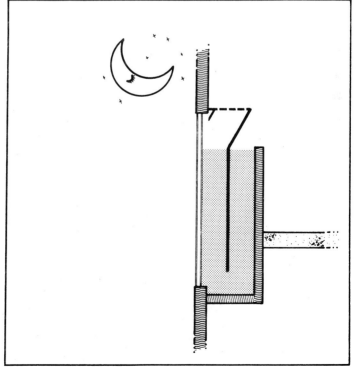

Fig. 7.18 U-tube convective collector, designed by Felix Trombe for heating multi-storied buildings: (**a**) daytime operation, (**b**) nighttime operation.

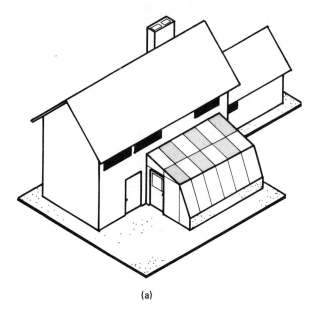

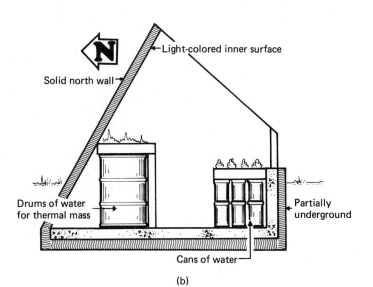

Fig. 7.19 Solar greenhouses: (**a**) attached, (**b**) free standing.

large heat losses and heating bills. To be as efficient as possible, a greenhouse must be oriented with its long axis east–west.

A free-standing solar greenhouse should have a solid and well-insulated north wall and roof. Building a small greenhouse partially or mostly underground will significantly reduce otherwise nearly uncontrollable heat losses.

Attaching a greenhouse directly against the south wall of a house reduces the heat losses for both structures because all or a portion of one wall is in contact with a warmer "outside" environment. If there is no south wall on the house, placing a greenhouse along a southeast wall for exposure to the morning sun, or constructing an L-shaped greenhouse along the southernmost corner of the house are possible compromises. Often the wall between the house and the greenhouse is insulated so that heat will not be lost at night from the house into the greenhouse. Also, having the greenhouse connected to the house automatically provides a location for an air-lock entry into the greenhouse, which is a big heat saver.

An attached greenhouse can be made to provide some solar heat to the house via open doors, windows, or vents that connect the two, and/or by a masonry wall or water wall between the house and greenhouse. Usually two openings between the house and the greenhouse, one low and one high, are recommended to encourage good airflow. In the summer outside vents can be opened also, a cooling convective current through the house and greenhouse can help to increase comfort.

Any solar greenhouse in all but the mildest of climates should have double glazing of plastic, fiberglass, or glass. One low vent and one high vent should be included at opposite ends of the greenhouse for cross ventilation. The entire structure should be tightly constructed so that infiltration around the glazing, entry, and vents is minimal. The perimeter, and preferably the perimeter and bottom of any floor and the growing beds, should be insulated with waterproof and rodent-proof insulation.

Replacement of some of the glazing with solid sections of roof or wall will have virtually no detrimental effects on the plants and can provide a significant reduction in the greenhouse's heat loss. The best method of reducing total heat loss, however, is by the use of nighttime insulation over the (reduced) glazed area. Tight-fitting rigid foam panels, while time consuming and tedious to use, are the most effective.

Figure 7.20 shows another method of insulating the large area of greenhouse glazing. Styrofoam beads are blown between two layers of glazing by a vacuum cleaner motor, creating a Beadwall of movable insulation for the greenhouse pictured. When sunlight is present the beads are moved back into storage tanks. The Beadwall has an insulating value about equal to that of a normally insulated wall.

Fig. 7.20 Solar greenhouse with Beadwall insulation. (Photograph courtesy of Zomeworks Corporation, Albuquerque, NM.)

What makes a solar greenhouse work is the inclusion of thermal mass inside the greenhouse to help collect and store solar energy and modulate temperature extremes. The thermal mass of the structure can be increased by including extensive concrete work with insulation on the outside and by incorporating concrete, or pools, bottles, or drums of water on the inside. Storage can be placed under growing beds so that it not only collects and stores heat, but supports the beds at a comfortable working height. Storage can also be stacked along and up the side of the north wall. Large, deep growing beds will also store solar heat. Sufficient mass should be incorporated to ensure that ventilation during the winter provides adequate carbon dioxide for the plants but is not needed to cool an overheated greenhouse. A full backup heating system must usually be included in case of an extended period of cloudy or unusually cold weather.

7.5 LIVING IN A PASSIVELY HEATED BUILDING

Passive solar heating can usually be incorporated into new structures without too much trouble or additional cost. Only careful design can ensure an attractive and workable solar building, however. With the exception of solar greenhouses, retrofitting for passive solar heating is usually difficult and not recommended.

Passively heated buildings are often quite small and open on the inside since they rely only on natural airflow for the circulation of heat. A passively heated building may have semiprivate sleeping lofts instead of enclosed bedrooms, and other rooms may be defined by low walls and different floor levels. This closeness and lack of privacy is considered cozy by some and unacceptable by others.

Since the thermal mass must be able to collect and store the solar heat, massive floors must be kept free of large insulating carpets and tall blocking furniture. Massive concrete or stone walls should not be paneled, and hanging pictures on rock-hard walls can sometimes prove difficult. Also, fading of furniture and pictures will occur because of their daily exposure to the sun.

Passive solar heating systems generally operate at lower temperatures than do active systems. Usually all the heat collected during the day is released to the building's interior at night. Therefore passive systems work best (as do all solar systems) in climates where cold weather is usually clear and sunny. A well-sized passive system can be effective in any part of the country, however. Passive systems can collect some heat even on hazy or bright, cloudy days, when many active systems may contribute nothing. Backup heating is required for passive systems in case of periods of extended bad weather. Often a wood stove is used instead of a furnace.

The daily placement and removal of nighttime insulation to reduce heat loss as well as manual damper maneuvering may be required. Aside from that and occasional window washings, a passive solar heating system is virtually maintenance free.

Even with carefully sized passive systems, no control over heat delivery or storage rate is possible and so sometimes a building may overheat (25–30° C or 77–80 F) during the afternoon and need venting, wasting solar heat, and may be very cold (10–15° C or 50–59° F) in the early morning. If the backup heating system is used in the early morning some of this expensive backup heat will be stored in the thermal mass, helping to build up afternoon overheating. Large temperature fluctuations can be minimized with additional thermal mass or with fans that improve storage efficiency. Potentially uncomfortable temperature swings of up to 10° C (18° F), depending on climate, must still be expected, however. Anyone who requires a constant temperature of 18–21° C (65–70° F) and would find it too troublesome to put on an extra sweater in the morning while removing the window insulation and stoking up the wood stove should probably not live in a passively heated house.

No solar heating system need be totally passive or totally active. A combination of passive and active solar features may provide the best performance.

REFERENCES

1. Balcomb, J. D., J. C. Hedstrom, and R. D. McFarland (1977). *Passive Solar Heating of Buildings.* Los Alamos, N. Mex.: Los Alamos Scientific Laboratory, LA-UR-77-1162.
2. Chahroudi, D., and S. Wellesley-Miller (1976). Thermocrete heat storage materials: applications and performance specifications. *Sharing the Sun* 8:245–259.
3. Mahene, D. (1978). Three solutions for persistent passive problems. *Solar Age* 3(9):20–23.
4. Mazria, E. (1979). *The Passive Solar Energy Book.* Emmaus, Pa.: Rodale Press.

BIBLIOGRAPHY

Anderson, B., and M. Riordan (1976). *The Solar Home Book.* Harrisville, N. H.: Cheshire Books.

Löf, G. O. G. (1978). A problem with passive. *Solar Age* 3:17–19.

McCullagh, J. C. (ed.) (1978). *The Solar Greenhouse Book.* Emmaus, Pa.: Rodale Press.

Morris, S. (1979). Natural convection: no moving parts. *Solar Age* 4(1):38–41.

Sunset Books (ed.) (1978). *Sunset Homeowners Guide to Solar Heating.* Menlo Park, Calif.: Lane.

Yanda, B. and R. Fisher (1978). *The Solar Heated Greenhouse.* Santa Fe, N. Mex.: John Muir Press.

White, J. W. (1978). Thermal blankets for greenhouses. *Solar Age* 3(3):31–33.

White J. W. et al. (1976). Energy conservation systems for greenhouses. In M. H. Jensen (ed.), *Proceedings of the Solar Energy—Fuel and Food Workshop.* Photocopied.

PROBLEMS

7.1 What is the fundamental difference between passive and active solar heating?

7.2 What is the difference between direct-gain and indirect-gain passive systems? Which type of system is an attached greenhouse? A roof pond?

7.3 Where should the backdraft dampers for a Trombe wall be placed and why? Suggest a simple homemade backdraft damper.

7.4 The solar heated portion of a house will measure 18 m × 6 m × 2.4 m high. How much direct-gain glazing is needed for this building? If direct solar radiation is to be stored in the concrete (2240 kg/m^3) floor and back wall, how thick must they be (minimum)?

7.5 The owners of the house described in problem 7.4 changed their minds and decided to install a solid-concrete indirect-gain wall in the house. Compare the glazing areas for the two types of systems and describe how they can be made to fit the building. How thick should the solid wall be if the same amount of thermal storage is to be provided as in the direct-gain construction?

7.6 You are building a passive Drumwall system. The floor area to be heated is 780 ft^2. How many 55-gal drums filled to 85% capacity will you use? How much will the water weigh?

7.7 Based only on information from Table 4.1, Thermal Properties of Materials, suggest why thinner solid storage walls and Trombe walls are often recommended for adobe than for brick and concrete.

7.8 Approximately three times as much glass area per unit area of heated interior space is required for an attached greenhouse as for a simple direct-gain system [4]. If the interior floor area to be heated is 50 m^2, what is the minimum recommended greenhouse glazing area? A 7500-kg floor and masonry wall are built between the house and the greenhouse. How many 1-l containers of water should be placed in the greenhouse as a first attempt to achieve acceptable temperature stability?

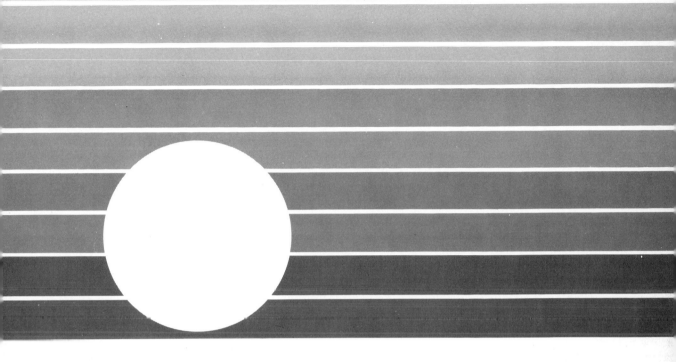

8 SOLAR-BUILDING DESIGN

Solar architecture differs fundamentally from presolar architecture. It represents a basic change in ways of thinking about shelter. It is true that all building styles throughout the centuries have been, to a great extent, influenced by the natural enviornment of the areas in which they flourished. Many building styles—notably those in Mediterranean and desert climate regions—have taken advantage of the thermal mass in thick earth walls for the storage of coolness and solar heat. Nevertheless, the new solar buildings represent a large shift in emphasis from what might be called *inward-looking* shelter to a more *outward-looking* approach. Solar buildings must still provide shelter from the elements. For example, to be effective they must be heavily insulated. Yet their orientation is outward in that they require sunlight for energy as green plants do. And the site, or surrounding land, must be considered in the design. It is not possible to ignore surrounding trees, hills, and structures, or to face a building in any direction arbitrarily.

Eventually some standard ways of dealing with bulky solar collectors, bulky heat-storage areas, and so forth will emerge. Monstrosities with collectors propped up on the roof or with cavelike interiors will no longer be acceptable. Until that time comes, good solar design will require a lot of deliberate, conscious effort on the part of the designer. Each building will be a new challenge. All of the bulky hardware and special design considerations must be integrated into one unified structure. At the same time, practical economic factors must be kept in mind. Radical new construction methods are usually out of the question owing to the high costs of labor, and even a building that appears to have just landed from outer space must still be bought by earthlings. In this chapter we attempt to outline the most important factors to be considered in solar building design and some practical solutions to special problems.

8.1 SOLAR BUILDING TYPES AND THEIR SPECIAL CONSIDERATIONS

Buildings fall into several classes. We will consider five: single family, multifamily, commercial, industrial, and institutional. Each type of building has a different function and each has a different use for solar heating or cooling; this must enter into its design. For example, factories must be functional. A little ugliness is usually tolerated, but excess cost is not. As a result, the only real justification for putting solar heating into a factory or warehouse is energy cost savings. If a slightly beastly looking collection system is available and if it can be shown to pay for itself at a rate that compares favorably with the rate of return of any other investment, the architect should recommend the system. On the other hand, the same collectors may be totally inappropriate for an office building, where an attractive solar system might have a value above what it saves in energy—that is, the solar system might lead to increased income for the company simply because it projects a forward-looking image to customers and the public. The five tables that follow list which factors are generally the most important for each class of building. Of course such lists can never be complete and the factors will vary depending on individual circumstances.

A solar installation for a home must be as simple and low in maintenance as possible (Table 8.1). A moderately large initial cost may be ac-

Table 8.1 **Design Considerations for Single-Family Homes**

Most Important Considerations	Less Critical Considerations
Visual harmony and appeal	Efficient space usage
Low mortgage payments	Perfect temperature stability
Eventual payback	Size of initial investment
Infrequent maintenance and adjustments	Low installation labor (in homeowner-installed systems)
Simplicity of system	
Unimpaired windows; view	
Low heat loss	

Table 8.2 **Design Considerations for Multifamily Residences**

Most Important Considerations	Less Critical Considerations
Visual harmony and appeal	Infrequent maintenance
Low initial cost	
Compact solar components	
Compliance with building codes	
Low maintenance costs	

ceptable if the monthly payments can be kept low and if the system can eventually pay for itself. The collector array should look professionally made and installed and be well integrated into the overall design so as not to draw undue attention to itself. While heat loss should be minimized, window area is still essential for livability. Passive systems and air-type active systems are desirable because of their low maintenance requirements. The temperature fluctuations of a passive system can usually be lived with.

Passive solar features are simple to add to a multifamily dwelling, but active systems can only be considered when there are clear economic advantages for doing so, such as a market of potential occupants interested in solar systems (Table 8.2). This is because systems suitable for this type of construction must be compact and efficient and thus tend to be expensive. Liquid-type systems are desirable for this use because of their compactness. Local building codes must be carefully studied since there may be restrictions on such things as foam insulation and plastic glazing materials as well as use of a single rock-type heat storage unit for two or more dwellings. Regular maintenance is feasible in apartments and condominiums, but high maintenance costs cannot be tolerated.

Collector arrays on commercial buildings (stores and offices) must be visually attractive and professional in appearance (Table 8.3). Because of the high cost of commercial space and real estate, the solar equipment chosen must be efficient and compact. Active-type systems should be

Table 8.3 **Design Considerations for Commercial Buildings**

Most Important Considerations	Less Critical Considerations
Visual harmony and appeal	Night temperature drop
Compact, efficient equipment	Direct return on investment
Continued solar access	
Low maintenance	
Good temperature stability and control	

Table 8.4 **Design Considerations for Industrial Buildings**

Most Important Considerations	Less Critical Considerations
Sufficient return on investment	Perfect temperature control
Adequate insulation	Bulkiness of equipment
	Physical appearance

chosen because of the importance of maintaining a constant, controllable temperature in sales and office environments. Large thermal storage units may not be essential in buildings unused during late-night hours, as some night temperature drop may be acceptable. Zoning of surrounding land should be checked to calculate potential future shading hazards, particularly in dense commercial areas. As long as solar heating remains a novelty, the monetary return on solar installations in commercial buildings may be more than just energy savings; it may also include such factors as good will, free advertising, and projecting a public-spirited image.

Economy is generally the number one consideration in industrial solar heating installations (factories, warehouses, etc.) (Table 8.4). Will a proposed system provide a return on investment comparable to that of alternative investments? In many cases special consideration for visual attractiveness is not justified, though by applying some creativity visual harmony can usually be achieved. Because many existing buildings lack adequate insulation, the greatest monetary return can often be gained simply by insulating and reducing infiltration losses. Many industrial buildings contain considerable thermal mass in their contents alone—a factor that can reduce the requirement for active heat storage.

Institutional solar installations (hospitals, schools, religious buildings) are similar to industrial ones in that economy and performance are primary considerations (Table 8.5). Lower returns on investments may be acceptable to nonprofit organizations and to those that are public-

Table 8.5 **Design Considerations for Institutional Buildings**

Most Important Considerations	Less Critical Considerations
Low initial cost	Infrequent maintenance
Reasonable payback period	
Acceptable appearance	
Economy of maintenance	
Good temperature control	
Resistance to vandalism	

spirited in the cause of energy savings. A special problem in many institutional buildings is susceptibility to vandalism. In such cases, collection areas should be high up and out of the way. Again, regular maintenance and adjustments may be feasible, but maintenance economy is important.

Buildings proposed for solar retrofitting are a special class in themselves. Retrofitting refers to the installation of a solar heating or cooling system in an existing building that was not designed to include a solar system. Unfortunately there are several problems involved with solar retrofitting of most existing buildings.

The first and most important problem is poor insulation. In general the older a building is, the less effectively it was insulated to begin with and the more draftiness it has developed over the years. But even newly constructed buildings often lack enough insulation to make solar heating economical. Often upgrading the insulation of an existing building and adding solar equipment requires such extensive remodeling that the entire project becomes too costly. In such cases the optimum energy savings can be achieved by just reinsulating and not installing solar heat. Ascertaining whether heat loss can be eliminated sufficiently is the first step in evaluating the feasibility of solar retrofitting.

The second step in evaluating solar retrofitting is determining where to place solar collection areas for the building. The best location for a solar greenhouse or solar collector array is on the south side of the building. Collectors perform best when facing true south and tilted from the ground at an angle equal to latitude + 15°. A tilt angle several degrees more or less than the optimum is acceptable. Many existing buildings will not have enough free space available on a south-facing wall or roof; some buildings may not even have a south-facing wall or roof. If a suitable wall or roof area is available it is often shaded by trees, shrubs, or surrounding buildings, or it is obstructed by existing doors, windows, gables, and so forth. Even if the south-facing roof space is free of obstructions it is often flat or too low in pitch for effective solar collection. A related problem with retrofitting is finding room for heat-storage units, large new ductwork, and other mechanical components.

A warehouse successfully retrofitted with a solar system is pictured in Fig. 8.1. The south wall of this brick, masonry, and wood commercial facility in Manchester, New Hampshire was retrofitted to function as a Trombe wall. The exterior south wall was darkened with a matte-black coating to increase its ability to absorb solar irradiation for heat production. Covered with Kalwall's Sunwall I® System, the system has proved overall to be a cost-effective net heat producer.

This brief coverage of the special considerations required for the installation of solar systems in different kinds of buildings indicates that a close relationship exists among solar architecture, economics, and the in-

Fig. 8.1 Retrofitted warehouse in Manchester, New Hampshire. (Photographs courtesy of Kalwall Corporation, Manchester, NH.)

tended use of a structure. Solar building design depends on so many factors that great care must be used in the initial planning stages. The experiences of architects, builders, and do-it-yourselfers have shown that solar projects that are not carefully thought out from the beginning have a great chance of (1) not working, (2) not selling, or (3) being far too costly.

8.2 MAKING BEST USE OF THE SITE

Site planning is always an important factor in architecture, but for solar buildings proper site planning is essential. First, a building must be oriented so that the collector array is facing south to maximize energy collection. Second, all present and possible future obstructions of sunlight must be anticipated and eliminated or designed around. Third, the building should be located to take advantage of any natural wind shelter that exists. Finally, the site should complement the design of the building and the building should be in visual harmony with its environment.

Solar collection will occur on the south side of the building (for buildings in the Northern Hemisphere) and, throughout much of the United States, the prevailing winds will be from the north, west, or east. This is fortunate for two reasons: (1) the building itself will prevent wind from striking the solar collection areas and (2) the designer can place the building so as to take advantage of planted or naturally occurring windbreaks that do not interfere with sunlight. Windbreaks to the north, east, and west of the building will not shade the collectors during the winter months.

The best wind protection is afforded by hills, solid structures, and multilayered vegetation—trees above and bushes below. Large groupings of trees alone are effective too. As indicated in Fig. 8.2, the protective zone provided by a good windbreak extends downwind by a distance many times the barrier's height. The height of the protective zone gradually decreases to ground level. Thus a structure should be located

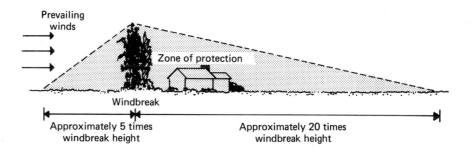

Fig. 8.2 Protection from wind.

as close as possible to the windbreak. Extensive plantings sheltering the north, west, and east sides of a building are effective against wind-induced heat loss.

For best solar collection, it is desirable to orient a building squarely in the true north–south direction (not magnetic north–south), with the solar collection areas on the south wall or on a south-sloping roof. On certain sites this creates problems because of the direction of a view, a street, or neighboring structures. In such cases a collector array can be faced as much as 15–20° east or west of true south without any substantial loss in performance. If there is a choice, it is better to face an array of active-type collectors toward the west since the outside air temperature tends to rise during any given day and it will be warmest outside when the collectors are operating at their peak efficiency. A passive collection array is sometimes oriented slightly east of south, however, to take advantage of early morning sun. This is because passively heated buildings drop in temperature during the night and heat is needed most in the morning hours. A slight performance penalty occurs with the east-of-south orientation.

Shading of collectors is one of the largest problems faced by the site planner. Not only are trees and structures on the property a threat, but trees and structures on neighboring properties are threats as well, especially in cities having small lot sizes. In northern latitudes the sun never rises very high in the sky in the winter months, so trees and buildings do not have to be very tall before they start to obstruct sunlight. In the central United States, for example, at 45° north latitude the sun is only 21.6° off of the horizon at solar noon on December 21. This means that 15 m (50 ft) south of a solar collector array, a tree that is just 6 m (20 ft) taller than the array will shade collectors from the noon sun. And noon is when the sun is highest—at 9:00 A.M. a tree 15 m away has to be only 2.7 m (9 ft) taller than the array to obscure the sun.

Figure 8.3(a–d) consists of four contour maps that may be used to determine whether there are serious shading problems on a particular site. The maps are for four different latitudes. Each map has a series of contour lines on it that represent different heights (in English units). One way to understand what the lines mean is to think of each of them as a fence. For example, the contour marked 10 can be thought of as a fence 10 ft tall. Any object, such as a tree, that is located anywhere along this imaginary 10-ft fence and is taller than the fence will shade point A at least part of the year. If a solar collector is at point A, it will be shaded by the tree some days of the year during parts of the day when it could be collecting solar energy.

The higher a solar collector is, the less troublesome the shading problem usually is. The heights marked on the map are not heights measured from the ground, but heights measured from the solar collector: If

MAKING BEST USE OF THE SITE 283

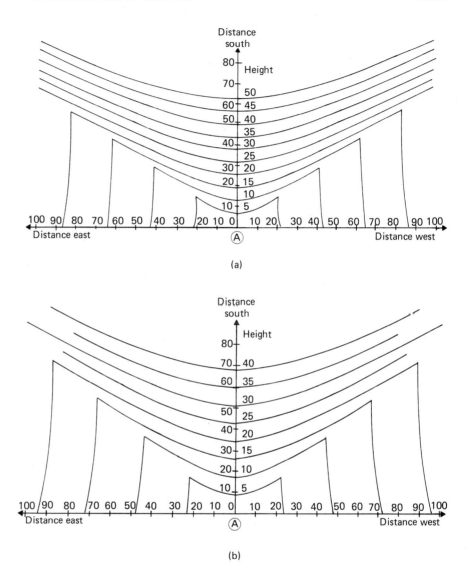

Fig. 8.3 Solar shading maps for collectors (point A) at (**a**) 28° north latitude, (**b**) 36° north latitude (continued)

the collector is 12 ft above the ground, we must add 12 ft to every contour on the map. The 5-ft contour line will become a 17-ft contour line and so on.

An object that is almost as tall as one of the contour lines is also a problem. The long curved portion of each line is based on the path of the sun on December 21, the day when the sun is lowest in the sky. If the

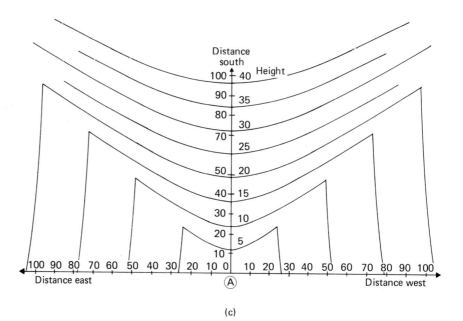

(c)

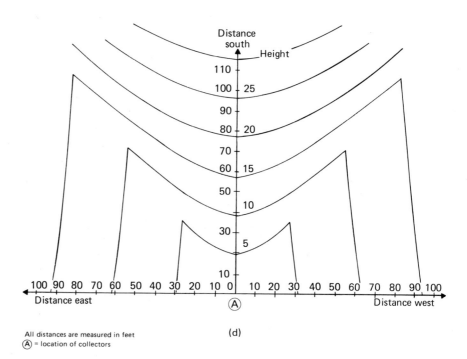

All distances are measured in feet
Ⓐ = location of collectors

(d)

Fig. 8.3 (continued) (**c**) 44° north latitude, (**d**) 52° north latitude.

line really were a fence and if we were watching from point A on December 21, we would see the sun just at the top of the fence throughout the day. Point A would not be totally shaded, but it would not receive all of the light it could because not all solar energy comes directly from the sun. The sky around the sun is a source of diffuse sunlight, so any object that obstructs this area will block some diffuse light. The nearly straight portions of the contours, which run north–south, are based on the time of day one hour after sunrise. If one of the contour lines were a fence and we were watching from point A, the sun would appear over the fence exactly one hour after it rose in the morning and it would disappear from our view exactly one hour before sunset. We would actually miss two hours of sunlight per day, but the only way to avoid this would be to have no obstructions on the east and west sides. In short, any object in front of a collector array will block some solar energy and if the object is taller than the height recommended on the map the shading problem may be serious.

The curves in Fig. 8.3 apply only to the shading of one specific point, point A. An array of solar collectors is an area, not a point. This means that the map should be checked for several test points on the collector array. A good way to check for shading hazards is to draw a site plan that shows each object located in front of the collector array and indicates the object's height. Then one point on the collector array is chosen as point A and the site plan is matched against the shading map. Next, another point on the array is chosen as point A and the process is repeated. By this procedure it is possible to find out which objects may shade which areas of the collector array and steps can be taken if necessary.

Some shading of a collection area is usually inevitable and can be tolerated as long as it does not increase. One problem is that landscapes change. Obviously, trees grow. Trees on the property can be controlled, but trees on neighboring properties cannot always be controlled, nor can buildings that might be constructed at some future date. The question of rights to sunlight is a legal question; steps are being taken in some areas to guarantee these rights. In any case, the designer must try to anticipate where future problems might arise and, if possible, design around these problems.

There are as many ideas of beauty as there are beholders. Nevertheless, some properties have a tendency to stick out like sore thumbs—particularly solar properties. Part of the reason for this may be poor site planning. A solar heated building is often quite different from neighboring buildings and these differences can be either emphasized or deemphasized by the designer. Difference and contrast are good, but incongruity is not. Trees and shrubbery, located where they will not shade the collection areas, can soften some of the starkness while providing protection from the wind. The building can be designed to go with the natural flow of the land. The trend now is to blend with the natural environment instead of fighting it; this applies to solar as well as any other type of construction.

8.3 HEAT LOSS AND SOLAR GAIN

Reducing a building's heat loss (or heat gain, in summer) is one of the best ways to minimize heating or cooling costs. In fact, upgrading the insulation and taking other steps to reduce heat loss can often be far more economical than adding solar collection capability. It is a waste of money to install expensive solar collection features in a poorly or even moderately insulated building when it is probable that a few simple and inexpensive measures to reduce heat loss will have an equal or greater effect on energy bills. Solar heating should only be considered for a building that has been adequately insulated

What is adequate insulation? The answer is not simple. A complete economic analysis of how much solar heating and how much insulation a specific building should have is rarely done.[1] Most authorities, however, recommend that the higher R values given in standard published guidelines be used for solar structures. Table 8.6 lists the levels of housing insulation recommended by three respected organizations—the National Association of Home Builders (NAHB), the Federal Housing Authority (FHA), and the Farmer's Home Administration (FmHA). (Degree days are discussed in Chapter 6, as is a simple method for calculating a building's heat loss.) Where there is a range of suggested R values, it is recommended that the upper figure be chosen.

An example given in the *Insulation Manual* [2] illustrates the effectiveness of proper insulation. An average house with minimally insulated walls and ceilings and single-glazed windows is compared with an energy-saving house that has the same design but more insulation and double-glazed windows. The energy-saving house has half the heat loss (and thus half the fuel costs) of the first house.

Insulation is by no means the only way to reduce heat loss. First, the shape of a building has a large bearing on heat loss. Two buildings with identical amounts of floor space can have different amounts of outer surface area (wall and roof area) exposed to the weather (Fig. 8.4). Heat transfer is closely related to surface area. A sprawling building with sev-

---

1. For instance, in one of the most highly regarded solar sizing methods, f-Chart [1], heat load is taken as a given quantity for a specific building and then the solar system is sized to match this heat load.

Table 8.6 **Recommended Insulation Levels for Standard Residential Construction (in English Units)**

	Degree Days* of Location	National Association of Home Builders Guidelines	Farmer's Home Administration Guidelines	Federal Housing Authority Proposed Standards
Walls	Under 1000	R-12	R-11	R-11
	1000–2500	R-12	R-11	R-11
	2500–4500	R-12–R-16	R-19	R-19–R-11
	4500–6000	R-12–R-16	R-19	R-19–R-11
	6000–7000	R-12–R-17	R-19	R-19–R-11
	Over 7000	R-12–R-18	R-19	R-19
Ceilings	Under 1000	R-19–R-21	R-19	R-19
	1000–2500	R-19–R-22	R-22	R-22–R-19
	2500–4500	R-19–R-30	R-30	R-30–R-22
	4500–6000	R-21–R-30	R-30	R-30
	6000–7000	R-21–R-34	R-38	R-30
	Over 7000	R-21–R-35	R-38	R-38
Floors	Under 1000	R-0–R-9	R-11	R-11–R-0
	1000–2500	R-0–R-15	R-11	R-11–R-10
	2500–4500	R-7–R-19	R-19	R-19–R-11
	4500–6000	R-13–R-19	R-19	R-19–R-11
	6000–7000	R-16–R-20	R-19	R-19–R-11
	Over 7000	R-16–R-22	R-19	R-19

Note: For metric equivalents, multiply degree days by 0.556 to get °C day, and R values by 5.68 to get W/m^2 °C.
*Degree days listed here do not include the heat-loss/degree-day factor included in the modified degree days used in Chapter 6.

eral wings is likely to have much more surface area exposed to the weather than a building with a compact design. One way to minimize heat loss is to design square or rectangular buildings, perhaps with two stories instead of one, and attempt to minimize the number of wings, protrusions, and so on. This will reduce both heat loss and building costs.

The best building shape for maximum solar gain and minimum heat loss is a rectangle with the elongated side facing toward the south. The long south side will collect solar energy in the winter and the short east and west sides will lower solar heat gain during the summer.

Infiltration of cold outside air (or leakage of warm inside air) is one of the largest sources of heat loss, often amounting to nearly half of the total losses. Cold air may enter from many places—baseboards, cracks around windows and doors, electrical outlets on outside walls, dryer vents, and light fixtures, to name a few. Some air infiltration is neces-

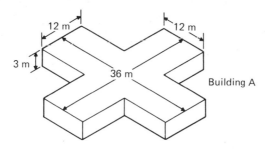

Gross floor space: 720 m²

Total wall and roof area: 1152 m²

(a)

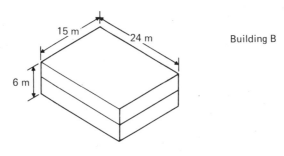

Gross floor space: 720 m²

Total wall and roof area: 828 m²

(b)

Conclusions:
Each building has the same amount of floor space. Building A has 40% more wall and roof area exposed to the weather.

Fig. 8.4 How building shape affects heat loss.

sary: a certain amount of ventilation is desirable or the air becomes stuffy. Air will always be exchanged when people enter and leave a building. But usually the problem is too much, not too little, infiltration. There are inexpensive and effective ways to reduce infiltration losses. All doors and windows should be weather-stripped. Weather-stripping tape, and silicone caulking of aerosol polyurethane foam should be used to seal cracks, wall outlets, and all other possible leaks. Revolving doors or air-lock entryways are very helpful. In large buildings or where substantial ventilation is needed, a *heat-exchange ventilator* can be used to exchange inside air for outside air without excessive heat losses (Fig. 8.5).

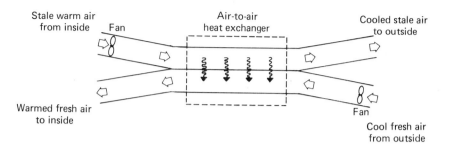

Fig. 8.5 Heat-exchange ventilator.

Heat loss can also be substantially cut if window area is reduced. Today double windows are almost universally used in new construction, at least in the northern regions of the United States. But even double windows lose more heat than walls. A wall might have an R value 5–10 (or more) times that of a double window. For winter heating purposes, double windows on the south side of the building may gain more heat than they lose because of the solar input, but on the north, east, and west sides this is not so. Many new solar heated buildings have only very small windows or even no windows on these walls, but these designs have drawbacks. Residents need windows for psychological and aesthetic reasons, and builders find that buildings without windows or with small windows do not sell well. When large windows are placed on the north, east, or west walls of a building, triple-glazed or double windows with storm windows can be used. Another possibility is the use of insulating shutters or insulating shades. If it seems unrealistic to expect the building occupants to close insulating shutters or shades when they should, these coverings can be made to operate automatically.

As mentioned previously, any south-facing window area is a definite plus as long as the solar heat can be retained once it is collected and the excess can be stored. These are the fundamentals of passive solar heating. Passive design features can be included in almost any new building, even one with an active solar heating system.

The direct-gain, earth-sheltered adobe home in Santa Fe, New Mexico shown in Fig. 8.6 was the first merchant-built structure to exclude a conventional heating system, normally required by building codes and lending agencies. Auxiliary heat is provided by two wood-fired adobe fireplaces. The house contains 41 m² (444 ft²) of south-facing vertical double glazing and has a floor-to-collector ratio of 3:1. Figure 8.7 shows a solar home located near the ocean at The Sea Ranch, California. In addition to passive, direct-gain storage, hot air is circulated by a fan from the top of the building around along the outside wall and through

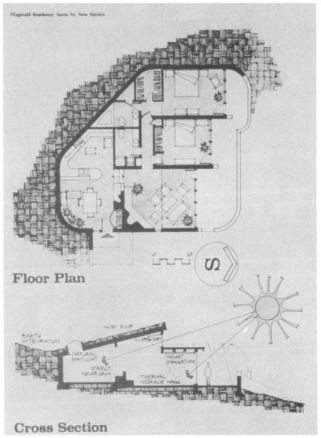

Fig. 8.6 Direct-gain, earth-sheltered adobe home in Santa Fe, New Mexico. (Photographs courtesy of David Wright and Dennis Andrejko, SEAgroup, Nevada City, CA.)

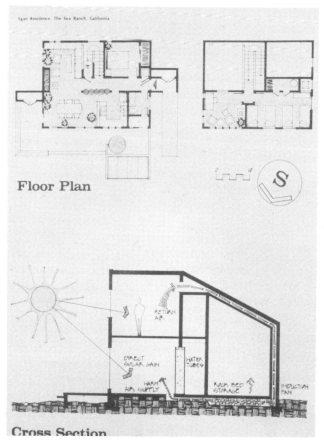

Fig. 8.7 Solar home combining active and passive components, The Sea Ranch, California. (Photographs courtesy of David Wright and Dennis Andrejko, SEAgroup, Nevada City, CA.)

291

a rock bed under the slab before being returned to the living space. A thermosiphon hot-water heater provides hot water for domestic uses and for the hot tub.

Perhaps the best way to realize the importance of good thermal design for buildings is to consider what size solar heating system would be needed to heat a building that has zero heat loss. No heating system of any kind would be needed. If the residents just opened the door on a summer day, they would have enough heat to last the winter. In fact they would have too much since human beings and their machines are constantly generating heat. Of course no such building is possible, but the example illustrates why weathertight, well-insulated structures are worth striving for.

8.4 PLACEMENT OF SOLAR COLLECTION AREAS AND SOLAR COMPONENTS

Solar heating and cooling introduce some new considerations into the building-design process. It is necessary to find room on a south-facing wall or roof for an expanse of glass or plastic that might be as large as one fourth or more of the building's floor area. Ideally this collection area will face directly south and be tilted at what can only be described as an architecturally inconvenient angle (50–60° from the horizontal). If the system is passive, there may be a lot of concrete or water containers to be designed around. If the system is active, a larger-than-normal utility room is needed somewhere in the building to house pumps, fans, and sizable heat-storage units in addition to a conventional furnace or boiler. If this task seems difficult for new construction, it may appear to be impossible for retrofitting.

Nevertheless, it has been shown that the solar heating system does not have to dictate the architectural style of a building. It certainly will have to be considered, but the hundreds of solar buildings now in existence come in all sizes, shapes, and styles. Collector arrays cannot simply be tacked onto a building or propped up on the roof without appearing totally out of place, but if carefully integrated into the overall design they can be an aesthetic asset.

Solar collector arrays, like any other external feature of a building, can be either accented or disguised. Many solar homes are contemporary in style because large planes and odd angles are part of that style and a large, tilted expanse of glass does not seem out of place. It can even become the home's chief design feature. The house shown in Fig. 8.8 might be considered an example of this "blatantly solar" approach. In this design no excuses are made for the fact that collectors perform best at a certain angle or that many are needed. Instead, the building is designed around the collector array. Figure 8.9 shows a somewhat more conventional home, with a small array of vertical collectors that appear to be a row of clerestory windows. This we might call the more "dis-

PLACEMENT OF SOLAR COLLECTION AREAS 293

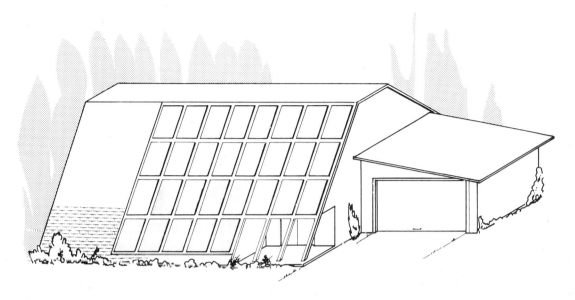

Fig. 8.8 Solar collection as a central feature of building design.

Fig. 8.9 Solar collection as a de-emphasized feature of building design.

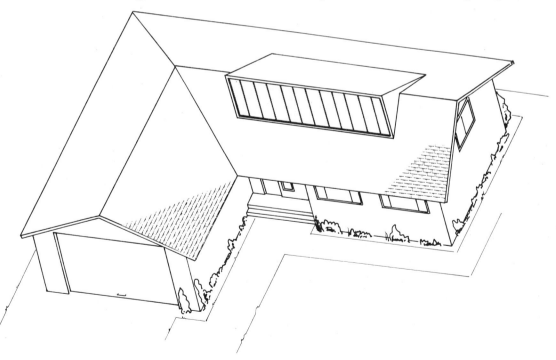

creetly solar" approach, one that is well suited for retrofitted installations and for some of the more traditional styles of architecture.

Collector tilt is important to the architect. In many cases there are distinct architectural advantages to installing a collector array vertically. A house with a vertical array of collectors is pictured in Fig. 8.10. Heat generated by the air-type collectors in this home is stored in a rock bin. This house contains both passive and active energy-conservation features.

With some systems, such as passive Trombe walls, the collection area has to be vertical. How does this affect performance, and what are the relative advantages of vertical and tilted collectors? It is commonly accepted that a system will collect the most energy if its solar collectors are tilted from the ground at an angle about equal to latitude + 15°. In the central United States at 40° north latitude, a collector array should face true south and be tilted up about 55°. Studies have shown that a tilt of 45–50° above the recommended tilt will reduce the yearly energy collection by about 12–15%. Thus a vertical collector array at 40° north latitude will collect about 85% of the energy it could collect if it were tilted. At higher latitudes the situation gets better. For example, in Canada at 55° north latitude a vertical collector is only about 7% less effective than one tilted at the ideal 70° angle.

Fig. 8.10 Attractive vertical collectors set off this mountain residence in Evergreen, Colorado. (Photograph courtesy of Richard L. Crowther, AIA, Crowther Architects Group, Denver, CO.)

Does 12–15% lower performance mean that an architect must always find a way to tilt the collectors? No. In fact, many feel that it is often best to mount an array vertically and then simply add another collector or two to make up the difference. There are several reasons for this. As already mentioned, many passive systems require vertical collection areas. Active-type collectors are generally heavy and structural support can sometimes be more difficult for tilted arrays. Also, a steeply pitched roof is not aesthetically pleasing on all buildings. A more substantive advantage of vertical collectors is that they are not subject to the same level of overheating as tilted collectors are. The sun is high in the sky in summer, so light will strike vertical collectors at a steeper angle. Protection from the summer sun can also be enhanced for vertical collectors if roof overhangs are installed (judiciously) above them. What is more, damage from hail and snow loading and water leakage are less of a problem with vertical collectors. Snow is an asset to vertical collectors because it increases the reflectivity of the ground: more light will enter. Finally, access and maintenance for large collector arrays is easier with vertical arrays. Thus the choice between vertical and tilted collection areas is not as simple as it might seem at first. There are advantages to both.

How about horizontal or low-pitched collectors? For example, does it make sense to retrofit with solar collectors an existing house roof that has a slope of 20°? No. It has been found that performance drops faster for underpitched collectors than for overpitched ones. At 40° north latitude a horizontal collector array will perform only about 85% as well as one tilted to 55°. A 20° tilt would be about equivalent to a vertical tilt (except that there would be none of the other advantages of vertical collectors). To calculate how performance changes as tilt changes see Chapter 6.

One effective method of reducing the number of collectors required on a building is to install light-colored or reflective surfaces near the array. For example, the area beneath an array of vertical collectors can be covered with reflective plastic sheeting or, if the collectors are near the ground, the ground can be landscaped with white gravel or other light-colored material. A body of water in front of a collector array can substantially improve collection. It has been estimated [3] that the use of fixed reflecting surfaces near a collector array can improve energy collection at times by as much as 50%.

In the sections that follow is a series of sketches showing some of the more common solutions to the problem of finding room for solar collection areas for both residential and commercial/industrial buildings. Some are suitable for retrofitting as well as for new construction.

8.4.1 Residential Buildings

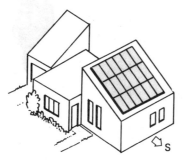

Solar Wall

The *solar wall* is commonly used on homes that are otherwise quite conventional in design. The landscape must be designed to avoid shading and some rooms should lack windows (although one or two of the collectors is sometimes replaced by a window). A retrofitted solar wall can require extensive remodeling of the existing building to accommodate ductwork or piping and may require the removal of trees and shrubs.

Solar Pitched Roof

The *solar pitched roof* is perhaps the most common approach to locating collector arrays. The roof is sloped at approximately the optimum solar collection angle (latitude + 15°). Hazards are overheating, hail, and snow loading. Access for maintenance may be difficult. This approach limits the range of architectural styles that can be used.

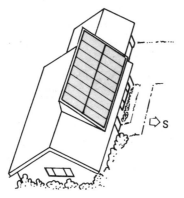

Solar Pitched Roof Extension

A *solar pitched roof extension* allows placement of collectors at optimum tilt, which need not match the roof slope. Many variations are possible. The extension is useful for adding a large number of collectors to an existing flat or low-pitched roof.

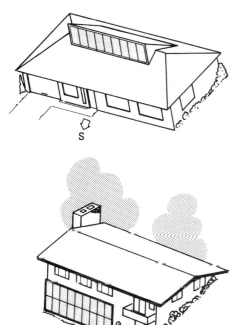

Solar Clerestory

A *solar clerestory* is useful where a relatively small number of collectors is needed and where the solar aspects of the design are to be secondary to other features. The clerestory is a simple and effective retrofitting style: collectors are up and out of the way and duct manifolds or pipes can be placed in the attic space without extensive remodeling.

Solar Subfacade

A *solar subfacade* is useful for buildings on steep hillsides with southern orientations. Collectors are placed at and/or below basement level. Careful insulation of collector array is important. This style leaves the south wall of the house free for window area, which leaves the view open and takes advantage of passive solar gain through the windows.

Solar Shed

A *solar shed* is a separate outbuilding that may be built solely for this purpose or onto an existing outbuilding such as a separate garage. The solar shed is extremely useful for some retrofitting situations and in other cases where insufficient collection area is available on the house. The shed may also hold heat-storage and mechanical components. Heat-carrying lines to the house should be as short as feasible and well insulated.

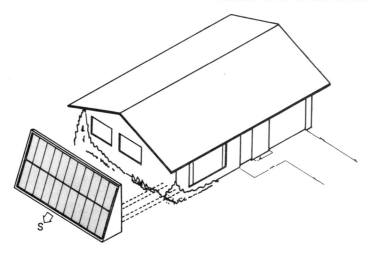

8.4.2 Commercial/Industrial Buildings

Solar Wall

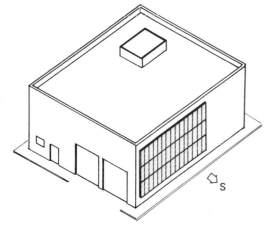

A solar wall is a simple solution to the problem of locating collection areas in industrial and many commercial buildings. Roofs are typically flat, and building a tilted array would require construction of a support frame. Reduction in performance due to vertical tilt is compensated for by the advantages of vertical collectors. If there is not enough collection area to overheat the building during the winter months, no thermal storage is required and a simple air-type system with fan (and in parallel to the backup heating system) may be all that is required.

Roof-Mounted Array

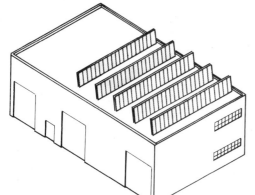

A roof-mounted array is useful where high solar fraction is desired and/or where solar heat is to be used for industrial purposes. Banks of collectors are oriented at optimum tilt and are spaced far enough apart so that the banks do not shade one another (shading can become a serious problem at high latitudes). Since collectors are typically quite heavy the roof must be reinforced adequately.

Three buildings with solar heating systems are shown in Fig. 8.11, 8.12, and 8.13. Nearly any part of a building can be attractively adapted to include a solar heating system. The active collection system is placed over the garage in the Soda Creek, Colorado home shown in Fig. 8.11. A wood stove is used for backup heat during the winter months. The liquid-type solar heated and solar-absorption-cooled residence at Colorado State University shown in Fig. 8.12 first began operation in 1974. In 1975 a virtually identical house, using instead an air-type solar heating system, was constructed only 50 m (165 ft) from this house. Both houses have 68 m² (730 ft²) of collection area. Comparison studies show that both systems perform well, and provide nearly 70% of the heating needs for the houses. The Raypak solar water heating system shown in

PLACEMENT OF SOLAR COLLECTION AREAS

Fig. 8.11 Home in Soda Creek, Colorado with active collection system on garage roof. (Photograph courtesy of Richard L. Crowther, AIA, Crowther Architects Group, Dever, CO.)

Fig. 8.12 House at Colorado State University. (Photograph courtesy of Richard L. Crowther, AIA, Crowther Architects Group, Denver, CO.)

Fig. 8.13 Banyan Street Manor high rise for the elderly in Hawaii. (Photgraph courtesy of Raypak, Inc., Westlake Villiage, CA.)

Fig. 8.13, located on the Banyan Street Manor high rise for the elderly, was the first to be designed as an integral part of a large building in Hawaii. The thermosiphoning system saves the building the equivalent of 470 barrels of oil per year. Imperfect building orientation did not preclude facing the collectors in the proper direction.

8.4.3 Mechanical Components

The mechanical components of active systems—pumps, fans, heat-storage units, heat exchangers, and so forth—are generally located in one central area of the building. For structural reasons (because of the weight of the heat-storage material) this area is usually on the ground floor or in the basement if there is one. The rock bin or water tank is typically supported by a reinforced-concrete slab. For many years there has been a misunderstanding regarding how much heat-storage material is required in active solar heating systems. For instance, one fairly common belief is that for a house with an air-type heating system it is necessary to "fill up the basement with rocks." As discussed in Chapter 4, it is generally not economical to store more heat than the amount needed for about one day's use. For rock storage this volume is about equal to that of a small room (e.g., a bathroom). This is not a tremendous volume, but it represents a lot of weight that must be adequately supported.

Sometimes it is difficult to find space for storage units and mechanical components. This is especially true for retrofitted solar installations. Not only must space be found in the existing building, but bulky equipment and storage material must be moved in. One solution for this problem is external storage, or an external room for mechanical components. For example, a rock storage bin may be buried outside the building near the mechanical room and connected to the heating system by insulated ducts. Or a solar shed may contain heat-storage material, fans, dampers, and other system equipment. When this approach is used it is essential to minimize the distance the connecting ducts or pipes must travel and to insulate them extremely well.

The ductwork needed for air-type heating systems is bulky but can generally be placed in an attic or crawl space. One exception is the manifold ducts, or headers, sometimes required for a solar wall. One header is needed at the top of the array and another at the bottom. These take up valuable floor space but, as shown in Fig. 8.14, the space between ducts can often be used for cupboards or shelving.

Fig. 8.14 Storage built into wall containing ductwork for solar collectors.

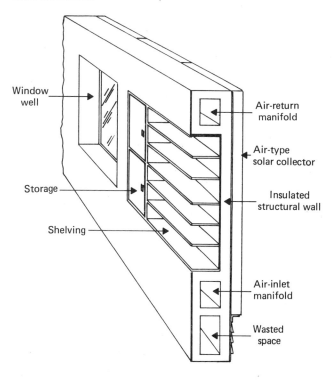

8.5 LEGAL FACTORS

Four legal considerations relate directly to the design and construction of solar buildings. These are

1. Sun rights. What happens when XYZ Corp. decides to build a skyscraper across the street from your solar heated building, blocking sunlight to your collectors?
2. Building codes. What is the city planning commission or building inspector going to object to?
3. Labor jurisdiction. What happens when the carpenters' union, the plumbers' union, and the roofers' union each insists that its members install your solar collectors?
4. Nuisance laws. Is it true that you can be sued because sunlight reflected from your collectors annoys a neighbor?

It is a good idea to be aware of these issues and to consider their effects during the building-design process.

Surprisingly, there is very little protection of *sun rights* under United States laws. Several states are considering legislation that would protect the owners of solar heated buildings from shading by neigboring properties, but few have actually done anything yet. The problem is the historical definition of property ownership, in which it is assumed that property lines extend from the center of the earth upward to the heavens. This concept of ownership has been around for a long time and its meaning in this country is that property owners can put anything in the air space above their land as long as it is not prohibited by local zoning laws. A limit of sorts was imposed by the United States Supreme Court in 1946 to protect airplanes from trespass suits. The Court ruled that the upper reaches of the air are to be viewed as a public highway and that property owners can control only the air space needed for reasonable use and enjoyment of their land. Except for this case, there has been little need in the past to protect rights to unobstructed air space.

Evidently there has been a need for such a law in England. In that country, protection may be afforded by the Doctrine of Ancient Lights, which states that property owners are entitled to continue receiving enough light from across neighboring land for reasonable use and enjoyment if they have been receiving it for the past 27 years. It is questionable whether this law would protect a new solar project and whether it would guarantee enough sunlight for solar buildings.

Now that a need exists, local governments may begin using their zoning powers specifically to protect solar properties. Restrictions could be placed on building heights, setbacks, fencing, street widths, and so forth. It is also possible for local governments to condemn air space by resorting to eminent-domain laws. New communities can be bound under restrictive covenants (rules agreed to when a property is purchased).

Because there is little protection under United States laws, any large solar project should be carefully evaluated regarding risk of future shading. If the risk appears great enough one possible legal means of protection is purchase of a *solar skyspace easement* from the neighboring property owner. An easement is an acquired right to use or enjoy another person's land. When a property is sold, any easements are transferred to the new owners.

A solar skyspace easement is a *negative easement:* the easement owner does not have the right to enter the property of the grantor but the grantor is prohibited from doing something with his or her land. In this case, the grantor would be prohibited from erecting new structures or allowing vegetation to grow up and cast shadows on the protected solar collectors. Such an easement must clearly define what constitutes solar skyspace. Several states now allow this type of protection (see Chapter 10).

Building codes are powerful means of restricting property owners and of specifying what can and cannot be done with a property. Generally building codes are drafted on a local level, so they vary widely from one community to the next. Some building codes place restrictions on the kinds of materials allowed for solar collectors and on how they can be mounted. Others set standards for the kinds of plumbing or ducting that may be used. Presently codes are not highly specific about solar heated buildings but, needless to say, government agencies such as the Department of Housing and Urban Development (HUD) and the Department of Energy (DOE), as well as model-building-code organizations such as the Building Officials Code Association, would like to see this change. Future building codes will be based on standards put out by organizations such as the American Society of Heating, Refrigeration, and Air-Conditioning Engineers (ASHRAE) and the National Bureau of Standards (NBS). Appendix E is a list of several organizations involved in drafting solar standards and codes and Appendix F is a table listing the major standards that have been published as of this writing, including a model building code put out by the International Association of Plumbing and Mechanical Officials. By studying these documents before a project is designed and built it may be possible to prevent the necessity of making expensive modifications later on when building codes are very likely to become more specific regarding solar properties.

As the number of solar buildings grows, the question of which unions have jurisdiction over which aspects of solar installation must be resolved. For example, the mounting of solar collectors is a problem since collectors can be considered part of the heating system, part of the roof, or part of the plumbing, depending on the type of collector and which union is being asked. Eventually jurisdiction agreements will be worked out among various unions. To give an example, one union of plumbers and pipe fitters (UAJAPPFFI) has negotiated an agreement with a union of sheet metal workers (SMWIA) [4]. For air-type systems, the collectors and all ducts will be installed by the sheet metal workers. For liquid-type systems with fluid pipes, collectors will be rigged by a crew made up of workers from both unions. The number of workers from each union will be proportional to respective union sizes. All pipes and valves will be installed by the plumbers, however. Labor unions take such questions of jurisdiction seriously, so until the issues are settled it is reasonable to expect a few problems on union jobs.

A question that has been raised is whether the glare from solar collectors can be viewed legally as a nuisance. Nuisance doctrines restrict the use of property in ways that interfere unreasonably with the rights of neighbors or the general public. Affected individuals have the right to initiate legal action. Glare from solar collection arrays may conceivably be a basis for lawsuits. One factor should always be considered in such cases however: a solar collector reflects no more light than a window.

8.6 THE DESIGN OF SOLAR BUILDINGS

Solar buildings draw energy from a nearby star, the sun. They differ in many respects from more conventional structures. In this chapter we have examined several of the factors involved in the design of such buildings. The list that follows summarizes the important phases of solar building design and demonstrates how these factors come together in the design process. If conventional buildings require a lot of careful planning, solar buildings require even more. Yet the rewards, in satisfaction as well as money saved, are well worth the effort involved.

1. Values specification
    a) Determination of the values of the people involved in the project
    b) Type of building
    c) Intended uses of the building
    d) Needs and desires of the owner and occupants
    e) Economic limitations

2. Site analysis
   a) Boundary survey
   b) Topographical map
   c) Map of existing objects and structures on the property
   d) Environmental impact study, if applicable
   e) Analysis of climate and how it will affect choice of solar features
3. System specification
   a) Choice of active, passive, or hybrid system
   b) Type of backup or auxiliary heat
   c) Heating and cooling modes desired
   d) Types of equipment needed
4. Building/site layout
   a) Building layout and site plan
   b) Insulation plan
   c) Solar-gain strategy
   d) Thermal-mass placement
   e) Building-elevation sketches
5. Project reanalysis
   a) Analysis of feasibility in light of initial work
   b) Rough calculation of heat load
   c) Rough calculation of system sizing
   d) Economic study
   e) Analysis of shading hazards
   f) Analysis of building codes
   g) Analysis of legal constraints
   h) Plan and outlook for financing
   i) Reworking of initial design if necessary
6. Detailed design
   a) Layout and sizing of solar components
   b) Architectural plans, specifications, and elevations
   c) Site plan
   d) Heating- and mechanical-systems plan
   e) Construction scheduling

REFERENCES

1. Beckman, W. A., S. A. Klein, and J. A. Duffie (1977). *Solar Heating Design by the F-Chart Method.* New York: John Wiley.

2. NAHB Research Foundation (1971). *Insulation Manual.* 2nd ed. Rockville, Md: NAHB Research Foundation.

3. Seitel, S. C. (1975). Collector performance enhancement with flat reflectors. *Solar Energy* 17: 291–295.

4. SMACNA (1978). Legal responsibilities. In *Fundamentals of Solar Heating.* Washington, D.C.: U.S. Government Printing Office, stock no. 061-000-00043-7.

BIBLIOGRAPHY

AIA Research Corp. (1976). *Solar Dwelling Design Concepts.* Washington, D.C.: U.S. Government Printing Office, stock no. 023-000-00334-1.

Anderson, B. (1976). *The Solar Home Book: Heating, Cooling, and Designing with the Sun.* Church Hill, N.H.: Cheshire Books.

Anderson, B. (1977). *Solar Energy: Fundamentals in Building Design.* New York: McGraw-Hill.

Barber, E. M., and D. Watson (1975). *Design Criteria for Solar Heated Buildings.* Guilford, Conn.: Sunworks.

Hudson Home Guides (ed.) (1978). *Practical Guide to Solar Homes.* New York: Bantam/Hudson.

Mazria, E. (1979). *The Passive Solar Energy Book.* Emmaus, Pa: Rodale Press.

Shurcliff, W. A. (1978). *Solar Heated Buildings of North America: 120 Outstanding Examples.* Church Hill, Harrisville, N.H.: Brick House.

Sunset Books (ed.) (1978). *Sunset Homeowner's Guide to Solar Heating.* Menlo Park, Calif.: Lane

Szokolay, S. V. (1975). *Solar Energy and Building.* New York: John Wiley.

Thomas, W., A. Miller, and R. Robbins (1978). *Overcoming Legal Uncertainties about Use of Solar Energy Systems.* Chicago: American Bar Foundation.

PROBLEMS

8.1 Design a retrofitted solar heating system for your house or apartment or that of a friend. Discuss why you chose the type of system, storage, and so forth you would use and where you would put the various components. Sketch the system.

8.2 Insulating your old house will cost $1800 cash but will save you $260 by the end of the year on your heating bill (a 20% savings). If the discount rate is 10% and energy inflation is 14%, how many years will it take until the present value of your savings exceeds the cost of the insulation? This figure is known as the *number of years until payback* or the *discounted payback period*.

8.3 Compare the discounted-payback period for problem 8.2 with the *simple payback period*, defined as the number of years until the cumulative savings equal the initial cost. Why is the simple payback period shorter than the discounted payback period?

8.4 For the house in problem 8.2, if a retrofitted solar heating system is installed now it will cost $10,000 cash and will provide 50% of the yearly heat needs, If the house is insulated first, the same system will be able to supply 60% of the reduced heat load. Your first yearly heating bill with no solar system would be $1300. Using only the payback period as your guide, decide which system you will invest in. (See inflating-discounting function, section 6.3.2.).

8.5 A 20-ft-tall row of trees extends along the road in front of a solar building at 44° north latitude. The row of trees is 40 ft from the building. Will the building be shaded by the trees during part of the year? What if the trees were 60 ft from the building? If the building were located at 28° north latitude?

8.6 To avoid being shaded by the trees, how high off the ground would the solar collectors for the 44°-north-latitude house have to be? How far away from the trees would the house have to be built to avoid shading?

8.7 The overhang on a solar building you are designing is to be sized so that no solar radiation can enter the 7-ft-long window at noon on 30 April (day 120). Calculate the overhang length for 30°, 35°, 40°, and 45° north latitudes (see Section 1.1.1.).

8.8 A factor, $\bar{R}$, that relates total radiation on a horizontal surface to radiation on a tilted surface is given by

$\bar{R}$ = beam factor + diffuse factor + ground = reflectance factor

The ground-reflectance factor is defined as

ground-reflectance factor = $\rho(1 - \cos s)/2$,

where ρ is ground reflectance (typically varying from 0.2 for bare ground to 0.7 for fresh snow) and s is the collector tilt. What is the effect of increasing ground reflectance to the maximum for a collector tilted at 60° from horizontal for a typical winter $\bar{R}$ value of 2 and a spring value of 1?

8.9 Your town does not have any sun-rights provisions in its codes and laws. List several methods you could use to ensure access to the sun

9 OTHER APPLICATIONS OF SOLAR ENERGY

To be put to work, solar energy must be transformed, or converted, into more useful forms of energy. Most of this book has been devoted to one specific transformation—light to heat. And we have concentrated on one very specific application of solar heat—the space heating of buildings. The sunlight-to-heat transformation is very simple and space heating is one of the best-developed solar technologies at this time. Solar space heating holds promise for greatly reducing the world's dependence on supplies of oil, coal, and uranium that are inherently limited. Solar technology is rapidly expanding into areas other than space heating, however, and it is important to be aware of this progress. In this chapter we briefly examine what is happening in several of these areas of research and review some of the applications that were only touched upon in earlier chapters.

9.1 TRADITIONAL APPLICATIONS

For years solar energy has been used to heat water and to cook and dry food. Perhaps the most widely used alternative application of solar technology is domestic water heating, which is similar to space heating in concept and has become at least as popular.

9.1.1 Solar Water Heating

As we all know, water exposed to sunlight will become warm. Three passive water heaters are shown in Fig. 9.1. There are several variations on this simple theme, mostly involving the inclusion of a glazing layer to reduce heat loss, tilting the collector toward the sun, and including pipe connections for easy filling and discharging. These types of collectors provide for solar heat collection and storage in one unit. While they work well on sunny days, they can lose tremendous amounts of heat during the night and on cloudy days. Nighttime insulation with a reflective inner surface can significantly reduce these losses while increasing gains during sunny periods by reflecting additional solar energy onto the collector/storage unit.

Solar water-heating systems usually resemble small liquid-type space-heating systems in both appearance and performance. In passive,

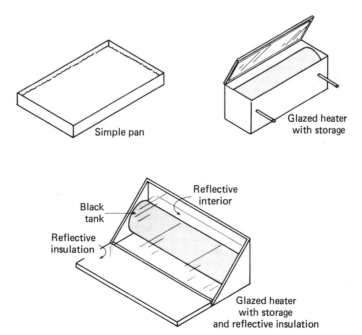

Fig. 9.1 Passive water heaters.

thermosiphoning types natural convection provides circulation, and in active types a pump circulates the water. As with all solar heating systems, a conventional backup system (in this case a water heater) is needed.

Both the passive and active water-heating systems must provide some hot-water storage. The storage tank should contain enough water for one and one-half to two days' use. Hot-water use of about 75 l (20 gal) per person per day can be assumed for sizing purposes. The tank should be insulated with a minimum of 9 cm ($3\frac{1}{2}$ in) of fiberglass insulation.

The collector area can be determined using the general rule of thumb that 0.012–0.024 m^2 of collector area should be provided per liter of storage volume (0.5–1 ft^2/gal), depending on location and use patterns [1, 2, 3, 4, 5]. For many locations, at the lower figure (0.012 m^2/l or 0.5 ft^2/gal) the system will provide a limited amount of hot water, perhaps 20 30% of the total load, but will not be oversized for summer heating.

The recommended collector tilt for solar water heating is an angle of latitude + 20° [1, 5, 6, 7]. Systems with collectors so tilted often provide summer hot-water needs but only a small fraction of winter needs. Since the hot-water load is essentially constant all year, a steeper tilt and larger collector area may be more economical and productive in some locations.

Because water remains in the collectors overnight, precautions against freezing must be taken. In areas where occasional freezing occurs, nighttime insulation or circulation of warm water through the collectors may be sufficient. In more severe climates, antifreeze can be used in the collectors and a heat exchanger installed between the collectors and the storage tank. An alternative is to drain the collectors at night.

All connecting pipes in the system should be as short as possible and should be insulated to minimize heat loss. Solar heated water can exceed 75° C (165° F), so it is wise to include a mixing valve in the system to prevent scalding (see Chapter 5).

Thermosiphoning Water Heaters

Thermosiphoning water heaters are the most common type of solar water heater (Fig. 9.2). Their operation is based on the same principles as thermosiphoning air collectors (see Chapter 7). As the water in the collector heats it becomes less dense and rises. Cold water from the bottom of the storage unit enters the bottom of the collector and heated water from the top returns to the storage tank.

The storage unit should be placed at least 0.3 m (1 ft) above the top of the collector surface. This placement will facilitate efficient daytime

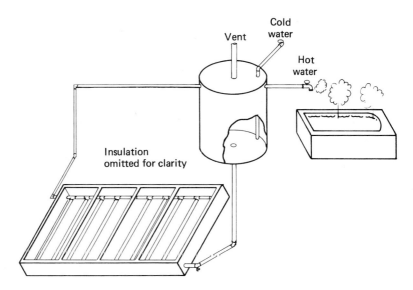

Fig. 9.2 Thermosiphoning water heater.

thermosiphoning action and will prevent reverse thermosiphoning and cooling of the water at night by allowing cool, dense water to settle in the collector. The hot-water storage tank can be placed inside an attic or on top of the roof in a false chimney. If it is impossible to place the storage tank above the top of the collector, a check valve may be installed to prevent reverse thermosiphoning.

Active Water Heaters

The setup of an active solar hot-water heater is very similar to that of an active liquid solar heating system (see Chapter 5). There are many possible variations in the system design. The active hot-water system pictured in Fig. 9.3 circulates potable water through the collectors and has

Fig. 9.3 Solar potable-hot-water heater with two tanks.

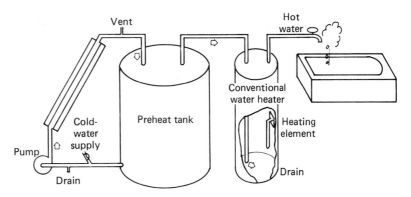

two storage tanks connected in series. The first and larger tank preheats the water, the hottest of which enters the second tank, which is a conventional water heater (and the backup heater). If the solar heated water is not hot enough to use, it is heated by the backup water heater to the proper temperature.

Some components of a solar hot-water system are shown in Fig. 9.4. In this closed-loop system a mixture of propylene glycol antifreeze and distilled water is used as the solar heat-transfer fluid. When solar heat is available the pump is automatically activated and the heated fluid flows through the heat exchanger, heating the potable water in the solar water tank. The solar tank provides solar preheated water directly to the cold side of the conventional water heater.

A retrofitted domestic hot-water installation is shown in Fig. 9.5. This Sunworks hot-water system illustrates one solution for an imperfectly tilted roof. The add-on collector bank is placed on racks. Collectors can also be built in to provide continuity in the house's appearance.

Fig. 9.4 Components of the Lennox LSHW2 Domestic Solar Hot Water System: (**a**) collector; (**b**) water tank with internal, double-walled heat exchanger; (**c**) control module and pump. (Photographs courtesy of Lennox Industries Inc., Dallas, TX.)

Fig. 9.5 Retrofitted domestic solar hot-water installation. (Photograph courtesy of Sunworks, Somerville, NJ.)

9.1.2 Swimming Pool Heating

One key to successful solar heating, whether it be for a house or a swimming pool, is reducing the heat load. For a swimming pool the heat load depends on the heat loss of the pool. Heat is lost by conduction to the ground (usually negligible), radiation to the sky, convection (especially when the wind blows), and evaporation of the water. Wind barriers can significantly decrease convection losses and evaporation losses can be stopped altogether by the use of a solar swimming pool cover (while the cover is in place). If an inflated cover is used, convection and radiation losses will also decrease because the top of the cover will be lower in temperature than the water. A cover will also help reduce chemical loss and keep the pool cleaner. A cover alone can increase the pool temperature 3–6° C (5–10° F) and can be expected to last three to five years.

In addition to or instead of a pool cover, solar collectors can be used to provide warmed water for a pool. Since swimming pool heating is a low-temperature solar energy application, glazed and insulated collectors are not needed. Often unglazed plastic or rubber collectors are used. Water from the pool is circulated directly through the collectors. If freezing is a possibility, provisions for draining the collectors must be

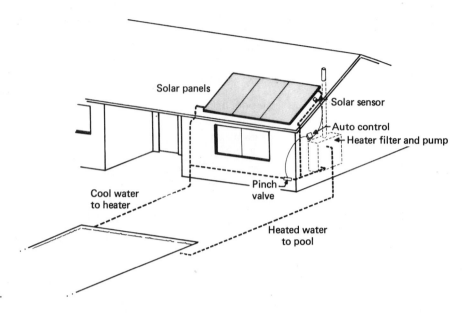

Fig. 9.6 Solar pool heater.

made. Figure 9.6 shows how a solar pool-heating system might be connected.

Heat for the pool at Oakbrook Townhouses in Thousand Oaks, California is provided by Raypak's swimming pool solar panels (Fig. 9.7).

Fig. 9.7 Swimming pool heating system at Oakbrook Townhouses, Thousand Oaks, California. (Photograph courtesy of Raypak, Inc., Westlake Village, CA.)

TRADITIONAL APPLICATIONS

Collectors are placed directly on the roof of a recreational building and also on a special rack designed to blend with the existing architecture. The system is so successful that the old gas connection has been shut off.

Usually it is recommended that the collector area equal 50–75% of the pool area, depending on geographical location and orientation of the system [8, 9, 10]. The collectors should be tilted at an angle within the range of latitude − 10° and latitude + 10°.

A small temperature rise through the collectors, often only 3–6° C (5–10° F), is desirable because it allows the collectors to operate very efficiently. This low temperature rise, which is sufficient to heat the pool, is often accomplished by establishment of a fast flow rate through the collectors. Solar collectors often raise pool temperatures 6–8° C (10–15° F).

9.1.3 Water Purification

The distillation of seawater or brackish or otherwise impure water for drinking can be accomplished with the use of a *solar still*. One of the simplest types of solar stills is the *glass-house*, or *glass-roof*, still. It consists of a long, narrow black tray with small troughs along the edges and a tent-shaped glass cover (Fig. 9.8). The water is heated by the sun and

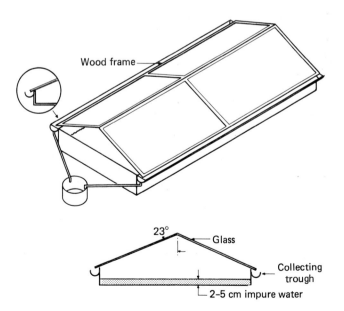

Fig. 9.8 Glass-house solar still.

evaporates onto the glass roof as a thin film. The distilled water slowly runs down the roof and is collected by the trough.

The glass-roof solar still can be made with plastic sheeting instead of glass. If untreated plastic is used, however, the evaporated water will collect as droplets on the plastic instead of as a thin film. These droplets reflect some of the sunlight away and can easily fall back into the tray to be evaporated again. Special treatment of the plastic surface is necessary to allow formation of the thin water film. The plastic sheeting is held in place by long, white, cylindrical sandbags, as shown in Fig. 9.9.

Usually water depth of 2–5 cm (1–2 in.) is best. Considerably more fresh water is collected per day in the summer than in the winter. Daily collection of fresh water can range from 7–8 l/m^2 (2.8–3.1 cups/ft^2) of still area in the summer to less than 1 l/m^2 (0.4 cups/ft^2) in the winter [11].

For the still to operate efficiently all air leaks that can lead to high heat losses must be plugged. The water in the tray should be replaced every few days. Small-scale solar stills for household-water distillation are often quite practical, but large-scale solar distillation has not been found to be economical except in certain remote areas.

9.1.4 Solar Cooking

Two basic methods for cooking with solar energy correspond to cooking with a conventional stove and oven. The solar stove is a concentrating parabolic reflector that supports the food at its focal point. The solar oven is an insulated box with a window top and side reflectors.

Parabolic solar cookers are usually constructed of aluminum or plastic and are about 1.2 m (4 ft) in diameter. Their position has to be read-

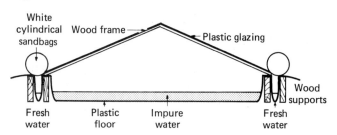

Fig. 9.9 Plastic solar still.

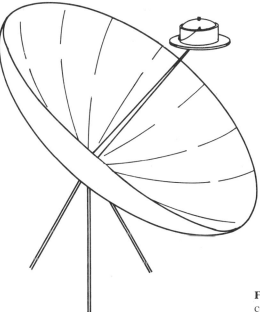

Fig. 9.10 Umbrella-type solar cooker.

justed every 15–20 min to follow the motion of the sun. Figure 9.10 shows a concentrating umbrella-type solar cooker developed by G. O. G. Löf [12].

Most solar ovens are basically well-insulated boxes with a glass or plastic top that admits solar radiation. Flat reflectors placed around the window increase the solar energy input and the temperatures reached inside the oven. Heat-storage material can be incorporated inside the box and used to preheat the oven. The hot storage material can also provide some heat for cooking if the sun should set or disappear behind a cloud. Solar ovens require repositioning every 30–45 min to follow the sun and can easily sustain normal baking temperatures during the day. A solar oven of the type developed by Maria Telkes [13] is shown in Fig. 9.11.

Solar cookers and solar ovens have several drawbacks. Tests showed that it took an average of 41 min for a solar cooker to bring 2 l (2 qt) of room-temperature water to a boil, while it took the solar oven an average of 112 min [14]. Cooking can be halted by the appearance of clouds, and wind can increase the heat-loss rate of the food. Also, with both solar cookers and solar ovens, cooking in the early morning or in the evening is impractical or impossible.

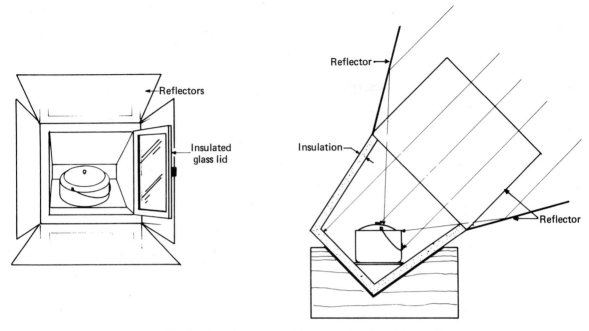

Fig. 9.11 Solar oven of the type developed by Telkes.

9.1.5 Agricultural Drying

Much agricultural drying is essentially the setting out of the material to be dried in the sun. Faster drying can be accomplished by use of specially constructed solar dryers, however. Usually such dryers have a dark absorbing surface with clear-plastic glazing and may have a fan to promote good air circulation.

Faster drying reduces contamination of the food by dirt, fungi, insects, and animals and exposure to wind and rain. Use of solar dryers also minimizes overheating and overdrying; usually the food so dried will taste better and may have a higher vitamin content than traditionally dried food. A solar dryer for crop drying is pictured in Fig. 9.12 and a small solar dryer for home use is shown in Fig. 9.13 (p. 320).

9.2 NEWER TOOLS AND APPLICATIONS

The key to unlocking the energy in sunlight is the development of cheap and efficient energy-conversion devices. As we saw in Chapter 1, an energy-conversion device takes in one form of energy and converts it to another useful form. Many machines and inventions are energy-conversion devices. Some examples are electric motors, batteries, lights, engines, and even solar collectors. To understand the various uses of solar

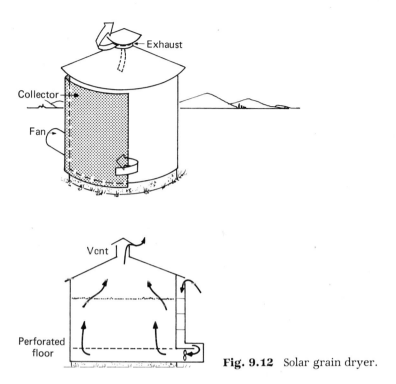

Fig. 9.12 Solar grain dryer.

energy we must also become familiar with some energy-conversion devices. Many of these devices have been around for a long time, but new designs and uses for them are still being developed.

9.2.1 Concentrating Collectors

Concentrating solar collectors are light-to-heat energy convertors, which have a wide variety of uses. One of the best ways to generate high temperatures from sunlight is to concentrate the light; that is, to use a lens or mirror to focus a large area of sunlight onto a small receiving surface. The *concentration ratio* is a measure of how large the area of collected sunlight is in relation to the area of the small receiver (Fig. 9.14).

There are two common misconceptions regarding concentrating collectors. The first is that high temperatures mean high efficiency. People assume that a lens or parabolic mirror is a better collector of solar energy than a flat-plate collector "because it collects such high temperatures." This is a mistaken assumption: solar collectors do not "collect temperature," they collect heat energy, and it is possible to collect tremendous amounts of heat energy at low temperatures. This misunderstanding shows a basic lack of knowledge about the differences between heat and temperature (see Chapter 1). The second misconception might

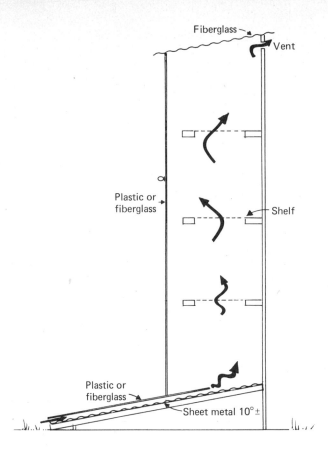

Fig. 9.13 Small solar dryer for home use.

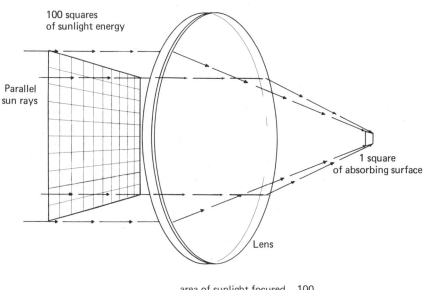

Fig. 9.14 Bending light rays to concentrate solar energy.

$$\text{concentration ratio} = \frac{\text{area of sunlight focused}}{\text{area of target}} = \frac{100}{1} = 100:1$$

be called "the burning-glass myth." It is assumed that a lens has the ability to "magnify light" just as it can magnify images and that a lens can start a fire because "it can magnify the solar energy." Of course, lenses cannot magnify energy, they can only focus it, as is shown in Fig. 9.15.

Fig. 9.15 The burning-glass myth.

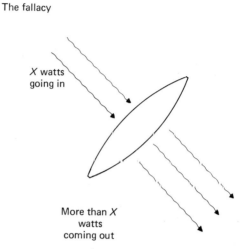

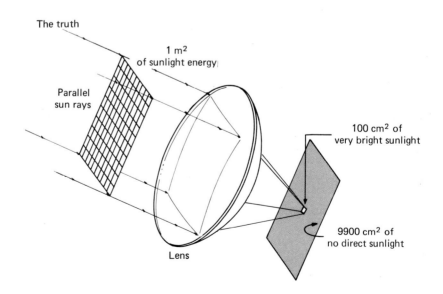

Figure 9.16 illustrates the construction of some of the basic types of concentrating collectors. The circular concentrator (Fig. 9.16 [a]) is used to generate extremely high temperatures. A concentration ratio of 1000 to 1 is possible, giving temperatures of over 3000° C. Daniels describes a plastic collector 1.8 m in diameter with an area of 2.7 m² [15]. It focused light onto a 0.015 m² target, giving a concentration ratio of 174:1 and producing a temperature of 500° C. A parabolic curve gives the most precise focus, but many other shapes can be used.

Fig. 9.16 Basic concentrating collectors: (**a**) circular collector, (**b**) circular Fresnel lens, (**c**) linear concentrator, (**d**) fixed-mirror concentrator.

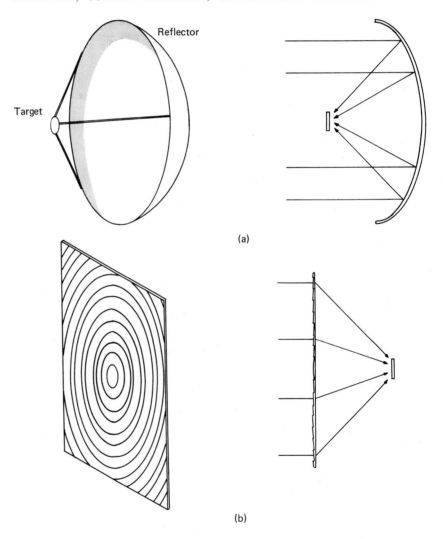

(a)

(b)

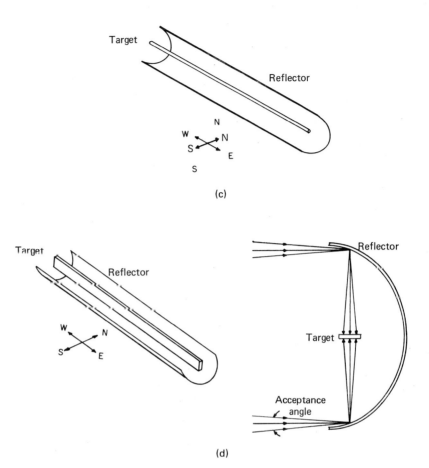

(c)

(d)

The circular Fresnel lens (Fig. 9.16 [b]) works like an ordinary convex lens but can be much thinner. It is often press-formed from plastic sheeting, so it can be inexpensively produced. It is difficult to make the grooves optically accurate, but fairly high concentration ratios are possible. The plastic lenses are susceptible to deterioration as a result of exposure to ultraviolet light and weather.

The linear concentrator (Fig. 9.16 [c]) focuses light onto a line, typically a black, fluid-filled tube serving as the absorber. Concentration ratios are not as high as for circular concentrators, but moderately high temperatures can be generated. Linear concentrators are commonly fixed in the east–west direction and are designed to follow only the changes in the sun's altitude during the day (east to west). This greatly simplifies the drive and support mechanisms required. A linear concentrator may have either a linear reflector (as shown) or a linear Fresnel lens.

It is not possible to design an effective concentrating collector that accepts light from any angle. All concentrating collectors must be moved to follow the sun. Concentrators can, however, be designed to accept light from a small range of angles (at the price of lowering the concentration ratio). Such collectors may be designed so that they need adjusting only once a day. One such design is the fixed-mirror concentrator shown in Fig. 9.16 (d). Pierce describes an experimental collector of this type that is 1.2 m × 2.4 m and has an acceptance angle of 10° [16]. The reflector has a cover of Tedlar stretched across it to reduce heat loss. For an outdoor-air temperature of 2° C and for a water temperature of 77° C the efficiency is 62%. Many other fixed or semifixed designs are possible [17].

All concentrating collectors use liquid heat-transfer fluids. Many have a liquid-filled tube as the absorbing surface. Since the absorber is very hot, the tube is usually enclosed by a larger glass tube (often containing a vacuum) to reduce radiative heat losses. For large concentrators reflectors tend to be much more economical than lenses. Fresnel lenses can be inexpensively pressed from plastic sheeting, but they are subject to the deteriorating effects of ultraviolet light and weather.

The reflectors for concentrating collectors are almost always *aluminized* plastic or glass; that is, glass, plastic sheeting, or plastic film that has been coated with a thin layer of aluminum. The mirror surfaces may be on the front or the rear of the glass or plastic. Front-surface mirrors must have a transparent coating or they will rapidly degrade when exposed to the elements. Aluminum coatings typically reflect 80–90% of the light they receive. Silver coatings are better, reflecting 90–98% of the light, but they tend to tarnish easily and come off.

If concentrators have one major disadvantage, it is that they have to track the sun as it moves. Specially geared drive motors or electric-eye devices enable them to do this. Concentrating collectors can only make use of the parallel rays of direct sunlight, so they must face the sun at all times. As indicated in Fig. 9.16, efforts have been made to design concentrators that do not require constant, precise alignment with the sun. In one such design, the absorbing surface is enlarged slightly. The reflector does not have to focus so precisely, and the larger collector will accept light from a wider range of angles. This solution has two disadvantages, however. First, increasing the absorber area also increases the heat loss. Second, the concentration ratio is decreased. Less light concentration means lower operating temperatures. Yet these disadvantages are often a small price to pay for simplifying collector design.

Because concentrating collectors are "blind" to nondirect sunlight, their performance is sometimes inferior to that of other types of collectors. On cloudy or hazy days a large part of the sunlight is diffuse or indirect. This diffuse light will not be received by concentrating collectors, but it can be absorbed by flat-plate and evacuated-tube-type collectors. Evacuated-tube collectors can also generate fairly high temperatures efficiently and they do not have to be motorized to follow the sun. For these reasons, tubular collectors are expected to take over some of the functions of concentrators in the future (see Chapter 3). For power generation and other very-high-temperature applications, concentrating collectors remain unsurpassed, however.

Concentrating collectors, because of their complexity and the high temperatures they generate, are more suitable for industrial applications than for home use. Almost any industrial process that requires heat can be adapted to run on solar energy. Concentrating collectors are also used to supply the high temperatures needed for solar heat engines and solar air conditioners. A special type of concentrating collector, the thermal power tower, will eventually be used to generate electricity.

9.2.2 Solar-Powered Heat Engines

Strictly speaking, a solar-powered turbine is not a solar energy-conversion device—it is a heat energy-conversion device that can convert heat from any source into mechanical energy. Nevertheless turbines, Stirling-cycle engines, and other heat engines are part of solar technology because solar heat is so easy to generate.

In Chapter 1 it was pointed out that some energy-conversion processes are irreversible. The conversion of sunlight to heat is one such process. Once light is converted to the random molecular motion of heat, it is not possible to convert all of the heat back to light. The conversion of mechanical energy to heat is also an irreversible process, as is the conversion of electrical energy to heat. No device that converts solar heat into mechanical energy or electricity can ever even approach 100% efficiency. In fact, there is a theoretical limit to how efficient any heat engine can be, a limit based on the temperature of the heat source and *heat sink* (the cooling source). The maximum efficiency of a heat engine, *Carnot efficiency*, is explained in Fig. 9.17.

Flat-plate solar collectors are generally not practical heat sources for engines because they cannot generate sufficiently high temperatures. For example, for a heat source of 80° C (176° F) and a heat sink of 10° C (50° F), the maximum possible efficiency of a heat engine (the Carnot efficiency) would be 19.8%. Real heat engines will operate at much less than the Carnot efficiency. (What is more, flat-plate collectors are not very efficient at 80° C.) To operate heat engines with solar heat, it is necessary to use concentrating collectors or at least evacuated-tube collectors.

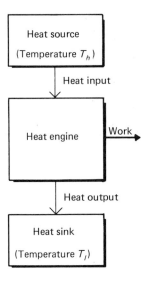

Maximum possible efficiency (Carnot efficiency):

$$N_c = \frac{T_h - T_l}{T_h}.$$

T_l, T_h must be absolute temperature, (i.e., K or R).

Fig. 9.17 Maximum efficiency of a heat engine.

Figure 9.18 describes two heat engines that can convert solar energy to mechanical energy. In the Rankine-cycle turbine (Fig. 9.18 [a]) water or another working fluid such as freon is heated at constant pressure in the solar collectors (or a nonboiling heat-transfer fluid is heated in the collectors and the heat is transferred to the water—this is better because it is not desirable to have the water boil in the collectors). At its boiling point water is vaporized at constant temperature and pressure. The steam is then superheated above the boiling point and expanded in the first stage of a turbine. The steam is then reheated with solar heat and expanded in the second stage of the turbine. Often there are three or more of these stages. Finally the steam is condensed and the heat is rejected from a cooling tower (or a radiator). The reheating and use of a multistage turbine increases overall efficiency. Other variations are possible that can increase it further. The Rankine cycle is the standard cycle used in present-day steam-turbine power plants.

The Stirling-cycle hot-air engine (Fig. 9.18 [b]) is a totally sealed engine containing air as the working fluid. In the ideal Stirling cycle, air is

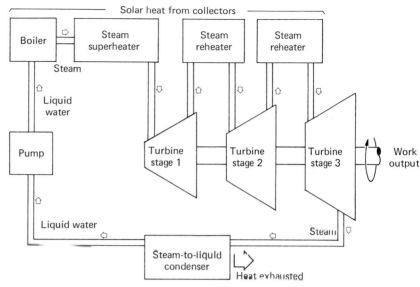

Fig. 9.18 Heat engines:
(**a**) Rankine-cycle turbine (continued)

(a)

heated at constant volume (using solar heat, in this case). As this is done, the air pressure rises. Next, the air expands, doing work, and is kept at constant temperature by the addition of more heat during the expansion. Third, the air is cooled at constant volume until it reaches the original temperature. Finally, work is done as the volume is reduced by the removal of heat at a constant temperature. The sketches show one form the engine can take; essentially one cylinder acts as two. The top part of the cylinder is continuously heated with solar heat and the bottom part is continuously cooled with cooling water or its equivalent. The upper and lower parts of the cylinder are separated by a sliding piston. The actual working piston is at the bottom. Once the solar energy has been transformed in a heat engine, it can be used to power any conventional type of machinery that requires a motor. For example, in a conventional air conditioner or heat pump a mechanical compressor is used that is powered by an electric motor. The compressor is used to increase the temperature and pressure of a vapor that has been heated and vaporized by the hot indoor air. The hot *refrigerant* then condenses and gives up heat from the room air to the outdoor air. The liquid then throttles through an expansion valve to reduce its temperature and pressure. The cycle continues as the indoor air heats and revaporizes the liquid. The part of the cycle that takes the most energy is the compressing of the re-

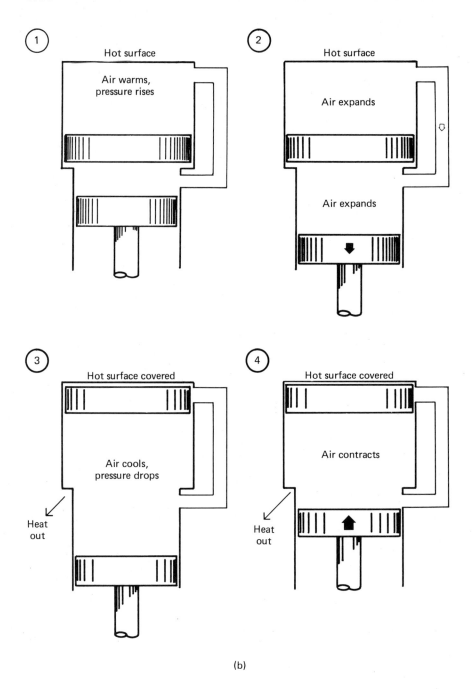

Fig. 9.18 (continued) (**b**) Stirling-cycle hot-air engine.

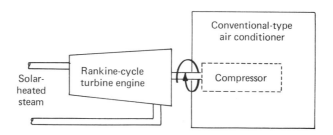

Fig. 9.19 Rankine-cycle air conditioner.

frigerant vapor. To build a solar-powered air conditioner, we must find a way to use solar energy to compress the refrigerant. One way to compress the refrigerant is to use a mechanical compressor powered by a heat engine. This has been done with a Rankine turbine: the method is called *Rankine-cycle solar air conditioning* (Fig. 9.19). This type of system requires a lot of expensive equipment, however, most of which cannot yet be purchased. Solar turbines themselves are complex and not very easy to come by.

In general, light-to-heat-to-mechanical-energy conversion by heat engines is not widely applied, except for agricultural pumping in remote areas. The reason is that the equipment needed tends to be too complex and costly for small-scale uses (for which solar cells are becoming increasingly practical). As large-scale solar-power generation is developed, however, heat engines will be increasingly used.

9.2.3 Solar Absorption and Adsorption Air Conditioners

An *absorption air conditioner* is a cooling device powered by heat. A conventional air conditioner requires electric power because it uses an electric compressor (a pump) to compress refrigerant vapor. To run a conventional air conditioner with solar energy, we could convert light to electricity and run the compressor with the electricity. An alternative would be to collect solar heat and use a heat engine to run the compressor. Or, instead of using a mechanical compressor, we can compress the refrigerant vapor chemically. This is possible because certain chemicals tend to attract other chemicals. For example, a strong solution of lithium bromide attracts water. Whenever water vapor is present lithium bromide tends to absorb the water, which compresses it. Thus if we were to use water as the working fluid of an air conditioner, we could use lithium bromide as the compressor. This is the principle of absorption air conditioning.

The absorption cooling cycle is an energy-conversion process. Heat energy is converted into the mechanical energy of "compressing" the refrigerant. Heat input is required because once the chemical absorbent has absorbed some refrigerant it loses its potency and needs to be regenerated, or purified. The absorbent is pumped to the generator where the refrigerant is boiled back out of it using a heat source that can be provided by solar energy. Concentrating collectors, evacuated-tube collectors, and even flat-plate collectors have been used to provide the heat for absorption coolers.

Figure 9.20 illustrates how the absorption cooling cycle works. The water/lithium bromide combination is the one most commonly used, but other refrigerant/absorbent combinations are possible, such as ammonia (as the refrigerant) and water (as the absorbent).

An absorption cooler has very few moving parts and almost any source of heat can be used for backup, or auxiliary heat necessary to produce cooling. Its big disadvantage is that for most efficient operation an absorption cooler should be run for long periods of time, not just intermittently. For this reason, buildings having absorption cooling systems

Fig. 9.20 The absorption cooling cycle.

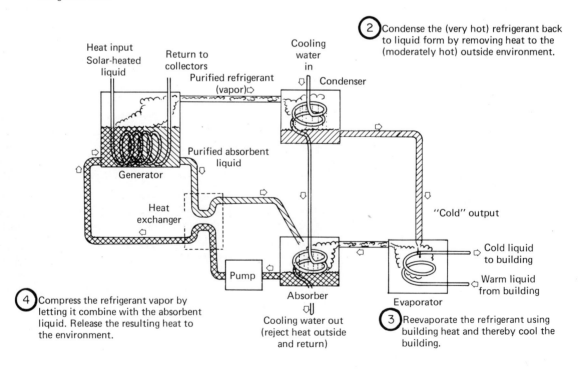

commonly have at least two storage tanks, one for storing solar heat and another for storing coolness. Absorption-type coolers (run with conventional sources of fuel) have been used in large buildings for many years, so several manufacturers have had experience with them. At least two manufacturers are marketing machinery for solar applications. Absorption cooling has not yet been widely used for home air conditioning because of the expense and because it works best with high-temperature solar collectors such as concentrating collectors or evaporated-tube-type collectors.

Adsorption cooling is similar to absorption cooling in that the refrigerant vapor is compressed by a substance. Adsorption involves using a solid such as silica gel instead of a liquid to compress the refrigerant vapor. Silica gel, like lithium bromide, attracts water. To be regenerated, the silica gel must be heated (by solar or conventional heat) to drive the water out. Adsorption cooling theoretically works well, but most solar cooling is of the absorption kind.

9.2.4 Solar Cells

The direct conversion of sunlight to electricity is accomplished with photovoltaic cells, or *solar cells.* There are several types of solar cells under development, but the kind most widely used at this time is the *silicon cell.*

Silicon is a *semiconductor*, meaning that it is neither a good nor a bad conductor of electricity (better than glass, for example, but worse than copper). A pure crystal of silicon can be *doped* with a few atoms of other substances to create either *p-type* or *n-type* silicon (Fig. 9.21). P-type silicon has been doped with *acceptor atoms* and n-type silicon has been doped with *donor atoms.* Each donor atom has an extra electron to donate that becomes free to move around in the crystal (although the donor atom itself is part of the crystal and cannot move). Each acceptor atom has an empty space that would accept an electron if one were available. This space is called a *hole.* Holes, like electrons, are free to move around in a crystal of silicon. A hole moves as an acceptor atom grabs an electron from a neighboring silicon atom. Now it is the silicon atom that has the hole and it can pass the hole along by grabbing an electron from another silicon atom. Silicon can have four kinds of charged particles in it: electrons and holes, which are free to move, and donors and acceptors, which are not.

A single crystal of silicon can be doped with donors in one area and with acceptors in another. Where the p region meets the n region we have a *pn junction* (Fig. 9.21). In the area around the pn junction there is a strong, permanent electric field that is vital to the operation of solar

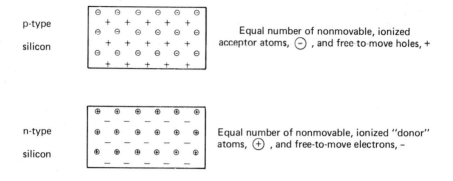

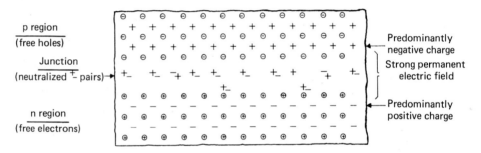

Fig. 9.21 Doped silicon and the pn junction.

cells. This field exists because some of the free holes from the p region were attracted to the free electrons in the n region and they combined at the junction, neutralizing one another and leaving an excess of negative charges in the p region and an excess of positive charges in the n region. The permanent electric field this creates is what makes a solar cell work. As a ray of sunlight enters the silicon and is absorbed, it breaks an electron free from an atom. This creates a free electron and a new hole. Because of the natural electric field near the pn junction, the electron and the hole are forced apart, and an electric potential (a voltage) is created. This is diagrammed in Fig. 9.22. Solar photons enter the silicon and are absorbed. Each absorbed photon creates an electron-hole pair. If the photon is absorbed near enough to the junction, the hole and electron are separated by the natural electric field at the junction. The hole goes to the p region and the electron to the n region. (If the electron-hole pair is not created near enough to the junction, the electron and hole will recombine, generating heat.) As the electrons and holes move apart, an

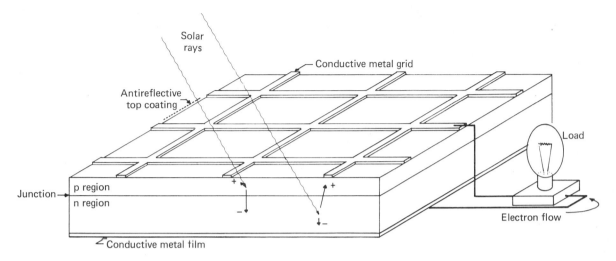

Fig. 9.22 The silicon solar cell

electric field or voltage is created, which causes electrons to flow though the wire (and recombine with holes near the metal grid).

The conventional silicon cell is made from a single crystal of very pure silicon. The silicon must be pure and regular in structure or the electron-hole pairs a light ray creates will immediately recombine and the cell will not work. Because of this, conventional solar cells are quite expensive and may never become cheaply available. Several other types of solar cells and some nonconventional silicon cells hold more promise of reducing the cost of solar electricity. One of these is the *cadmium sulfide cell*, which is less efficient than the silicon cell but can be produced at fairly low cost. Another is the *gallium arsenide* cell, which is still quite expensive but can have twice the efficiency of the silicon cells[7]. Some of these alternative cells and their efficiencies are listed in Table 9.1.

Table 9.1 **Solar-Cell Efficiencies**

Type	Efficiency (Percentage)
Conventional silicon	12–18%
Cadmium sulfide	5–8%
Gallium arsenide	16–25%
Ribbon-grown silicon	10–15%
Ovshinsky silicon	Approximately 10%

The conventional (and costly) way to make silicon cells is to crystallize long bars of silicon and then saw off thin, round wafers. The wafers are so thin that the saw cuts are similar in thickness to the wafers themselves, and much of the bar is wasted. A newer way to grow silicon crystals is the *ribbon-growth*, or *edge-defined-film growth* (EFG) method. In this process, molten silicon is pulled through a die, takes the shape of a thin ribbon, and crystallizes as it hardens. The sawing process is thus eliminated and cells can be made in larger sizes. It is believed that this can eventually reduce the cost of silicon cells.

Another new silicon cell is the *Ovshinsky cell*, which is made from amorphous silicon, or silicon that does not have the regular crystalline structure. One layer is undoped and another layer is doped with fluorine and hydrogen. This development is claimed to reduce the silicon-material cost by a factor of 500. There is some controversy as to whether such large cost reductions are possible, but it is believed that substantial reductions can be achieved.

The major problem with solar cells is that they remain very expensive in relation to the amount of power they can generate. The Department of Energy (DOE) has a goal of $500 per *peak kilowatt*[1] by 1986, but the present cost range is $10,000–$20,000. This cost is put into perspective by the fact that a typical United States household consumes 1–10 kW or more of electricity continuously. Fair technological advances are being made, but even faster progress will be required to make solar cells economical for power generation within the near future.

One approach to cost reduction has been to reduce the number of cells required and increase their efficiency by focusing sunlight onto the cells. A concentrating collector of some kind is used to focus a large area of sunlight onto a small area of solar cells. To avoid overheating the cells, a cooling liquid is circulated behind them. An expensive, sun-tracking mounting is required for the concentrator, but sun tracking results in maximum energy collection. Also, the heat collected by the cooling water is a useful by-product. Sandia Laboratories has experimented with concentrator-photovoltaic systems. In one system, plastic Fresnel lenses were used to concentrate sunlight onto the cells. Concentrator systems are intrinsically expensive, but it is believed that they may make photovoltaic cells economical, at least sooner than full arrays will become feasible.

---

1. A peak kilowatt means that the array of solar cells could generate one kilowatt of electricity under the best conditions—for example, near the middle of the day. To generate one kilowatt around the clock would require a much larger array of cells and an expensive battery system, to store the energy.

9.2.5 Wind Power

Many of our sources of energy are actually solar energy "in disguise." Two examples are fossil fuels and hydroelectric power. A lot of research is being done on wind power and other sources of energy derived from solar energy.

Wind energy is provided by air that has been set in motion by solar energy. Wind energy harnessed by wind turbines (windmills) can provide power in remote areas and can pump drinking and irrigation water, aerate water, dry crops, and more.

There is much current interest in using solar wind energy to generate electricity. Small, home-sized generating systems can be hooked up to the power grid and feed power into the distribution system when they are generating more electricity than the home can use. In such systems the utility grid thus becomes a storage device. The homeowner receives credit for the power fed into the utility. The rate the utility pays for electricity received is lower than the homeowner must pay when he or she wishes to retrieve some of this "stored" electricity. The difference in the rates compensates the utility for costs associated with providing the service.

Figure 9.23 shows an experimental, vertical-axis Darrieus wind turbine at Sandia Laboratories, Albuquerque. The 16.7 m (55 ft) diameter

Fig. 9.23 Experimental Darrieus wind turbine at Sandia Laboratories, Albuquerque, New Mexico. (Photograph courtesy of Lori Anderson, Seattle, WA.)

turbine can produce 60 kW [18]. It is not quite as efficient as regular windmill-type turbines with horizontal propellers, but it is lighter and simpler in construction. The blades are shaped as airfoils to reduce blade-bending stresses. The goal of the experimental program is to develop and build a small, utility-sized, 1-MW system.

Some have predicted that by the year 2020 wind energy will be the most beneficial of all the types of solar energy. The truth of this prediction will depend on how much progress is made in other energy-conversion areas.

9.2.6 Biomass

Biomass is another indirect form of solar energy. It is the product of the storage of solar energy in the photosynthesis process. It includes dead and living plant material, animal and municipal wastes, and even coal, natural gas, and oil, since they are believed to be the result of plant and animal decomposition. Biomass usually means renewable biomass, however, which does not include the fossil fuels. Biomass can be used to produce heat and electricity as well as fuels, sugars, alcohols, and other chemicals.

Burning wood or other plant materials for heating homes or for generating heat or electricity for industry is the use of solar energy through biomass. Widespread use of biomass in this manner may, however, interfere with the production of food.

Anaerobic digestion, or the conversion by bacteria of animal wastes and plant material to *biogas* (mostly methane and carbon dioxide) and fertilizer is another use of renewable biomass. It is possible for a farm to meet a significant portion of its energy needs with the anaerobic digestion products from swine or cow manure and hay.

The production of alcohols from biomass may become more important in the future if our country continues to suffer from periodic gas crises. Gasohol, 10% alcohol mixed with gasoline, is already in use in many areas and may become more common. One study has predicted that biomass will be the largest single application for solar energy in the future, contributing over 50% of the solar energy used in the year 2020 [19].

9.3 FUTURE TOOLS AND APPLICATIONS

If the preceding overview of solar related energy-conversion technology indicates anything at all, it should be that there is an endless number of possible ways to collect and use solar energy. In the three or four decades in which solar research has been somewhat active, it is likely that we have only scratched the surface. The following sections explore some of the proposed uses of solar energy and some of the experimental tools that may someday be used for not yet thought of applications of solar energy.

9.3.1 Solar Power Towers

It is generally agreed that for the rest of this century the major contributions of solar technology will be in the areas of building space heating, space cooling, and water heating. For the longer stretch, however, it is believed that electrical power will be generated increasingly from solar energy. Most scenarios do not include large numbers of rooftops covered with solar cells (though this is not an impossibility). Instead, large central solar power plants are pictured.

Solar heat-to-electricity conversion in *power towers*, or central receiver plants, is one of the most promising areas of research. A power tower is a large receiving tower surrounded by many acres of mirrors or heliostats. Figure 9.24 is an artist's conception of Solar One, the first solar receiver electric power plant. When completed in late 1981, Solar One is expected to produce up to 10 MW of electric power for Southern California Edison Company. A field of 1818 computer-directed mirrors will track the sun, reflecting solar radiation to the 90 m (300 ft) high cen-

Fig. 9.24 Solar One, the first electric solar receiver power plant. (Photograph courtesy of Southern California Edison Company.)

tral tower. The solar heat will be used to produce superheated steam that will run a conventional steam-turbine generator.

Each mirror surrounding a power tower has its own motorized mount and is individually controlled so as to reflect continously the sun's rays onto a collection area on the receiving tower. Sunlight from perhaps hundreds of acres is focused onto one small area, heating it tremendously. The heat can be used to fire a conventional steam power plant.

There are many advantages to this approach. It is expensive because of all the mirrors and tracking devices, but even so it is perhaps the most economical solar power concept yet proposed. Power towers can be built today and they are being tested on a large scale. DOE has a Solar Thermal Central Power Program, whose goal is to guide power-tower development to the point of commercial viability by the mid-1980s. Sandia Laboratories in New Mexico has built a large, 5 MW test facility. Several corporations are working on heliostat and receiver designs such as the one pictured in Fig. 9. 24.

High solar concentration ratios (1000:1 or more) are possible, which means that the small receiver can be highly efficient at high temperatures (since it only has a small surface from which to lose heat). It is estimated that a 100-MW plant (a respectable size) could be built using 0.8 km^2 (0.3 mi^2) of reflector area [20].

9.3.2 Ocean Thermal Energy Conversion

The ocean is an extremely large "solar collector." It is also a low-temperature solar collector since it is such a large thermal mass. But there are ways to harness its energy. One way would be to use ocean currents to turn large, underwater turbines. A method that is receiving more attention is *ocean thermal energy conversion* (OTEC). An OTEC power plant would be a large floating ship or barge that would use a heat engine such as a turbine to generate power from the temperature difference between surface water and deep water. Warm, surface water at, say, 25° C (77° F) would provide the heat to run a boiler that would boil a working fluid such as freon. The cooling water would come from far beneath the surface, where the temperature might be 5–7° C (40–45° F).

DOE and other organizations believe strongly in the potential of OTEC because of the tremendous amount of energy stored in the oceans. There are problems to be overcome, however. The major problem is poor efficiency. The temperature difference between surface water and deep water is not large enough to give a Carnot efficiency of more than 5–8%. The energy may be "free," but the equipment will definitely not be. Tremendously large heat exchangers would be required because the temperature differences are so small. There are materials problems because ocean water is extremely corrosive. Finally, there is the problem of transporting the energy once it is produced. OTEC plants must be located in southern latitudes (where the water is warm), so the cost of transportation must be added to the cost of production.

9.3.3 Solar Satellites

One imaginative scheme for generating electricity from sunlight is the *solar satellite* or orbiting solar power station. Electricity would be generated in the satellite by either a thermal cycle (similar to the power tower), or by solar cells. It would be converted to microwaves and beamed in a tight beam by an antenna approximately 1 km (0.6 mi) in diameter to a fixed receiving area on earth. The receiver on earth would have to be considerably larger than the satellite's antenna to capture most of the microwave energy and convert it back to electricity.

The satellite would be in a *geosynchronous* orbit around the earth, meaning that it would remain constantly over one location on the earth. Otherwise the receiver would be out of range at times and the satellite would have to be aimed constantly to follow the receiver. The solar collection area on the satellite would be huge—50–150 km^2 (20–60 mi^2)—and could be constructed very simply since very little structural strength would be required in the zero-gravity space environment. A satellite of this size could generate nearly 10 GW (10,000 MW) of power.

The primary advantage of an orbiting power station would be increased energy collection. Because of its high orbit the solar satellite would receive direct sunlight nearly 24 hours a day. It would not be bothered by cloudy weather, and all of the sunlight would be direct, not diffuse. Power from any one satellite could be beamed almost anywhere in an entire hemisphere of the earth. The primary disadvantage is the cost of (1) transporting materials into space and (2) building and operating two collection areas, one in space to capture the sunlight and another on earth to capture the microwaves.

9.3.4 Other Solar Energy Converters

In addition to the devices, methods and applications we have examined so far, many other techniques for the conversion of solar energy are being explored. Some techniques are new and others are further developments of old ideas. The *heat-pipe* concept, for example, was patented in 1917 by General Motors but is now being reexamined as a way to simplify flat-plate solar heat-collection systems. The list that follows summarizes several of these somewhat exotic areas of research.

1. A *thermic diode* is a liquid-type flat-plate solar collector with built-in heat storage [21]. The absorber plate is in front and the flat storage tank behind. A special check valve allows liquid to flow from the absorber to the storage unit by thermosiphoning but does not allow reverse thermosiphoning at night that would result in heat loss.

2. A *heat pipe* is a device for transferring heat that does not require the pumping of fluids [22]. One end of the heat pipe (tube) is filled with a liquid and heat is applied to this end, vaporizing the liquid. The vapor travels to the other (cooler) end and condenses, releasing heat. The heat pipe has been suggested as a way to transfer solar heat passively and effectively from a solar absorber plate to the inside of a building.

3. A *photochemical flat-plate collector* is a device that would make use of a light-absorbing chemical reaction [23]. A chemical such as nitrosyl chloride would absorb light energy at fairly low temperatures and break up into other compounds. These would be moved to another place and recombined into nitrosyl chloride, releasing heat at high temperatures. This process could result in flat-plate collectors that would be highly efficient at low outdoor temperatures and that could generate high temperatures.

4. A *solar hydrogen generator* is a device that would use solar energy to decompose water directly into hydrogen and oxygen [24]. The hydrogen could then be burned to provide heat or it could power a conventional fuel cell to produce electricity. An electrode material such as titanium dioxide would absorb sunlight and act as a catalyst to split water into hydrogen and oxygen.

5. A *wet-type solar cell* is a unique photovoltaic cell for converting light to electricity. It has a semiconductor electrode and a platinum electrode in an electrolyte solution [25]. When light strikes the boundary between the semiconductor electrode and the solution, a voltage is generated. The hope is to develop efficient solar cells that would be lower in cost than existing cells.

6. A *thermoelectric generator* generates electricity from solar heat [2]. It focuses light onto special high-temperature *thermocouples*. (Thermocouples are devices formed by joining electrically dissimilar metals to semiconductors.) When one junction is heated and another is cooled, a voltage is generated. This generator cannot yet be made very efficient.

7. A *thermionic generator* generates electricity from high-temperature solar heat [2]. A metal surface in a vacuum is heated until electrons evaporate. A cooler metal receiving plate is placed very near to the hot surface and collects the electrons. High-temperature collectors are required but the thermionic elements could be relatively inexpensive.

8. With a *thermochemical energy collector,* high-temperature heat from concentrating solar collectors is used to cause a reversible chemical reaction (e.g., the disassociation of sulfur trioxide into sulfur dioxide and oxygen). This reaction stores the heat. Later the two products can be made to recombine, releasing heat. Thus solar heat could be collected at one central location and used at other locations [26].

9. A *nitinol engine* is a heat engine in which a metal alloy of nickel and titanium is the working element [27]. The metal changes its shape in response to temperature changes. Low-temperature heat is thus converted into mechanical energy.

New and unique energy-conversion techniques are being conceived daily—many of them related to solar energy. Some will be known only to readers of obscure scientific journals and others will become commonplace during the coming decades. Despite claims to the contrary, our rapidly expanding world population will require constant advances in technology, particularly new ideas for collecting energy and making it useful.

REFERENCES

1. Morse, R. N., and D. J. Close (1977). Solar water heating. In R. C. Jordan and B. Y. H. Liu (eds.), *Applications of Solar Energy for Heating and Cooling of Buildings.* New York: ASHRAE.

2. Daniels, F. (1964). *Direct Use of the Sun's Energy.* New York: Ballantine, pp. 75–81.

3. Morse, R. N. (1964). Water heating by solar energy. In *Proceedings of the United Nations Conference on New Sources of Energy,* vol. 5. New York: United Nations, p. 62.

4. Farber, E. A. (1964). The use of solar energy for heating water. In *Proceedings of the United Nations Conference on New Sources of Energy,* vol. 5. New York: United Nations, p. 24.

5. Anderson, B., and M. Riordan (1976). *The Solar Home Book: Heating, Cooling and Designing with the Sun.* Harrisville, N. H.: Cheshire Books.

6. Beck, E. J. *et al.* (1977). *Solar Heating of Buildings and Domestic Hot Water.* Washington, D. C.: U.S. Department of Commerce, NTIS AD-A054 601.

7. National Solar Heating and Cooling Information Center (1977). *Solar Hot Water and Your Home.* Rockville, Md.: National Solar Heating and Cooling Information Center.

8. D'Alessandro, B. (1979). Swimming with the sun. *Solar Age* 4 (4):30.
9. Lucas, T. (1975). *How to Build a Solar Heater.* New York: Ward Ritchie.
10. SolarVision (1977). *Solar Age Catalog.* Port Jervis, N.Y.: SolarVision, p. 128.
11. Gomella, C. (1964). Use of solar energy for the production of fresh water. In *Proceedings of the United Nations Conference on New Sources of Energy,* vol. 6. New York: United Nations, p. 157.
12. Löf, G. O. G., and D. A. Fester (1964). Design and performance of folding umbrella-type solar cooker. In *Proceedings of the United Nations Conference on New Sources of Energy,* vol. 5. New York: United Nations, p. 347.
13. Telkes, M., and S. Andrassy (1964). Practical solar cooking ovens. In *Proceedings of the United Nations Conference on New Sources of Energy,* vol. 5. New York: United Nations, p. 394.
14. Nutrition Division, Food and Agriculture Organization of the United Nations (1964). Report on tests conducted using the Telkes solar oven and the Wisconsin solar stove over the period July to September 1959. In *Proceedings of the United Nation Conference on New Sources of Energy,* vol. 5. New York: United Nations, p. 353.
15. Daniels, F. (1964). *Direct Use of the Sun's Energy.* New York: Ballantine.
16. Pierce, N. T. (1978). Low-cost concentrators. *Solar Age* 3 (2):18.
17. Meinel, A. P., and M. P. Meinel (1977). *Applied Solar Energy: An Introduction.* Reading, Mass.: Addison-Wesley.
18. SRI International (1978). *A Comparative Evaluation of Solar Alternatives: Implications for Federal Research, Development and Demonstrations,* vols. 1 and 2. Washington, D.C.: U.S. Department of Energy.
19. Alich, J. A., and J. G. Witwer (1976). Agricultural and forestry wastes as an energy resource. *Sharing the Sun* 7:146.
20. Skinrood, A. C. (1976). Barstow: prototypical power tower. *Solar Age* 3 (6):24.
21. Buckley, S. (1976). Thermic diode solar panels: a brief summary. *Sharing the Sun.* 2:1.
22. Kusianovich, J. (1977). Heat pipes. *The Solar Age Catalog.* Port Jervis, N.Y.: SolarVision.
23. Carlsson, B., and G. Wettermark (1978). The Photochemical heat pipe. *Solar Energy* 21(2):87.
24. Maruska, H. P., and A. K. Ghosh (1978). Photocatalytic decomposition of water at semiconductor electrodes. *Solar Energy* 20(6):443.
25. Kamat, P. V. *et al.* (1978). Thermophotochemical cells for solar energy conversion. *Solar Energy* 20(2):171.
26. Chubb, T. A. *et al.* (1976). Application of chemical engineering to large scale solar energy. *Sharing the Sun* 7:28.
27. Banks, R., and M. Wahlig (1976). Nitinol engine development. *Sharing the Sun* 7:29.

BIBLIOGRAPHY

Solar Water Heating

Anderson, B., and M. Riordan (1976). *The Solar Home Book: Heating, Cooling and Designing with the Sun.* Harrisville, N. H.: Cheshire Books.

Beck, E. J. *et al.* (1977). *Solar Heating of Buildings and Domestic Hot Water.* Washington, D.C.: U.S. Department of Commerce NTIS AD-A054 601.

Daniels, F. (1964). *Direct Use of the Sun's Energy.* New York: Ballantine, pp. 75–81.

Farber, E. A. (1964). The use of solar energy for heating water. In *Proceedings of the United Nations Conference on New Sources of Energy,* vol. 5. New York: United Nations, p. 24.

Lucas, T. (1975). *How To Build a Solar Heater*. New York: Ward Ritchie.

Morse, R. N. (1964). Water heating by solar energy. In *Proceedings of the United Nations Conference on New Sources of Energy*, vol. 5. New York: United Nations, p. 62.

Morse, R. N., and D. J. Close (1976). Solar water heating. In R. C. Jordan and B. Y. H. Liu (eds.), *Applications of Solar Energy for Heating and Cooling of Buildings*. New York: ASHRAE.

National Solar Heating and Cooling Information Center [n.d.]. *Solar Hot Water and Your Home*. Rockville, Md.: National Solar Heating and Cooling Information Center.

Silverstein, M. (1977). Solar hot water—looking before you leap. *Solar Age Catalog*. Port Jervis, N.Y.: SolarVision p. 122.

SolarVision (1979). *Solar Age* 4(2).

Swimming Pool Heating

ASHRAE (1978). *Draft—Methods of Testing to Determine the Thermal Performance of Liquid Type Solar Collectors for Heating of Swimming Pools*. New York: ASHRAE, standard 96P.

Czarnecki, J. T. (1963). A method of heating swimming pools by solar energy. *Solar Energy* 7(1):3.

D'Alessandro, B. (1979). Swimming with the sun. *Solar Age* 4(4):30.

deWinter, F. (1974). *How To Design and Build A Solar Swimming Pool Heater*. New York: Copper Development Association.

Lucas, T. (1975). *How To Build a Solar Heater*. New York: Ward Ritchie

Root, D. E., Jr. (1959). A simplified engineering approach to swimming pool heating. *Solar Energy* 3(1):60.

SolarVision (1977). *Solar Age Catalog*. Port Jervis, N.Y.: SolarVision, p. 128.

Water Purification

Daniels, F. (1964). *Direct Use of the Sun's Energy*. New York: Ballantine, pp. 75–81.

Gomella, C. (1964). Use of solar energy for the production of fresh water. In *Proceedings of the United Nations Conference on New Sources of Energy*, vol.6. New York: United Nations, p. 157.

Howe, E. D. (1964). Solar distillation research at the University of California. In *Proceedings of the United Nations Conference on New Sources of Energy*, vol. 6. New York: United Nations, p. 239.

Sales Section, United Nations (1970). *Solar Distillation as a Means of Meeting Small-Scale Water Demands*. New York: United Nations, no. E.70.11.B.1.

Talbert, S. G., J. A. Eibling, and G. O. G. Löf (1970). *Manual on Solar Distillation of Saline Water*. Washington, D.C.: Office of Saline Water, U.S. Department of Interior.

Solar Cooking

Bernard, R. (1978). Easy to build solar cookers. *Solar Age* 3(2):14.

Daniels, F. (1964). *Direct Use of the Sun's Energy*. New York: Ballantine.

Löf, G. O. G., and D. A. Fester (1964). Design and performance of folding umbrella-type solar cooker. In *Proceedings of the United Nations Conference on New Sources of Energy*, vol. 5. New York: United Nations, p. 347.

Nutrition Division, Food and Agriculture Organization of the United Nations (1964). Report on tests conducted using the Telkes solar oven and the Wisconsin solar stove over the period July to September 1959. In *Proceedings of the United Nations Conference on New Sources of Energy*, vol. 5. New York: United Nations, p. 353.

Telkes, M., and S. Andrassy (1964). Practical solar cooking ovens. In *Proceedings of the United Nations Conference on New Sources of Energy*, vol. 5. New York: United Nations, p. 394.

Agricultural Drying

Brace Research Institute (1976). Survey of solar agricultural dryers. *Sharing the Sun* 7:7.

Daniels, F. (1964). *Direct Use of the Sun's Energy*. New York: Ballantine.

Löf, G. O. G. (1964). Use of solar energy for heating purposes: solar drying. In *Proceedings of the*

United Nations Conference on New Sources of Energy, vol. 5. New York: United Nations, p. 248.

Schoenau, G. J., and R. W. Besant (1976). The potential of solar energy for grain drying in western Canada. *Sharing the Sun* 7:33.

Concentrating Collectors

Daniels, F. (1964). *Direct Use of the Sun's Energy*. New York: Ballantine.

Pierce, N. T. (1978). Low-cost concentrators. *Solar Age* 3:2.

Meinel, A. P., and M. P. Meinel (1977). *Applied Solar Energy: an Introduction*. Reading, Mass.: Addison-Wesley.

Solar-Powered Heat Engines

Daniels, F. (1964). *Direct Use of the Sun's Energy*. New York: Ballantine.

Meinel, A. P., and M. P. Meinel (1977). *Applied Solar Energy: an Introduction*. Reading, Mass.: Addison-Wesley.

Wark, K. (1977). *Thermodynamics*. 3rd ed. New York: McGraw-Hill.

Solar Absorption and Adsorption Air Conditioners

Duffie, J. A., and W. A. Beckman (1974). *Solar Energy Thermal Processes*. New York: John Wiley.

ISES (1976). Solar heating and cooling of buildings. *Sharing the Sun* 3:27.

NSF/RANN (1976). *Proceedings, NSF/RANN Workshop on Solar Cooling for Buildings*. Washington, D.C.: U. S. Government Printing Office, stock no. 3800-00189.

Solar Cells

Daniels, F. (1964). *Direct Use of the Sun's Energy*. New York: Ballantine.

Hovel, H. J. (1976). *Solar Cells*. In R. K. Willardson and A. C. Beer (eds.), Semiconductors and Semimetals, vol. 11. New York: Academic Press.

ISES (1976). Photovoltaics and materials. *Sharing the Sun* 6.

Meinel, A. P., and M. P. Meinel (1977). *Applied Solar Energy: an Introduction*. Reading, Mass.: Addison-Wesley.

SolarVision (1978). *Solar Age* 3(6).

Wind Power

Burke, B., and R. N. Meroney (1977). *Energy from the Wind: an Annotated Bibliography*. Fort Collins, Colo.: Engineering Research Center, Colorado State University.

Katzenburg, R. (1978). Plugging in the wind. *Solar Age* 3(6):23.

Lundergan, C. D. (ed.) (1977). *Sandia Laboratories Energy Programs*. Albuquerque, N. Mex.: Sandia Laboratories, SAND77-0034.

SRI International (1978). *A Comparative Evaluation of Solar Alternatives: Implications for Federal Research, Development and Demonstrations*, vols. 1 and 2. Washington, D.C.: Department of Energy.

Wind Power Digest, published quarterly.

Biomass[2]

Alich, J. A., and J. G. Witwer (1976). Agricultural and forestry wastes as an Energy Resource. *Sharing the Sun* 7:146.

Auerbach, L. (1976). A homesite power unit—methane generator. *Solar Age* 1(6):24.

Boersma, L., E. Gasper, and B. P. Warkentin (1978). Methods for the recovery of nutrients and energy from swine manure. In *Solar 78 Northwest Conference Proceedings*. Salem, Oreg.: Oregon Department of Energy, p. 212.

Bogan, R. H. (1976). Fuels and chemicals from the sun through bioconversion. *Sharing the Sun* 7:2.

Lingappa, B. T., and Y. Lingappa (1979). Family-size methane generators. *Solar Age* 4(3):16.

---

2. For further information, write to Biomass Energy Institute, Inc., P.O. Box 129, Postal Station C, Winnipeg, Manitoba, R3M 3S7, Canada.

Lipinsky, E. S. (1976). Field crops as a future source of fuels and chemical feedstocks. *Sharing the sun* 7:104.

Merrill, R., and T. Gage (1978). *Energy Primer—Solar, Water, Wind and Biofuels.* rev. ed. New York: Dell.

Ostrovski, C. M. et al. (1976). A feasibility study of bio-gas production in individual farms in southwestern Ontario. *Sharing the Sun* 7:129.

Robertson, E. E. (1976). Perpetually renewable biomass prospects. *Sharing the Sun* 7:157.

Solar Power Towers

ISES (1976). Solar thermal and ocean thermal. *Sharing the Sun* 5:323–391.

Lundergan, C. D.(ed.) (1977). *Sandia Laboratories Energy Programs.* Albuquerque, New Mex.: Sandia Laboratories, SAND77-0034

Skinrood, A. C. (1976). Barstow: prototypical power tower. *Solar Age* 3(6):24.

Ocean Thermal Energy Conversion

ISES (1976). Solar thermal and ocean thermal. *Sharing the Sun* 5:392–548.

Meyer, R. A. (1978). Ocean thermal energy conversion. *Solar Age* 3(6):20.

Solar Satellites

Glaser, P. E. (1976). The status of satellite solar power development. *Sharing the Sun* 1:1.

Harron, R. J. (1978). The solar power satellite as a viable energy production alternative. In *Proceedings of the 1978 Conference on Technology for Energy Conservation.* Rockville, Md.: Information Transfer, Inc.

Other Solar Energy Converters

Banks, R., and M. Wahlig (1976). Nitinol engine development. *Sharing the Sun* 7:360.

Buckley, S. (1976). Thermic diode solar panels: a brief summary. *Sharing the Sun* 2:1.

Carlsson, B., and G. Wettermark (1978). The photochemical heat pipe. *Solar Energy* 21(2):87.

Chubb, T. A. et al. (1976). Application of chemical engineering to large scale solar energy. *Sharing the Sun* 7:364.

Daniels, F. (1964) *Direct Use of the Sun's Energy.* New York: Ballantine.

ISES (1976). Photovoltaics and materials. *Sharing the Sun* 6.

Kamat, P. V. et al. (1978). Thermophotochemical cells for solar energy conversion. *Solar Energy* 20(2):171.

Kusianovich, J. (1977). Heat pipes. In Solar Vision (ed.), *The Solar Age Catalog.* Port Jervis, N. Y.: SolarVision.

Maruska, H. P., and A. K. Ghosh (1978). Photocatalytic decomposition of water at semiconductor electrodes. *Solar Energy* 20(6):443.

PROBLEMS

9.1 How many square meters of collector area and how much storage volume is needed for solar water heating for a family of four?

9.2 One family has a constant demand for hot water from early morning to early evening. Another family uses the same amount of water during the day, but only in the early morning and in the evening. Which family requires the larger solar water-heating system and why?

9.3 A thermic diode solar panel contains a thermosiphon water collector and water storage in one unit. The thermic check valve that allows one-way flow only is shown below. Explain how this valve works.

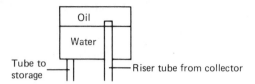

9.4 In Phoenix, Arizona, a 36 m² uncovered swimming pool may lose 0.3 cm of water per summer day due to evaporation. The heat of vaporization of water is 2.26 MJ/l. Compare the energy lost due to evaporation with the total daily radiation absorbed by the pool ($\alpha = 0.3$) if the daily radiation is 28.2 MJ/m².

9.5 A simple 2 m² solar still produces 7 l of water per day. If the total daily incident radiation is 26 MJ/m², what is the overall efficiency of the still? (See also problem 9.4.)

9.6 The theoretical concentration ratio of a 2 m² solar cooker (Fig. 9.10) is 90:1. The cooker takes 93 min to boil a liter of water in solar radiation of 800 W/m². How much energy was theoretically available to boil the water? How much energy did it really take if the water was initially at 20° C and all heat losses are ignored? Why are these two values so different?

9.7 The actual concentration ratio for most concentrating collectors is only 30%–70% of the theoretical value. List several reasons for this. (Hint: consider the actual path of a single light ray.)

9.8 A linear concentrator may run at 40% efficiency while producing a temperature of 300° C. A flat-plate collector (with the same area) running beside the concentrator at 40% efficiency may produce a temperature of only 60° C. How is this possible?

9.9 Suggest a simple tracking control device for an east–west linear parabolic collector that has two photovoltaic cells and a shading device. Why are two cells necessary?

9.10 What is the theoretical maximum efficiency of the three-stage Rankine steam generator shown in Fig. 9.18 if the superheated steam is heated to 250° C and the turbines expel the steam at 100° C?

9.11 The maximum coefficient of performance, *COP*, for an absorption cooler is defined as

$$COP_{max} = \frac{T_e(T_g - T_o)}{T_g(T_o - T_e)},$$

where

T_e = evaporating temperature,

T_g = generator temperature,

T_o = environmental temperature.

All temperatures are K or R. The actual refrigeration efficiency is given by

$$n = \frac{COP_{actual}}{COP_{max}}.$$

The evaporating temperature is 13° C, the generator temperature is 95° C, the environmental temperature is 35° C, and the actual *COP* is 0.53. What is the maximum *COP* and the actual refrigeration efficiency for this cooler?

9.12 Under full-sun conditions a typical photovoltaic cell will produce approximately 0.4 V with a maximum power output of 0.8 A. Cells must be connected in series and parallel to produce the desired voltage and current. In series, voltage is additive but current is not. In parallel, current is additive but voltage is not. How many cells are required and how should they be connected to produce up to 3.2 A and 3.2 V?

9.13 Photovoltaic systems are sized to meet the electric load. The number of panels needed can be estimated as

$$\text{Number of panels} = 1.2 \times \frac{\text{daily load (in ampere-hr)}}{\left(\begin{array}{c}\text{hours of peak}\\\text{sun per day}\end{array}\right) \times \left(\begin{array}{c}\text{one panel current}\\\text{(in amperes)}\end{array}\right)}$$

The extra factor of 20% accounts for system losses. How many 1.6 A panels are needed to meet a constant load of 3 A for a location that averages five peak-sun hours per day?

10 FACTORS AFFECTING THE FUTURE OF SOLAR ENERGY USE

10.1 GOVERNMENT Predictions of solar energy use for the United States in the year 2000 range from 5% to 25% of the total energy consumed. To reach the energy independence corresponding to the higher figure, government support and funding for solar energy projects must continue and increase.

Historically the federal government has spent or foregone receipt of large sums of money in order to increase energy production. Virtually all types of energy production have been and are continuing to be subsidized. Because of subsidized energy production, conventional energy sources appear to be less expensive than they really are. It is difficult for the relatively new solar energy industry to compete with other powerful energy industries benefitted by special long-standing legislation.

In the past the oil industry has directly received the majority of benefits from federal incentives, an estimated $77.2 billion. The largest portion of the funds, $40 billion, was for tax reductions for intangible expenses such as drilling and the oil depletion allowance. The remaining funds went for various incentives and subsidies, and a relatively small amount went for research [1]. Figure 10.1 shows the percentages of government subsidies received by the various energy industries through 1976. Since 1976 tax incentives for solar energy development have increased. Continued large subsidies for traditional energy sources make it difficult for newer and less subsidized alternative energy sources to compete economically, however.

Many important recommendations related to public policy regarding energy were forthcoming from the 55th American Assembly on Improving Energy Efficiency held in 1978 [2]. The assembly specifically recognized the need for removal of tax advantages granted to fossil-fuel and nuclear-energy industries. The assembly recommended that, until such tax advantages are revoked, energy conservation efforts, solar energy, and other renewable resources receive similar tax advantages and benefits according to their potential.

Fig. 10.1 Beneficiaries of government incentives to increase energy production, through 1976.

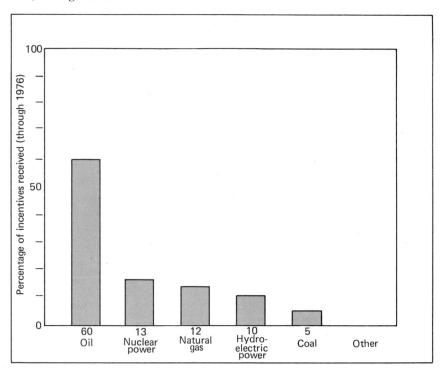

Originally taxes were designed to cover only government expenses. Now, however, additional items and activities are taxed, thereby allowing the government to accomplish social goals by redistributing wealth and the tax load. Most tax laws discriminate in some way, but such discrimination is usually justified as promoting the public welfare. For example, to support healthy state and local governments, interest paid on state and municipal bonds is not taxed. Donations to charities are not taxed because the charities provide services that the government would otherwise have to pay for. Also, rapid depreciation is allowed for pollution-control devices because a cleaner environment is better for everyone.

Similarly tax incentives supporting solar energy use show support for a cleaner environment, energy conservation, and the movement away from importing foreign gas and oil. Also, land that would have been used for refineries, power plants, and transmission lines can be used for other purposes. Since a tax incentive might influence decisions about whether to install a solar system, the incentive helps to improve local economies by encouraging additional solar-system building.

Some tax benefits do already exist for the use of solar energy, other alternative energies, and for energy-conservation measures. The National Energy Act of 1978 and the oil company windfall-profit-tax legislation of 1980 provide for a tax credit of up to $4000 (40% of the first $10,000 spent on the system) for the installation of certain solar, wind, or geothermal equipment on a home. A credit of 25% of a homeowner's outlay for energy-conservation devices (insulation, storm windows, etc.) up to a maximum of $300 is also allowed. A 25% business investment-tax-credit for solar energy use in a commercial building or industrial process is also often allowed. The act also provides for federal loans and loan guarantees to assist in the financing of solar systems.

Many states have also enacted tax incentives to encourage the use of solar energy. In many states, solar energy systems are exempt from property tax. The exemption means that the building and property the system is attached to will be taxed as if the system were not there. Since a solar energy system is usually quite expensive and the financial benefits arising from its presence do not accrue for many years, the existence or lack of a property-tax exemption for solar systems may be a deciding factor for many potential solar homeowners.

Some states also allow a tax credit or deduction on the state income tax for solar energy use. California is among the leaders in enacting favorable solar energy legislation and now allows homeowners a 55% tax credit for solar systems.

A deduction on state income tax for a solar energy system will result in a decrease in taxes paid. State taxes will be decreased by the amount of the allowed deduction multiplied by the state tax rate. A tax credit is the most effective tax incentive and reduces taxes by the amount of the credit. A summary of solar energy legislation by state appears in Appendix G. The government can also support solar energy use by providing low-interest loans, loan guarantees, loan insurance, and direct grants for demonstration projects or research.

Department of Energy [DOE] funds for solar energy and related technologies are divided among solar heating and cooling, electricity generation (including photovoltaic, thermal, wind, and ocean thermal energy conversion [OTEC]), and biomass (direct or indirect use of solar energy stored in plants through the photosynthesis process) programs. In the future, increased funding for central electricity-generation research is expected since the industrialized nations of the world seem to be moving toward electricity as the main energy-distribution method. Continued and increased support for all types of solar-technology development is also expected.

Solar energy use has strong popular support, which may influence government funding decisions. The problem is that many people expect too much too soon. They seriously underestimate the development costs and the time necessary to bring new products to consumers. Government funding for research into new solar demonstration projects in which existing technologies are used will, however, be necessary only until these technologies can compete economically with long-subsidized conventional energy sources.

10.2 PEAK LOAD

The hoped-for increase in the use of existing solar technologies (mainly space and water heating and some electricity generation) may have a significant effect on the projected use of conventional energy sources. Public utilities must have enough reserve capacity to provide energy during the times of highest, or peak, use. Often, meeting peak-load demands requires the use of auxiliary equipment that is expensive to operate. The utilities would prefer near-constant demand the year around. Unfortunately, extremely cold or hot weather usually precludes this. Solar systems, which decrease the energy demand during mild weather but may fail to operate during the coldest weather, may increase peak loads and hamper the utility's load-management efforts.

Solar heated and cooled buildings should not have too great an effect on gas utilities or fuel-supply dealers. Heating oil is delivered periodically at the site of the solar heated building. Since solar heated buildings have decreased overall demands for heat, they should serve to lower the total use of heating oil and gas and ease peak-load demand somewhat. The severest peak-load problems are anticipated for the electric utilities.

In the illustrations that follow, a utility's generating plants are represented by the men and the electricity demand (load) by the weights the men must hold.

Average load   Peak load

Case I Under average conditions the load is not great and most of the man's strength is not used. Under peak conditions, however, all the man's strength is needed. A very strong man is required to carry the peak load and, though he uses his maximum strength only rarely, he demands a high wage because he must have that reserve strength available.

Average load   Peak load

Case II In this example, the load is always nearly constant and the peak load does not vary much from the average load. A big strong man is not needed to carry these loads and the utility has fired him and hired a weaker man at a lower wage.

The utility's goal is the situation in case II, where the load is ideally matched to the generating capacity. Here, all the customers' needs are met at the lowest possible cost to the utility. To achieve this goal, the utility must sometimes force a decrease in the peak load by charging lower rates for electricity used at an off-peak time or by imposing prohibitively high rates for electricity used during the peak period.

Most utilities are *summer peaking*, which means the highest loads occur during the summer months. Solar hot water and air conditioning will reduce these peaks, helping the utilities to realize the situation presented in case II.

Many utilities are *winter peaking*, however. Customers whose solar heating systems may not work during extremely bad weather and who rely on backup electric heat may cause an increase in the peak load and put the utilities into the situation presented in case I. To meet that increased demand, the utilities may have to use expensive nuclear, gas, oil, or coal plants. To charge the same rate for all users all the time may be unfair to those people who use a relatively constant amount of conventional energy in their homes.

Rate structures devised to promote energy conservation and charge solar customers for the load-management problems the utilities feel solar customers cause are being investigated and implemented. Unless new proposed rates are confiscatory or illegal, they are usually approved. Since electric utilities are essentially regulated monopolies, the customers must either pay the new rates or do without electricity.

The utilities hope that low off-peak rates and high peak rates will reduce their peak loads by encouraging customers to use electricity during off-peak times. Such rates can vary with historical use patterns (low rates during traditionally low-use hours and high rates during traditionally high-use hours) or can vary directly with demand (if demand is high, the rate is high and if demand is low, the rate is also low).

A new type of discriminatory rate called a *demand/energy rate* was allowed and then repealed in Colorado in 1976 [3]. With this rate, the cost for electricity was based on the maximum load during any 15-min period as well as on the total kWh used. Even though a solar heated building may have used much less electricity than a conventionally heated building, its electric bill might have been nearly the same as if it had been heated solely by electric heat because much of its consumption would have been at peak times.

Since demand for electricity is often summer peaking, demand/energy rates can benefit users of solar air conditioners. Most solar energy systems decrease summer peak loads on utilities because water is heated solely by solar energy. This has led some to suggest that solar-system buildings should be granted a favorable rate for electricity by utilities that experience summer peaking.

In 1978 in Columbia, Missouri, a utility proposed a retroactive rate increase for solar buildings. The utility was not sure how much solar buildings contributed to peak load, and so proposed that the owners of solar heated buildings each pay $200 (half the metering cost), which would then be applied to the retroactive surcharge. The request was denied. It was pointed out that Columbia's utility experiences summer peaking and the solar buildings help to reduce summer peak load. In a letter to Columbia's mayor, Alan Starr of the Economic Regulatory Administration wrote, "Even without incentives to install storage devices, a solar-heated house backed up by resistance heat, probably does not cause a more severe peaking problem than its major competitor—an electric heat pump" [4].

Singling out solar buildings with rates specifically designed for them is perhaps unfair since they represent a very small minority of buildings in most utility districts. Unfairly high rates may severely inhibit the growth of solar communities. Do solar buildings really add disproportionately to peak loads of local utilities? If so, a fair but perhaps higher rate should be determined for solar buildings.

10.3 SOLAR ENERGY AND UTILITIES

Many utilities are now entering the solar energy business with direct system sales and leases or low- or no-interest loans for the installation of insulation, storm windows, and solar systems. Utilities can afford the initial investment and seem willing to participate in such programs.

The rights and duties of utilities regarding solar energy use were defined somewhat in the National Energy Act passed by Congress in 1978. Under the act, electric and gas utilities must provide customers with information about effective energy conservation and solar measures. The utilities must also inspect a customer's residence if requested and make recommendations about which conservation and solar energy devices would be economical. The utilities must also provide a list of lenders, suppliers, and contractors and offer to arrange for financing and installation.

Under this law, utilities are prohibited from installing or making loans for such systems themselves. Utilities that were engaged in such activities prior to the enactment of the law are exempted, however. Exemptions for utilities not previously engaged in such activities would probably not be difficult to obtain.

Utilities are prohibited from including the costs of the solar and conservation-device installations in the general utility rates. The solar and conservation customers pay back these costs through a surcharge on their monthly utility bills.

Whether utilities should be allowed to lease or sell solar energy systems continues to be an ethical as well as a political question. As Robert B. Reich of the Federal Trade Commission has said, "The utilities have access to every home and thereby are going to have an advantage over every other contractor" [5]. A utility would probably buy from one or two favored manufacturers or contractors. This unavoidable situation would be unfair to small, independent contractors and manufacturers who would not be able to market their systems as effectively as the utility. Though information on all systems available be presented, if the utility sells a particular kind of system that fact may carry a lot of weight with the uninformed homeowner.

Utility involvement in the solar energy and energy-conservation industry must be viewed solely as a convenience to the homeowner. The homeowner does not have to shop around for the best deal (and the best deal may not be available from the utility). Anyone interested in solar energy should get price quotes and economic-performance predictions from several contractors and manufacturers before deciding which system to install. The opinion of the utility should definitely not be considered the last word.

The installation of energy-conservation devices reduces the total amount of energy used and thus the utility's total revenue. Since the utility must still meet its fixed costs and earn a profit above these costs, increased utility rates must be expected.

REFERENCES

1. Cone, B. W. *et al.* (1978). An analysis of federal incentives used to stimulate energy production: an executive summary. In *Solar 78 Northwest Conference Proceedings.* Salem, Oreg.: Oregon Department of Energy.

2. American Assembly (1978). *Improving Energy Efficiency: the Role of Public Policy, Final Report of the 55th American Assembly.* New York: Arden.

3. Colorado Public Utility Commission (1976). Home Builders Association of Metropolitan Denver, Public Service Company of Colorado, Decision no. 87573. Denver, Colo.: Colorado Public Utility Commission.

4. Anonymous (1978). Gazette: news from the solar scene—Missouri city rejects rate hikes for solar users. *Solar Age* 3 (8): 13.

5. Bossong, K. (1978). A spokesman for a public interest group says no. *Solar Age* 3 (1): 23.

BIBLIOGRAPHY

Anonymous (1979). More on the solar tax incentives, solar times. *Solar Age* 4 (1): 14.

Bainbridge, D., and M. Hunt (1978). California's new solar tax credit. *Solar Age* 13 (6): 29-31.

Balcomb, J. D. D. (1978). A question of balance, the Department of Energy solar program. *Solar Age* 3 (5): 12-17.

Batt, S. H. (1978). The future for active solar systems. *Solar Age* 13 (1): 21-22.

Carter, J. (1978). A utility chart and solar California style. *Solar Age* 3 (5): 27-29.

Dolan, T. M. *et al.* (1978). The impact of widespread solar development on energy supply utilities and heating fuel dealers. In *Solar Energy for Pacific Northwest Residential Heating*. Seattle, Wash.. U.S. Department of Energy, Region X

Frank, A. (1978). From Washington—the Carter administration—promises, promises. *Solar Age* 3 (8): 4.

Henderson, H. (1978). Economics: a new look at energy is in progress. *Solar Age* 3 (8): 18-21.

Kassler, H. S. (1978). The domestic policy review: progress or promises. *Solar Age* 3 (10): 12-15.

Mills, G. (1977). Demand electric rates: a new problem and challenge for solar heating. *ASHRAE Journal* 19 (1): 42.

Thomas, W. A. (1978). The legal situation, the needs, the possible areas for legislation. *Solar Age* 1 (6): 11.

Thomas, W. A., A. S. Miller, and R. L. Robbins (1978). Role of government overcoming legal uncertainties about use of solar energy systems. Chicago: American Bar Foundation.

APPENDIXES

Conversion Factors

METRIC PREFIXES

giga	G	10^9
mega	M	10^6
kilo	k	10^3
milli	m	10^{-3}
micro	μ	10^{-6}
nano	n	10^{-9}

LENGTH

1 ft = 0.3048 m
1 in = 2.54 cm
1 mi = 1.6093 km

AREA

1 ft^2 = 0.092903 m^2
1 in^2 = 6.45 cm^2
1 mi^2 = 2.59 km^2

VOLUME	1 ft^3 = 0.02832 m^3
	1 ft^3 = 28.32 l
	1 gal = 3.78 l
	1 m^3 = 1000 l
	1 gal = 0.133 ft^3
MASS	1 lb = 0.4535 kg
FLOW RATE	1 ft^3/min = 0.472 l/sec
	1 gal/min = 0.0631 l/sec
	1 gal/min ft^2 = 0.6791 l/sec m^2
	1 ft^3/min ft^2 = 5.08 l/sec m^2
	1 lb/hr = 0.000126 kg/sec
	1 lb/hr ft^2 = 0.001356 kg/sec m^2
ENERGY	1 Btu = 1.05506 kJ
	1 thm = 10^6 Btu
	1 kWh = 3.6 MJ
	1 kWh = 3413 Btu
POWER	1 Btu/hr = 0.293 W
	1 W = 1 J/sec
ENERGY FLUX	1 Btu/hr ft^2 = 3.155 W/m^2
	1 Btu/hr ft^2 °F = 5.68 W/m^2 °C
	1 Btu/hr ft °F = 1.73 W/m °C
TEMPERATURE	°F = °C × 9/5 + 32
	°C = (°F − 32) × 5/9
	K = °C + 273
	R = °F + 460

B Derivation of the Economics for Collector Sizing

The sizing method used in Chapter 6, based on that developed by John C. Ward of Colorado State University (see Reference 10, Chapter 6), is the result of the following mathematical derivation. While knowledge of calculus is not required for use of the method as presented in Chapter 6, an understanding of partial derivatives and their use is necessary here.

Through performance simulation Ward found that the performance of a solar heating system over the year can be expressed as

$$\mathcal{F} = a + b \, ln\left(\frac{AS_{jan}k}{L_{jan}}\right),$$

where

$\mathcal{J}$ = the fraction of yearly heat load provided by solar energy,
a, b = collector-dependent constants found to be related to system type as follows:

Type	a	b
Liquid, flat plate	0.819	0.281
Liquid, evacuated tube	0.871	0.346
Air, flat plate	0.918	0.432

A = collector area,
S_{jan} = January solar radiation on a horizontal surface,
k = area-performance correction factor,
L_{jan} = January heat load.[1]

k accounts for decreased system performance due to nonoptimum orientation, tilt, and/or the presence of a heat exchanger between collectors and storage. k is assumed to remain constant with respect to area. For example, for a 10 m² evacuated-tube system, if

$S_{jan} = 250$ M J/m², $L_{jan} = 5600$ MJ, and $k = 1$,
$\mathcal{J} = 0.871 + 0.346 \; ln(10 \cdot 250 \cdot 1/5600) = 0.592$,

meaning that this system should provide 59.2% of the yearly heat and hot-water load for this building.

Since we wish to determine the optimum area, or the area that will result in the best possible economic return for the system, economics and calculus must enter into the picture. The total-life-cycle cost of the system, C_T, can be written as

$$C_T = (AC_a + C_b)(F_2 - T) + (1 - \mathcal{J}) \, LEF_1,$$

where

C_T = total lifetime costs associated with the solar heating system (in today's dollars),

---

[1]. Equation reprinted from *Sharing the Sun! Solar Technology in the Seventies* by permission of the American Section of the Solar Energy Society, 205B McDowell Hall, University of Delaware, Newark, DE 19711.

A = collector area,
C_a = collector-area-dependent initial cost,
C_b = area-independent initial cost,
F_2 = economic factor, including discounted mortgage costs (down payment, interest, and capital payments), tax credit or deduction and discounting and inflating property tax, and insurance and maintenance costs (for commercial systems, income-tax deduction for depreciation, insurance, and maintenance costs are included),
T = fraction of first cost refunded as a federal tax credit, given by

$$T = l + \frac{m}{AC_a + C_b},$$

where $(AC_a + C_b)$ are as defined earlier and l and m are constants determined by the system type and cost,

$\mathcal{J} = a + b \ln (AS_{jan}{}^k)/(L_{jan})$,
L = total yearly heat load,
E = delivered-backup-fuel price,
F_1 = economic factor for discounted and inflated energy costs.

The system's original cost is represented by $(AC_a + C_b)$, and LEF_1 represents the cost of providing conventional heat. Since part of the heat is provided by solar energy, the cost of conventional backup heat used is only $(1 - \mathcal{J})LEF_1$.

The variables used in these equations are also discussed in Chapter 6. Some extra understanding of how they are determined can be gained by reviewing that discussion.

What collector area leads to the lowest total costs (maximum savings)? The derivative of the cost curve with respect to area will equal zero at this minimum point:

$$0 = \frac{\partial C_T}{\partial A} = \frac{\partial [(AC_a + C_b)(F_2 - T) + (1 - \mathcal{J})LEF_1]}{\partial A}.$$

A partial solution is

$$0 = C_a F_2 - C_a T - AC_a \frac{\partial T}{\partial A} - C_b \frac{\partial T}{\partial A} - LEF_1 \frac{\partial \mathcal{J}}{\partial A}.$$

Rearranging, we find that

$$0 = C_a(F_2 - T) - (AC_a + C_b)\frac{\partial T}{\partial A} - LEF_1 \frac{\partial \mathcal{J}}{\partial A}.$$

Solving for $\partial T/\partial A$, we start with $T = l + m/(AC_a + C_b)$. T is continuous but cannot be differentiable at all points because the values of l and m change depending on the type of system (residential or commercial) and the system cost. The values of l and m relating to the federal tax credit only are as follows:

	l	m	Valid System Cost
Residential	0.4	0	$\leq \$10{,}000$
	0	4000	$\geq \$10{,}000$
Commercial	0.15–0.25	0	All

If we ignore the problem of differentiability at all points for a moment, we see that the T function is differentiable and

$$\frac{\partial T}{\partial A} = \frac{-mC_a}{(AC_a + C_b)^2}$$

Solving for $\partial \mathcal{J}/\partial A$, since S_{jan}, L_{jan}, and k are all constants, we are left with $\partial \mathcal{J}/\partial A = b/A_{\text{opt}}$. If we substitute these derivatives into our equation, we have

$$0 = \partial C_T/\partial A = C_a(F_2 - T) + \frac{(AC_a + C_b)mC_a}{(AC_a + C_b)^2} - \frac{bLEF_1}{A_{\text{opt}}}.$$

We will abbreviate $(AC_a + C_b)$, which is the initial system cost, to SC. Continuing, we find that

$$0 = C_a(F_2 - T + m/SC) - bLEF_1/A_{\text{opt}}$$
$$= C_a(F_2 - l) - bLEF_1/A_{\text{opt}}$$

or

$$A_{\text{opt}} = bLEF_1/C_a(F_2 - l).$$

To verify that this area truly corresponds to the minimum cost, we make a quick check of the second derivative, $\partial^2 C_T/\partial A^2 = bLEF_1/A_{\text{opt}}^2$. It is positive, ensuring a minimum cost at A_{opt}.

Now we must back up to the problem with T. T is clearly continuous and therefore differentiable for any set of l, m values. The problem arises because the l, m values are valid only for certain system costs. The minimum cost for one set of l, m values will often lie in an invalid system cost range (e.g., for $[l, m] = [0.4, 0]$ leads to an SC of $\$12{,}000$, while $[l$,

$m] = [0, 4000]$ leads to an SC of \$9,000, both invalid for the l and m values used). The two cost curves must cross at a single point that must correspond to the true minimum cost. Since the valid SC ranges are contiguous, this point will always be the borderline between the two ranges. This situation is illustrated in Fig. A.1. A T function in which only the federal tax credit is considered will always have a borderline system cost of \$10,000. If other tax credits are included and all minimum costs are associated with invalid system costs, the true minimum will similarly be found at one of the borderline system costs.

In Fig. 6.6, the check for a valid system cost is accomplished by calculating the system cost as $(AC_a + C_b)$ or

$$\left(\frac{BLEF_1}{F_2 - l}\right) + C_b$$

using different l values until a valid system cost is found. This system cost is then used to calculate backward to find the area. The l value and its associated m value are used later for the calculation of the tax-credit amount. New l and m values can be used to reflect changes in the federal credits or to include state credits.

Fig. A.1 Actual optimum system cost for a residential solar heating system with seemingly invalid optimum costs.

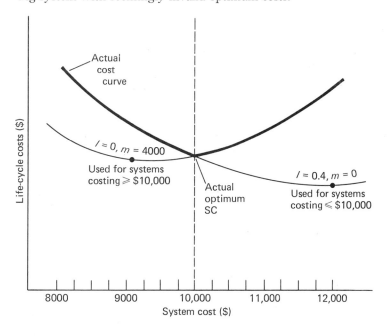

Whether the solar system results in a net savings or loss for the investor is checked also. A *positive solar savings* indicates the system is a good investment and has saved money. A *negative solar savings* indicates that the system is not as economical as had been expected.

Solar savings = C_T(no solar) − C_T(with solar)

or

solar savings = $LEF_1 - (AC_a + C_b)(F_2 - T) - (1 - \mathcal{J})LEF_1$.

Rearranging, we find that

solar savings = $\mathcal{J} LEF_1 - F_2(AC_a + C_b) + T(AC_a + C_b)$.

This economic check is the last step of the sizing procedure.

C Table of Design Conditions

			Metric Units			English Units		
State	City	Latitude	January Modified DD*	Yearly Modified DD*	S_{jan} (MJ/m²)	January Modified DD*	Yearly Modified DD*	S_{jan} (Btu/ft²)×10³
AK	Adak	51.9	—	—	81.34	—	—	7.167
	Annette	55.0	422	3044	62.59	759	5479	5.515
	Barrow	71.3	782	6417	0	1408	11551	0
	Bethel	60.8	588	4181	34.07	1059	7526	3.000
	Bettles	66.9	757	4970	3.53	1362	8946	0.310
	Big Delta	64.0	686	4338	16.20	1235	7808	1.420
	Fairbanks	64.8	755	4542	10.60	1359	8176	0.933

Source: Data adapted from Cinquemani, V., J. R. Owenby, Jr., and R. G. Baldwin (1978). *Input for Solar Systems.* Washington, D.C.: Department of Energy.

*The degree-day listings in this table include a factor relating measured degree days to actual heat load. Solar radiation measurements are given in MJ/m² or (Btu/ft²)×10³. One MJ = 1,000,000 J.

(continued)

			Metric Units			English Units		
State	City	Latitude	January Modified DD*	Yearly Modified DD*	S_{jan} (MJ/m^2)	January Modified DD*	Yearly Modified DD*	S_{jan} (Btu/ft^2)×10^3
	Gulkana	62.2	710	4413	25.60	1277	7944	2.260
	Homer	59.6	514	3939	42.78	925	7090	3.770
	Juneau	58.4	513	3589	40.92	923	6461	3.605
	King Salmon	58.7	559	4048	51.52	1007	7286	4.538
	Kodiak	57.8	487	4024	52.51	877	7243	4.628
	Kotzebue	66.9	674	5079	3.00	1214	9142	0.264
	McGrath	63.0	726	4588	20.40	1306	8258	1.790
	Nome	64.5	579	4536	10.5	1042	8165	0.924
	Summit	63.3	622	4550	19.70	1120	8190	1.730
	Yakutat	59.5	504	3799	35.31	908	6839	3.112
AL	Birmingham	33.6	323	1406	248.6	582	2531	21.90
	Mobile	30.7	223	833	291.4	401	1499	25.67
	Montgomery	32.3	275	1122	264.5	495	2019	23.30
AR	Fort Smith	35.3	384	1589	261.6	691	2860	23.05
	Little Rock	34.7	391	1658	257.3	704	2985	22.67
AZ	Phoenix	33.4	212	767	359.3	381	1381	31.66
	Prescott	34.7	375	1934	357.5	676	3481	31.50
	Tucson	32.1	219	866	386.7	393	1559	34.07
	Winslow	35.0	441	2077	346.4	793	3739	30.52
	Yuma	32.7	152	499	385.6	274	899	33.98
CA	Arcata	41.0	—	—	186.0	—	—	16.39
	Bakersfield	35.4	269	1081	269.6	483	1945	23.76
	China Lake	35.7	—	—	319.9	—	—	28.19
	Daggett	34.9	271	1089	337.1	489	1961	29.70
	El Toro	33.7	—	—	333.2	—	—	29.36
	Fresno	36.8	302	1311	231.0	544	2359	20.36
	Long Beach	33.8	168	794	326.4	302	1429	28.76
	Los Angeles	33.9	164	899	325.8	295	1619	28.71
	Mount Shasta	41.3	421	2546	197.3	757	4582	17.38
	Needles	34.8	208	706	346.5	375	1271	30.53
	Oakland	37.7	251	1438	249.0	452	2589	21.94
	Point Mugu	34.1	—	—	326.2	—	—	28.74

(continued)

TABLE OF DESIGN CONDITIONS

State	City	Latitude	Metric Units			English Units		
			January Modified DD*	Yearly Modified DD*	S_{jan} (MJ/m^2)	January Modified DD*	Yearly Modified DD*	S_{jan} $(Btu/ft^2) \times 10^3$
	Red Bluff	40.2	304	1329	200.6	546	2392	17.68
	Sacramento	38.5	305	1406	210.0	549	2530	18.50
	San Diego	32.7	155	745	343.3	279	1341	30.25
	San Francisco	37.6	256	1504	249.0	461	2707	21.94
	Santa Maria	34.9	222	1509	300.4	400	2717	26.47
	Sunnyvale	37.4	—	—	259.5	—	—	22.87
CO	Colorado Springs	38.8	454	2607	313.3	818	4692	27.61
	Denver	39.8	434	2398	295.6	781	4316	26.04
	Eagle	39.7	535	3092	265.3	962	5566	23.38
	Grand Junction	39.1	505	2380	278.4	910	4284	24.53
	Pueblo	38.3	425	2120	314.6	765	3816	27.72
CT	Hartford	41.9	529	2697	168.0	952	4854	14.80
DC	Washington-Sterling	39.0	486	2386	201.3	874	4295	17.73
DE	Wilmington	39.7	470	2270	201.0	846	4086	17.71
FL	Apalachicola	29.7	182	673	300.0	327	1211	26.44
	Daytona Beach	29.2	119	446	337.2	215	803	29.71
	Jacksonville	30.5	172	656	316.6	310	1181	27.90
	Miami	25.8	26	102	372.0	47	183	32.78
	Orlando	28.6	97	362	351.6	175	652	30.98
	Tallahassee	30.4	202	773	308.4	363	1391	27.17
	Tampa	28.0	100	355	355.6	181	639	31.33
	West Palm Beach	26.7	41	148	351.8	74	266	30.99
GA	Atlanta	33.7	347	1531	252.5	624	2755	22.25
	Augusta	33.4	297	1259	264.2	535	2267	23.28
	Macon	32.7	269	1108	270.5	483	1994	23.84
	Savannah	32.1	239	965	279.6	430	1737	24.64
HI	Barbers Point	21.3	0	0	424.9	0	0	37.44
	Hilo	19.7	0	0	393.9	0	0	34.72
	Honolulu	21.3	0	0	415.1	0	0	36.57
	Lihue	22.0	0	0	388.1	0	0	34.19

(continued)

			Metric Units			English Units		
State	City	Latitude	January Modified DD*	Yearly Modified DD*	S_{jan} (MJ/m^2)	January Modified DD*	Yearly Modified DD*	S_{jan} $(Btu/ft^2) \times 10^3$
IA	Burlington	40.8	499	2352	203.8	898	4233	17.95
	Des Moines	41.5	530	2514	204.3	954	4525	18.00
	Mason City	43.2	554	2779	194.8	997	5003	17.16
	Sioux City	42.4	535	2552	200.0	962	4593	17.63
ID	Boise	43.6	490	2560	170.7	882	4608	15.04
	Lewiston	46.4	440	2296	119.5	793	4132	10.53
	Pocatello	42.9	506	2757	189.7	911	4963	16.71
IL	Chicago	41.8	498	2417	178.4	896	4350	15.72
	Moline	41.5	511	2421	188.3	919	4357	16.59
	Springfield	39.8	478	2238	205.7	860	4029	18.13
IN	Evansville	38.1	436	2009	202.0	784	3616	17.80
	Fort Wayne	41.0	491	2474	160.1	883	4454	14.11
	Indianapolis	39.7	463	2246	174.3	834	4043	15.36
	South Bend	41.7	507	2576	146.3	912	4636	12.89
KS	Dodge City	37.8	441	2097	290.8	793	3775	25.62
	Goodland	39.4	457	2413	277.7	823	4344	24.47
	Topeka	39.1	472	2156	239.6	849	3881	21.11
	Wichita	37.7	434	1948	275.8	782	3506	24.30
KY	Lexington	38.0	427	2030	192.0	769	3654	16.92
	Louisville	38.2	431	2039	191.9	777	3670	16.91
LA	Baton Rouge	30.5	223	826	276.2	401	1486	24.34
	Lake Charles	30.1	205	741	256.3	369	1333	22.58
	New Orleans	30.0	199	724	293.6	359	1304	25.87
	Shreveport	32.5	273	1072	268.2	491	1929	23.63
MA	Boston	42.4	482	2439	167.3	867	4391	14.74
MD	Baltimore	39.2	445	2148	206.5	801	3866	18.19
	Patuxent River	38.3	—	—	214.0	—	—	18.85
ME	Bangor	44.8	—	—	159.9	—	—	14.09
	Caribou	46.9	579	3315	147.5	1043	5967	13.00
	Portland	43.7	527	2927	158.4	948	5269	13.96
MI	Alpena	45.1	542	3158	127.4	976	5685	11.22
	Detroit	42.4	515	2616	146.8	926	4709	12.94
	Flint	43.0	528	2806	134.8	950	5051	11.88

(continued)

TABLE OF DESIGN CONDITIONS

			Metric Units			English Units		
State	City	Latitude	January Modified DD*	Yearly Modified DD*	S_{jan} (MJ/m²)	January Modified DD*	Yearly Modified DD*	S_{jan} (Btu/ft²)×10³
	Grand Rapids	42.9	539	2827	130.0	970	5088	11.46
	Houghton	47.2	—	—	85.93	—	—	7.573
	Sault Sainte Marie	46.5	572	3338	114.3	1029	6009	10.07
	Traverse City	44.7	546	3068	109.3	983	5522	9.635
MN	Duluth	46.8	582	3244	136.7	1048	5840	12.05
	International Falls	48.6	619	3340	125.1	1115	6012	11.03
	Minneapolis-St. Paul	44.9	570	2839	163.2	1025	5110	14.38
	Rochester	43.9	562	2863	167.8	1012	5153	14.79
MO	Columbia	38.8	455	2091	212.0	820	3763	18.96
	Kansas City	39.3	494	2251	227.9	889	4051	20.08
	Springfield	37.2	432	1983	240.5	777	3570	21.19
	Saint Louis	38.8	444	2017	220.7	799	3631	19.45
MS	Jackson	32.3	281	1137	265.1	506	2047	23.36
	Meridian	32.3	284	1181	261.8	512	2125	23.07
MT	Billings	45.8	475	2583	171.0	855	4650	15.07
	Cut Bank	48.6	479	2861	141.5	862	5149	12.47
	Dillon	45.3	521	3136	185.2	939	5645	16.32
	Glasgow	48.2	562	2912	136.5	1011	5242	12.03
	Great Falls	47.5	464	2574	147.9	836	4634	13.04
	Helena	46.6	484	2724	147.6	870	4903	13.00
	Lewistown	47.1	473	2856	147.7	852	5140	13.02
	Miles City	46.4	517	2654	160.8	931	4778	14.17
	Missoula	46.9	508	2941	109.7	914	5293	9.666
NC	Asheville	35.4	386	1947	253.9	695	3504	22.37
	Cape Hatteras	35.3	302	1350	241.2	544	2430	21.25
	Charlotte	35.2	351	1591	253.0	632	2864	22.29
	Cherry Point	34.9	—	—	266.2	—	—	23.46
	Greensboro	36.1	400	1879	251.7	721	3383	22.17
	Raleigh-Durham	35.9	376	1737	244.1	676	3127	21.51
ND	Bismarck	46.8	565	2901	164.2	1017	5221	14.47
	Fargo	46.9	595	3011	146.0	1071	5419	12.86
	Minot	48.3	561	2979	135.0	1009	5362	11.89

(continued)

		Metric Units				English Units		
State	City	Latitude	January Modified DD*	Yearly Modified DD*	S_{jan} (MJ/m^2)	January Modified DD*	Yearly Modified DD*	S_{jan} (Btu/ft^2)×10^3
NE	Grand Island	41.0	506	2457	232.7	911	4423	20.50
	North Omaha	41.4	531	2524	223.0	956	4544	19.65
	North Platte	41.1	488	2552	243.6	879	4594	21.46
	Scottsbluff	41.9	475	2591	237.7	856	4663	20.95
NH	Concord	43.2	526	2815	161.7	947	5067	14.24
NJ	Lakehurst	40.0	—	—	197.0	—	—	17.35
	Newark	40.7	479	2313	194.1	862	4163	17.10
NM	Albuquerque	35.1	435	2019	357.6	783	3635	31.51
	Clayton	36.5	425	2237	338.3	764	4027	29.81
	Farmington	36.8	474	2400	332.3	853	4320	29.28
	Roswell	33.4	402	1783	368.2	724	3209	32.44
	Truth or Consequences	33.2	360	1577	393.2	649	2839	34.65
	Tucumcari	35.2	394	1838	354.9	709	3308	31.27
	Zuni	35.1	447	2417	347.0	805	4350	30.57
NV	Elko	40.8	501	2892	242.4	901	5205	21.36
	Ely	39.3	486	2958	288.3	874	5324	25.40
	Las Vegas	36.1	319	1286	344.1	574	2315	30.32
	Lovelock	40.1	502	2689	282.8	904	4840	24.92
	Reno	39.5	450	2643	281.6	810	4757	24.81
	Tonopah	38.1	474	2589	322.9	852	4661	28.45
	Winnemucca	40.9	464	2698	242.9	836	4856	21.41
	Yucca Flats	37.0	—	—	335.4	—	—	29.55
NY	Albany	42.8	532	2717	160.6	958	4890	14.15
	Binghamton	42.2	531	2903	135.7	956	5226	11.96
	Buffalo	42.9	538	2910	122.8	968	5238	10.82
	Massena	44.9	569	2991	137.6	1024	5384	12.13
	New York (Central Park)	40.8	473	2254	176.0	851	4057	15.51
	New York (La Guardia airport)	40.8	474	2282	192.7	854	4108	16.98
	Rochester	43.1	528	2793	128.2	951	5027	11.29
	Syracuse	43.1	517	2689	135.5	930	4841	11.94

(continued)

TABLE OF DESIGN CONDITIONS 373

State	City	Latitude	Metric Units			English Units		
			January Modified DD*	Yearly Modified DD*	S_{jan} (MJ/m^2)	January Modified DD*	Yearly Modified DD*	S_{jan} (Btu/ft^2)×10^3
OH	Akron-Canton	40.9	504	2614	150.7	907	4706	13.27
	Cincinnati-Covington	39.1	442	2130	176.1	795	3834	15.52
	Cleveland	41.4	491	2558	136.6	884	4604	12.04
	Columbus	40.0	472	2370	161.6	849	4266	14.24
	Dayton	39.9	470	2320	172.1	847	4176	15.17
	Toledo	41.6	497	2543	153.0	894	4578	13.48
	Youngstown	41.3	501	2643	135.5	902	4757	11.94
OK	Oklahoma City	35.4	397	1678	281.8	714	3020	24.83
	Tulsa	36.2	400	1671	257.4	719	3008	22.68
OR	Astoria	46.2	374	2618	110.7	673	4712	9.756
	Burns	43.6	487	2845	172.4	876	5121	15.19
	Medford	42.4	435	2438	143.2	783	4388	12.61
	North Bend	43.4	312	2318	154.3	562	4172	13.59
	Pendleton	45.7	425	2178	122.5	765	3920	10.79
	Portland	45.6	412	2369	109.1	742	4265	9.610
	Redmond	44.3	448	2761	172.8	807	4970	15.22
	Salem	44.9	401	2399	116.8	723	4318	10.29
PA	Allentown	40.7	500	2529	185.6	901	4552	16.35
	Erie	42.1	537	2973	121.6	966	5352	10.71
	Harrisburg	40.2	480	2319	188.4	865	4174	16.60
	Philadelphia	39.9	466	2236	195.4	839	4024	17.21
	Pittsburgh	40.5	476	2465	149.3	856	4437	13.16
	Wilkes-Barre-Scranton	41.3	502	2609	160.1	904	4696	14.11
PR	San Juan	18.4	0	0	466.4	0	0	41.09
RI	Providence	41.7	493	2592	178.1	887	4666	15.69
SC	Charleston	32.9	258	1061	261.8	464	1910	23.07
	Columbia	34.0	301	1284	268.0	541	2312	23.61
	Greenville-Spartanburg	34.9	348	1564	256.7	627	2815	22.62
SD	Huron	44.4	554	2742	171.7	997	4936	15.13
	Pierre	44.4	544	2729	186.5	980	4913	16.43

(continued)

			Metric Units			English Units		
State	City	Latitude	January Modified DD*	Yearly Modified DD*	S_{jan} (MJ/m²)	January Modified DD*	Yearly Modified DD*	S_{jan} (Btu/ft²)×10³
	Rapid City	44.1	490	2688	190.8	883	4838	16.81
	Sioux Falls	43.6	554	2757	187.4	997	4963	16.51
TN	Chattanooga	35.0	371	1690	221.8	667	3042	19.55
	Knoxville	35.8	369	1698	218.4	664	3057	19.24
	Memphis	35.1	366	1556	240.2	660	2801	21.16
	Nashville	36.1	380	1698	203.9	685	3057	17.97
TX	Abilene	32.4	326	1291	325.0	587	2323	28.64
	Amarillo	35.2	399	1857	337.8	718	3342	29.77
	Austin	30.3	239	859	304.1	430	1546	26.80
	Brownsville	25.9	111	322	321.1	200	579	28.30
	Corpus Christi	27.8	150	460	315.9	271	828	27.84
	Dallas	32.9	301	1132	289.0	541	2038	25.47
	Del Rio	29.4	222	753	337.2	399	1355	29.71
	El Paso	31.8	328	1324	395.8	590	2383	34.88
	Fort Worth	32.8	310	1178	283.3	557	2120	24.96
	Houston	30.0	206	709	271.7	370	1276	23.94
	Kingsville	27.5	—	—	321.0	—	—	28.28
	Laredo	27.5	148	433	337.3	266	780	29.73
	Lubbock	33.7	373	1648	362.7	672	2967	31.96
	Lufkin	31.2	252	959	279.3	453	1727	24.61
	Midland-Odessa	31.9	328	1296	380.4	590	2333	33.52
	Port Arthur	30.0	208	751	281.4	374	1351	24.79
	San Angelo	31.4	285	1108	338.4	514	1994	29.82
	San Antonio	29.5	223	776	315.0	401	1397	27.76
	Sherman	33.7	357	1416	279.3	643	2549	24.61
	Waco	31.6	276	1018	292.9	497	1832	25.81
	Wichita Falls	34.0	351	1400	303.3	633	2520	26.72
UT	Bryce Canyon	37.7	525	3422	321.5	945	6159	28.32
	Cedar City	37.7	468	2551	310.4	842	4591	27.35
	Salt Lake City	40.8	492	2568	224.8	886	4623	19.81
VA	Norfolk	36.9	376	1724	238.6	676	3104	21.03
	Richmond	37.5	406	1876	222.3	731	3377	19.59
	Roanoke	37.3	417	2027	232.4	751	3648	20.48

(continued)

TABLE OF DESIGN CONDITIONS

			Metric Units			English Units		
State	City	Latitude	January Modified DD*	Yearly Modified DD*	S_{jan} (MJ/m^2)	January Modified DD*	Yearly Modified DD*	S_{jan} (Btu/ft^2) × 10^3
VT	Burlington	44.5	548	2890	135.6	987	5202	11.94
WA	Olympia	47.0	426	2734	94.6	767	4922	8.333
	Seattle-Tacoma	47.5	411	2564	92.1	740	4615	8.113
	Spokane	47.6	495	2753	110.8	890	4955	9.765
	Whidbey Island	48.4	—	—	99.5	—	—	8.770
	Yakima	46.6	483	2497	128.4	870	4495	11.32
WI	Eau Claire	44.9	581	2951	158.9	1046	5311	14.00
	Green Bay	44.5	553	2910	158.8	995	5238	13.99
	La Crosse	43.9	545	2666	169.3	981	4798	14.92
	Madison	43.1	548	2837	181.3	987	5106	15.97
	Milwaukee	43.0	535	2810	168.7	963	5072	14.86
WV	Charleston	38.4	420	2037	175.3	756	3667	15.45
	Huntington	38.4	418	2029	185.1	752	3653	16.31
WY	Casper	42.9	486	2831	240.4	874	5095	21.18
	Cheyenne	41.2	465	2832	269.4	836	5098	23.74
	Rock Springs	41.6	543	3216	258.6	977	5789	22.78
	Sheridan	44.8	495	2799	182.1	892	5039	16.04

D Inflating-Discounting Factors

This appendix contains tables of Inflating-Discounting Factors for 5, 10, 15, 20, 25, and 30 years. The Inflating-Discounting Factor accounts both for inflating yearly payments (or savings) and for discounting these payments to their cumulative present value.

The first year's payment (or savings) is multiplied by the Inflating-Discounting Factor to give the cumulative present value of all future payments. The factor is discussed in Section 6.3.2. See Section 6.1 for a review of present value calculations.

The Inflating-Discounting Factor, $F(N, i, d)$ is a function of N, the number of years, i, the yearly inflation rate, and d, the yearly discount rate after taxes.

$$F(N, i, d) = \frac{1}{d-i}\left[1 - \left(\frac{1+i}{1+d}\right)^N\right] \quad \text{if } i \neq d$$

or

$$F(N, i, d) = \frac{N}{1+i} \quad \text{if } i = d.$$

$F(N, i, d)$ can be looked up in these tables as F (year, row, column).

Inflating-Discounting Factors for Five Years

i \ d	0	0.01	0.02	0.03	0.04	0.05	0.06	0.07	0.08	0.09	0.10	0.11	0.12	0.13	0.14	0.15	0.16	0.17	0.18	0.19	0.20
0	5.000	4.853	4.713	4.580	4.452	4.329	4.212	4.100	3.993	3.890	3.791	3.696	3.605	3.517	3.433	3.352	3.274	3.199	3.127	3.058	2.991
0.01	5.101	4.950	4.807	4.669	4.538	4.413	4.292	4.177	4.067	3.961	3.860	3.763	3.669	3.580	3.493	3.410	3.331	3.254	3.180	3.109	3.040
0.02	5.204	5.050	4.902	4.761	4.626	4.497	4.374	4.256	4.143	4.035	3.931	3.831	3.735	3.643	3.555	3.470	3.388	3.309	3.234	3.161	3.091
0.03	5.309	5.150	4.999	4.854	4.716	4.584	4.457	4.336	4.220	4.109	4.003	3.900	3.802	3.708	3.617	3.530	3.447	3.366	3.288	3.214	3.142
0.04	5.416	5.253	5.098	4.950	4.808	4.672	4.542	4.418	4.299	4.185	4.076	3.971	3.870	3.774	3.681	3.592	3.506	3.424	3.344	3.268	3.194
0.05	5.526	5.358	5.199	5.047	4.901	4.762	4.629	4.501	4.379	4.262	4.151	4.043	3.940	3.841	3.746	3.655	3.567	3.482	3.401	3.323	3.247
0.06	5.637	5.466	5.302	5.146	4.996	4.853	4.717	4.586	4.461	4.342	4.227	4.117	4.011	3.909	3.812	3.719	3.629	3.542	3.459	3.379	3.301
0.07	5.751	5.575	5.407	5.246	5.093	4.947	4.807	4.673	4.545	4.422	4.304	4.191	4.083	3.979	3.879	3.784	3.691	3.603	3.518	3.436	3.357
0.08	5.867	5.686	5.514	5.349	5.192	5.042	4.898	4.761	4.630	4.504	4.383	4.268	4.157	4.050	3.948	3.850	3.755	3.665	3.577	3.493	3.413
0.09	5.985	5.799	5.623	5.454	5.293	5.139	4.992	4.851	4.716	4.587	4.464	4.345	4.231	4.122	4.018	3.917	3.821	3.728	3.638	3.552	3.470
0.10	6.105	5.915	5.734	5.561	5.395	5.238	5.087	4.942	4.804	4.672	4.545	4.424	4.308	4.196	4.089	3.986	3.887	3.792	3.700	3.612	3.528
0.11	6.228	6.033	5.847	5.669	5.500	5.338	5.183	5.036	4.894	4.759	4.629	4.505	4.385	4.271	4.161	4.056	3.954	3.857	3.764	3.673	3.587
0.12	6.353	6.153	5.962	5.780	5.606	5.441	5.282	5.131	4.986	4.847	4.714	4.586	4.464	4.347	4.235	4.127	4.023	3.924	3.828	3.736	3.647
0.13	6.480	6.275	6.080	5.893	5.715	5.545	5.382	5.227	5.079	4.936	4.800	4.670	4.545	4.425	4.310	4.199	4.093	3.991	3.893	3.799	3.708
0.14	6.610	6.400	6.199	6.008	5.826	5.651	5.485	5.326	5.173	5.028	4.888	4.755	4.627	4.504	4.386	4.273	4.164	4.060	3.960	3.863	3.770
0.15	6.742	6.527	6.321	6.125	5.938	5.760	5.589	5.426	5.270	5.121	4.978	4.841	4.710	4.584	4.464	4.348	4.237	4.130	4.027	3.929	3.834
0.16	6.877	6.656	6.445	6.244	6.053	5.870	5.695	5.528	5.368	5.215	5.069	4.929	4.795	4.666	4.543	4.424	4.310	4.201	4.096	3.995	3.898

Note: i = inflation rate; d = discount rate.

Inflating-Discounting Factors for Ten Years

d \ i	0	0.01	0.02	0.03	0.04	0.05	0.06	0.07	0.08	0.09	0.10	0.11	0.12	0.13	0.14	0.15	0.16	0.17	0.18	0.19	0.20
0	10.000	9.471	8.983	8.530	8.111	7.722	7.360	7.024	6.710	6.418	6.145	5.889	5.650	5.426	5.216	5.019	4.833	4.659	4.494	4.339	4.192
0.01	10.462	9.901	9.383	8.903	8.459	8.046	7.664	7.308	6.976	6.667	6.379	6.110	5.858	5.622	5.400	5.193	4.997	4.814	4.641	4.478	4.324
0.02	10.950	10.354	9.804	9.295	8.825	8.388	7.983	7.606	7.256	6.930	6.625	6.341	6.075	5.826	5.593	5.374	5.169	4.976	4.794	4.623	4.462
0.03	11.464	10.831	10.248	9.709	9.210	8.748	8.319	7.921	7.550	7.205	6.884	6.584	6.303	6.041	5.795	5.565	5.349	5.146	4.955	4.775	4.606
0.04	12.006	11.335	10.716	10.144	9.615	9.126	8.672	8.251	7.859	7.495	7.155	6.838	6.543	6.266	6.007	5.765	5.537	5.323	5.123	4.934	4.756
0.05	12.578	11.865	11.209	10.603	10.042	9.524	9.043	8.598	8.184	7.798	7.440	7.105	6.793	6.502	6.229	5.974	5.734	5.509	5.298	5.100	4.913
0.06	13.181	12.425	11.728	11.085	10.492	9.943	9.434	8.962	8.525	8.118	7.739	7.386	7.057	6.749	6.462	6.193	5.940	5.704	5.482	5.273	5.077
0.07	13.816	13.014	12.275	11.594	10.965	10.383	9.845	9.346	8.883	8.453	8.053	7.680	7.333	7.008	6.705	6.422	6.156	5.908	5.674	5.455	5.248
0.08	14.487	13.635	12.851	12.129	11.462	10.846	10.277	9.749	9.259	8.805	8.382	7.989	7.622	7.280	6.961	6.662	6.383	6.121	5.875	5.644	5.428
0.09	15.193	14.289	13.458	12.692	11.986	11.334	10.731	10.172	9.655	9.174	8.728	8.313	7.926	7.565	7.228	6.914	6.619	6.344	6.085	5.843	5.615
0.10	15.937	14.979	14.097	13.286	12.537	11.847	11.208	10.618	10.070	9.562	9.091	8.652	8.244	7.864	7.509	7.177	6.867	6.577	6.305	6.050	5.811
0.11	16.722	15.705	14.770	13.910	13.117	12.386	11.710	11.085	10.507	9.970	9.472	9.009	8.578	8.177	7.803	7.453	7.127	6.822	6.536	6.267	6.016
0.12	17.549	16.470	15.479	14.567	13.727	12.953	12.238	11.577	10.965	10.398	9.872	9.383	8.929	8.505	8.111	7.743	7.399	7.077	6.776	6.494	6.230
0.13	18.420	17.275	16.225	15.259	14.369	13.550	12.793	12.094	11.447	10.848	10.292	9.776	9.296	8.850	8.434	8.046	7.684	7.345	7.028	6.732	6.454
0.14	19.337	18.124	17.010	15.987	15.045	14.177	13.376	12.637	11.953	11.319	10.732	10.188	9.681	9.210	8.772	8.363	7.982	7.625	7.292	6.980	6.688
0.15	20.304	19.017	17.837	16.752	15.755	14.836	13.989	13.207	12.484	11.815	11.195	10.620	10.085	9.589	9.126	8.696	8.294	7.919	7.568	7.240	6.932
0.16	21.322	19.957	18.707	17.558	16.502	15.529	14.633	13.806	13.042	12.335	11.680	11.073	10.509	9.985	9.498	9.044	8.621	8.226	7.857	7.511	7.188

Note: i = inflation rate; d = discount rate.

Inflating-Discounting Factors for Fifteen Years

i \ d	0	0.01	0.02	0.03	0.04	0.05	0.06	0.07	0.08	0.09	0.10	0.11	0.12	0.13	0.14	0.15	0.16	0.17	0.18	0.19	0.20
0	15.000	13.865	12.849	11.938	11.118	10.380	9.712	9.108	8.559	8.061	7.606	7.191	6.811	6.462	6.142	5.847	5.575	5.324	5.092	4.876	4.675
0.01	16.097	14.851	13.738	12.741	11.845	11.039	10.311	9.654	9.057	8.516	8.023	7.574	7.163	6.786	6.441	6.124	5.831	5.561	5.312	5.081	4.867
0.02	17.293	15.926	14.706	13.614	12.634	11.754	10.960	10.244	9.595	9.007	8.473	7.986	7.541	7.135	6.762	6.420	6.105	5.815	5.547	5.300	5.070
0.03	18.599	17.098	15.759	14.563	13.492	12.530	11.664	10.883	10.177	9.538	8.958	8.430	7.949	7.509	7.107	6.738	6.399	6.087	5.799	5.533	5.288
0.04	20.024	18.375	16.906	15.596	14.423	13.372	12.426	11.575	10.807	10.111	9.481	8.908	8.387	7.912	7.477	7.079	6.714	6.378	6.069	5.783	5.519
0.05	21.579	19.767	18.156	16.719	15.436	14.286	13.254	12.325	11.488	10.731	10.046	9.425	8.860	8.345	7.875	7.445	7.051	6.689	6.357	6.050	5.767
0.06	23.276	21.285	19.517	17.942	16.536	15.278	14.151	13.138	12.225	11.402	10.657	9.982	9.369	8.812	8.303	7.839	7.413	7.024	6.665	6.336	6.032
0.07	25.129	22.942	21.000	19.273	17.733	16.357	15.125	14.019	13.024	12.127	11.317	10.584	9.913	9.314	8.764	8.262	7.803	7.382	6.996	6.641	6.315
0.08	27.152	24.748	22.616	20.722	19.035	17.529	16.182	14.974	13.889	12.912	12.030	11.233	10.511	9.856	9.260	8.717	8.220	7.767	7.351	6.969	6.618
0.09	29.361	26.718	24.377	22.300	20.451	18.802	17.329	16.010	14.826	13.761	12.802	11.935	11.151	10.440	9.794	9.206	8.670	8.180	7.731	7.320	6.942
0.10	31.772	28.867	26.297	24.017	21.991	20.187	18.575	17.134	15.842	14.681	13.636	12.694	11.842	11.070	10.370	9.733	9.153	8.623	8.139	7.696	7.289
0.11	34.405	31.212	28.389	25.888	23.667	21.691	19.929	18.354	16.943	15.678	14.539	13.514	12.587	11.749	10.990	10.300	9.672	9.100	8.577	8.099	7.661
0.12	37.280	33.770	30.669	27.925	25.491	23.327	21.399	19.677	18.137	16.757	15.516	14.400	13.393	12.483	11.659	10.911	10.231	9.612	9.048	8.532	8.059
0.13	40.417	36.559	33.155	30.144	27.475	25.105	22.996	21.114	19.432	17.926	16.574	15.353	14.233	13.274	12.380	11.569	10.833	10.164	9.553	8.996	8.487
0.14	43.842	39.602	35.863	32.560	29.634	27.039	24.730	22.673	20.836	19.193	17.719	16.395	15.204	14.129	13.158	12.279	11.481	10.756	10.097	9.495	8.945
0.15	47.580	42.920	38.815	35.191	31.984	29.141	26.615	24.366	22.359	20.566	18.959	17.517	16.220	15.052	13.997	13.043	12.179	11.394	10.681	10.031	9.437
0.16	51.660	46.539	42.032	38.055	34.540	31.426	28.662	26.203	24.011	22.053	20.302	18.731	17.319	16.049	14.903	13.868	12.931	12.081	11.309	10.606	9.965

Note: i = inflation rate; d = discount rate.

Inflating-Discounting Factors for Twenty Years

i \ d	0	0.01	0.02	0.03	0.04	0.05	0.06	0.07	0.08	0.09	0.10	0.11	0.12	0.13	0.14	0.15	0.16	0.17	0.18	0.19	0.20
0	20.000	18.046	16.351	14.877	13.590	12.462	11.470	10.594	9.818	9.129	8.514	7.963	7.469	7.025	6.623	6.259	5.929	5.628	5.353	5.101	4.870
0.01	22.019	19.802	17.885	16.221	14.771	13.503	12.391	11.411	10.546	9.779	9.096	8.487	7.941	7.451	7.009	6.610	6.249	5.920	5.620	5.347	5.096
0.02	24.297	21.780	19.608	17.727	16.092	14.665	13.417	12.320	11.353	10.498	9.739	9.063	8.460	7.919	7.432	6.994	6.597	6.238	5.911	5.613	5.340
0.03	26.870	24.009	21.546	19.417	17.571	15.965	14.562	13.332	12.250	11.296	10.450	9.700	9.031	8.433	7.896	7.414	6.978	6.585	6.227	5.902	5.605
0.04	29.778	26.524	23.728	21.317	19.231	17.419	15.840	14.459	13.247	12.181	11.238	10.403	9.661	8.998	8.406	7.874	7.395	6.963	6.572	6.216	5.893
0.05	33.066	29.362	26.186	23.453	21.093	19.048	17.269	15.717	14.358	13.164	12.112	11.182	10.356	9.622	8.966	8.379	7.851	7.376	6.947	6.558	6.205
0.06	36.786	32.568	28.958	25.857	23.185	20.874	18.868	17.122	15.596	14.258	13.082	12.044	11.125	10.310	9.583	8.934	8.352	7.829	7.358	6.932	6.545
0.07	40.996	36.190	32.084	28.564	25.536	22.922	20.659	18.692	16.977	15.476	14.160	13.001	11.977	11.070	10.263	9.545	8.902	8.325	7.807	7.339	6.916
0.08	45.762	40.284	35.612	31.613	28.180	25.222	22.665	20.448	18.519	16.834	15.359	14.063	12.920	11.910	11.014	10.217	9.506	8.870	8.298	7.784	7.320
0.09	51.160	44.913	39.594	35.050	31.156	27.806	24.916	22.414	20.242	18.349	16.694	15.243	13.967	12.841	11.844	10.959	10.172	9.468	8.838	8.272	7.762
0.10	57.275	50.150	44.093	38.926	34.506	30.710	27.442	24.617	22.169	20.039	18.182	16.556	15.129	13.872	12.762	11.779	10.905	10.126	9.430	8.806	8.245
0.11	64.203	56.074	49.174	43.299	38.279	33.977	30.277	27.086	24.325	21.928	19.841	18.018	16.421	15.017	13.779	12.685	11.714	10.851	10.081	9.392	8.774
0.12	72.052	62.778	54.917	48.232	42.531	37.651	33.463	29.856	26.740	24.040	21.693	19.647	17.857	16.287	14.906	13.687	12.608	11.650	10.798	10.036	9.355
0.13	80.947	70.364	61.406	53.801	47.322	41.787	37.042	32.963	29.445	26.402	23.761	21.463	19.456	17.699	16.156	14.797	13.596	12.532	11.589	10.745	9.992
0.14	91.025	78.949	68.741	60.086	52.723	46.442	41.066	36.451	32.477	29.045	26.072	23.489	21.237	19.269	17.544	16.027	14.689	13.506	12.457	11.524	10.692
0.15	102.444	88.665	77.032	67.181	58.813	51.684	45.591	40.368	35.877	32.005	28.656	25.750	23.222	21.016	19.086	17.391	15.900	14.583	13.418	12.383	11.462
0.16	115.380	99.660	86.404	75.192	65.680	57.587	50.680	44.767	39.691	35.320	31.545	28.276	25.436	22.962	20.800	18.906	17.241	15.775	14.479	13.330	12.310

Note: i = inflation rate; d = discount rate.

Inflating-Discounting Factors for Twenty-Five Years

i \ d	0	0.01	0.02	0.03	0.04	0.05	0.06	0.07	0.08	0.09	0.10	0.11	0.12	0.13	0.14	0.15	0.16	0.17	0.18	0.19	0.20
0	25.000	22.023	19.523	17.413	15.622	14.094	12.783	11.654	10.675	9.823	9.077	8.422	7.843	7.330	6.873	6.464	6.097	5.766	5.467	5.195	4.948
0.01	28.243	24.752	21.832	19.375	17.298	15.532	14.024	12.729	11.611	10.641	9.796	9.056	8.405	7.830	7.320	6.865	6.457	6.092	5.762	5.463	5.192
0.02	32.030	27.929	24.510	21.644	19.229	17.184	15.444	13.954	12.674	11.568	10.607	9.769	9.035	8.388	7.817	7.309	6.856	6.451	6.086	5.758	5.460
0.03	36.459	31.633	27.622	24.272	21.459	19.085	17.072	15.356	13.885	12.620	11.525	10.574	9.743	9.014	8.372	7.803	7.298	6.848	6.444	6.081	5.753
0.04	41.646	35.958	31.245	27.322	24.038	21.277	18.943	16.961	15.269	13.817	12.566	11.482	10.540	9.716	8.993	8.355	7.790	7.288	6.839	6.437	6.075
0.05	47.727	41.014	35.470	30.867	27.028	23.810	21.098	18.803	16.851	15.182	13.749	12.512	11.440	10.506	9.689	8.971	8.338	7.776	7.277	6.830	6.430
0.06	54.865	46.933	40.401	34.994	30.498	26.740	23.585	20.923	18.666	16.743	15.097	13.682	12.459	11.398	10.473	9.662	8.950	8.321	7.763	7.266	6.822
0.07	63.249	53.869	46.164	39.804	34.531	30.137	26.458	23.364	20.750	18.530	16.636	15.002	13.615	12.406	11.356	10.439	9.636	8.929	8.304	7.749	7.255
0.08	73.106	62.003	52.906	45.417	39.224	34.079	29.784	26.183	23.148	20.580	18.396	16.530	14.929	13.548	12.353	11.314	10.406	9.609	8.907	8.286	7.735
0.09	84.701	71.550	60.800	51.974	44.693	38.660	33.639	29.440	25.912	22.936	20.412	18.264	16.425	14.846	13.483	12.301	11.272	10.372	9.582	8.886	8.269
0.10	98.347	82.762	70.051	59.639	51.071	43.990	38.112	33.210	29.103	25.648	22.727	20.248	18.133	16.322	14.764	13.417	12.249	11.230	10.339	9.556	8.864
0.11	114.413	95.935	80.897	68.606	58.516	50.197	43.308	37.578	32.791	28.774	25.388	22.523	20.086	18.005	16.220	14.683	13.353	12.197	11.189	10.306	9.529
0.12	133.334	111.419	93.621	79.104	67.213	57.431	49.350	42.645	37.058	32.382	28.452	25.134	22.321	19.926	17.878	16.119	14.602	13.288	12.146	11.148	10.272
0.13	155.620	129.624	108.552	91.398	77.377	65.868	56.381	48.529	42.001	36.552	31.983	28.137	24.885	22.124	19.769	17.753	16.020	14.523	13.225	12.094	11.106
0.14	181.871	151.032	126.078	105.803	89.263	75.714	64.570	55.366	47.732	41.375	36.053	31.594	27.829	24.641	21.930	19.615	17.630	15.921	14.444	13.161	12.044
0.15	212.793	176.208	146.655	122.686	103.168	87.211	74.112	63.316	54.382	46.959	40.766	35.578	31.213	27.527	24.401	21.739	19.463	17.509	15.824	14.366	13.098
0.16	249.214	205.816	170.815	142.475	119.438	100.639	85.236	72.567	62.105	53.430	46.209	40.173	35.109	30.842	27.232	24.166	21.552	19.313	17.389	15.728	14.288

Note: i = inflation rate; d = discount rate.

Inflating-Discounting Factors for Thirty Years

d \ i	0	0.01	0.02	0.03	0.04	0.05	0.06	0.07	0.08	0.09	0.10	0.11	0.12	0.13	0.14	0.15	0.16	0.17	0.18	0.19	0.20
0	30.000	25.808	22.396	19.600	17.292	15.372	13.765	12.409	11.258	10.274	9.427	8.694	8.055	7.496	7.003	6.566	6.177	5.829	5.517	5.235	4.979
0.01	34.785	29.703	25.589	22.235	19.481	17.203	15.307	13.716	12.372	11.230	10.253	9.411	8.682	8.046	7.489	6.997	6.562	6.174	5.827	5.515	5.233
0.02	40.568	34.389	29.412	25.374	22.076	19.363	17.116	15.241	13.667	12.335	11.202	10.232	9.395	8.670	8.037	7.482	6.992	6.558	6.171	5.825	5.513
0.03	47.575	40.042	34.002	29.126	25.163	21.919	19.246	17.028	15.176	13.618	12.299	11.175	10.211	9.379	8.658	8.028	7.475	6.987	6.554	6.168	5.822
0.04	56.085	46.878	39.529	33.624	28.846	24.955	21.765	19.131	16.942	15.111	13.569	12.262	11.147	10.190	9.363	8.646	8.018	7.468	6.981	6.550	6.165
0.05	66.439	55.164	46.201	39.029	33.254	28.571	24.751	21.612	19.017	16.856	15.046	13.520	12.225	11.119	10.169	9.347	8.633	8.009	7.460	6.976	6.545
0.06	79.058	65.225	54.270	45.541	38.541	32.891	28.302	24.549	21.461	18.904	16.771	14.982	13.472	12.188	11.091	10.147	9.331	8.621	8.000	7.453	6.970
0.07	94.461	77.462	64.050	53.404	44.900	38.065	32.537	28.037	24.351	21.313	18.792	16.687	14.918	13.423	12.151	11.063	10.126	9.315	8.608	7.990	7.446
0.08	113.283	92.367	75.922	62.914	52.563	44.276	37.601	32.190	27.778	24.157	21.166	18.681	16.603	14.855	13.375	12.115	11.035	10.104	9.298	8.596	7.980
0.09	136.308	110.545	90.353	74.435	61.813	51.746	43.668	37.147	31.851	27.523	23.965	21.022	18.572	16.520	14.792	13.327	12.078	11.007	10.083	9.282	8.583
0.10	164.494	132.735	107.916	88.413	73.000	60.748	50.953	43.076	36.704	31.518	27.273	23.776	20.879	18.464	16.438	14.729	13.279	12.041	10.979	10.061	9.265
0.11	199.021	159.843	129.313	105.392	86.545	71.613	59.716	50.182	42.499	36.271	31.192	27.027	23.590	20.738	18.356	16.356	14.667	13.231	12.005	10.951	10.040
0.12	241.333	192.981	155.400	126.034	102.965	84.744	70.272	58.715	49.433	41.937	35.848	30.873	26.786	23.407	20.599	18.250	16.275	14.605	13.184	11.968	10.922
0.13	293.199	233.508	187.225	151.152	122.891	100.632	83.007	68.976	57.745	48.705	41.389	35.435	30.561	26.549	23.227	20.462	18.146	16.195	14.543	13.137	11.932
0.14	356.787	283.085	226.067	181.734	147.089	119.875	98.387	81.331	67.722	56.803	47.997	40.855	35.031	30.254	26.316	23.050	20.326	18.042	16.116	14.482	13.089
0.15	434.745	343.743	273.489	218.986	176.494	143.199	116.979	96.226	79.714	66.508	55.890	47.308	40.334	34.635	29.954	26.087	22.875	20.192	17.939	16.037	14.421
0.16	530.312	417.960	331.394	264.377	212.243	171.488	139.473	114.198	94.144	78.151	65.332	55.003	46.637	39.825	34.249	29.660	25.862	22.703	20.060	17.838	15.958

Note i = inflation rate; d = discount rate.

E Solar-System Standards Organizations

A number of organizations are involved in solar standards. These organizations have developed different kinds of standards: for the materials that go into the manufacture of collectors and other components; for the testing of collectors and storage; and for installation of systems in houses and commercial buildings.

Standards differ from codes in that complying with standards is voluntary, whereas codes are legally enforceable in the jurisdictions that adopt them. Manufacturers and installers usually comply with standards to provide a more saleable product or service; to meet the requests of a general clientele who are standards conscious; or to comply with the requirements of granters such as the Department of Housing and Urban Development (HUD) and the Department of Energy (DOE), or loan guarantors such as the Federal Housing Authority (FHA), the Farmer's Home Administration (FmHA), and the Veterans Administration (VA).

Acronym	Name and Location	Activity
ANSI	American National Standards Institute 1430 Broadway New York, NY 10018 (212) 354-3300	Coordinates activities of other standard-setting organizations.
ARI	Air-Conditioning & Refrigeration Institute 1815 N. Fort Myer Dr. Arlington, VA 22209 (703) 524-8800	Formed Air Conditioning & Refrigeration Institute Foundation (ARIF) for certifying collectors under NBS contract.
SEIA	Solar Energy Industries Association Suite 800 1001 Connecticut Ave. NW Washington, DC 20036 (202) 293-2981	Formed Solar Energy Research and Education Foundation (SEREF) for administering the above certification program under contract to DOE.
ASHRAE	American Society of Heating, Refrigerating and Air Conditioning Engineers 345 E. 47th St. New York, NY 10017 (212) 644-7853	Developed methods of testing solar collectors and thermal storage, ASHRAE standards 93-77 and 94-77.
ASTM	American Society for Testing & Materials 1916 Race St. Philadelphia, PA 19103 (215) 299-5476	Developed performance tests for materials and components.

NBS	National Bureau of Standards Bldg. 225-A-114 Washington, DC 20234 (202) 921-3285	Government laboratories that initiate standards and testing procedures such as interim performance criteria for DOE and intermediate minimum-property standards for HUD.
UL	Underwriters Laboratories, Inc. 207 East Ohio St. Chicago, IL 60611 (312) 642-6969	Drafts safety standards for collectors.
ASME	American Society of Mechanical Engineers United Engineering Center 345 E. 47th St. New York, NY 10017 (212) 644-7722	Develops temperature and pressure standards for piping and containers under pressure.
SMACNA	Sheet Metal and Air Conditioning Contractors National Association, Inc. 8224 Old Courthouse Rd. Tyson's Corner Vienna, VA 22180 (703) 790-9890	Developed installation standards for air collectors.
IAPMO	International Association of Plumbing & Mechanical Officials 5032 Alhambra Ave. Los Angeles, CA 90032 (213) 223-1417	Issued Uniform Solar Energy Code, which has been adopted by several jurisdictions as a plumbing code for installing liquid collectors.

F Published Standards, Codes, and Performance Criteria

The list that follows contains publication and purchasing information for works containing standards, codes, and performance criteria.[1]

ASHRAE Standard 93-77, *Methods for Testing the Thermal Performance of Solar Collectors.* 1977. ASHRAE, 345 E. 47th St., New York, NY 10017, $8.35.[2]

---

1. Data from National Solar Heating and Cooling Information Center. Prices are subject to change.
2. ASHRAE Standards 93-77 and 94-77 are available together for the reduced price of $10.35.

ASHRAE Standard 94-77, *Methods of Testing Thermal Storage Devices Based on Thermal Performance.* 1977. ASHRAE, 345 E. 47th St., New York, NY 10017, $6.35.

Florida Solar Energy Center Test Methods and Minimum Standards for Solar Collectors. November 1976. Florida Solar Energy Center, 300 State Rd., 401, Cape Canaveral, FL 32920, free of charge.

Heating and Air Conditioning Systems Installation Standards for One and Two Family Dwellings and Multifamily Housing Including Solar. 3rd ed. February 1977. SMACNA, 8224 Old Courthouse Rd., Tyson's Corner, Vienna, VA 22180, $10 for individuals and $6 for architectural and engineering firms, educational institutions, public libraries, government agencies and departments.

Interim Performance Criteria for Solar Heating and Cooling Systems in Dwellings. 1975. U.S. Government Printing Office,[3] catalog no. SDC13.6/2:504, $1.90.

Interim Performance Criteria for Solar Heating and Cooling Systems in Commercial Buildings. November 1976. National Technical Information Center,[4] order no. PB 262 114, $5.50.

Intermediate Minimum Property Standards Supplement—Solar Heating and Domestic Hot Water Systems. 1977. U.S. Government Printing Office, catalog no. 4930.2, $12.

Intermediate Standards for Solar Domestic Hot Water Systems/Hud Initiative. July 1977. National Technical Information Center, order no. PB 271 758, $8.

Provisional Flat Plate Solar Collector Testing Procedures. October 1977. National Technical Information Center, order no. PB 272 500, $5.25.

Uniform Solar Energy Code. 1976. International Association of Plumbing and Mechanical Officials, 5032 Alhambra Ave., Los Angeles, CA 90032, $6.

---

3. Superintendent of Documents, U.S. Government Printing Office, Washington, D.C. 20402.
4. National Technical Information Center, 5285 Port Royal Rd., Springfield, VA 22151.

G Summary of State Solar Legislation

Solar legislation is changing constantly. For the most current legislation for your state, contact your state energy office or taxing authority. Up-to-date information can also be obtained from the National Solar Heating and Cooling Information Center (telephone [800] 523-2929 or [800] 462-4983 in Pennsylvania and [800] 523-4700 in Alaska and Hawaii).

SUMMARY OF STATE SOLAR LEGISLATION

State	Bill Number	Description	Contact
AK	CH. 94, Laws of 1977	Allows an individual income-tax credit of 10% ($200 maximum) of cost of installing a solar system in the taxpayer's personal residence. Expires 12/31/82.	State Department of Revenue, Income Tax Division, Pouch SA, State Office Bldg., Juneau, AK 99811 (907) 465-2326
	CH. 179, Laws of 1978	Creates the Alaska Renewable Research Corporation. Most of its budget is used for venture capital for new businesses involved in renewable resources. Grants are also available.	Alaska Renewable Research Corporation, Box 1647, Juneau, AK 99802 (907) 465-4616 Suite One 313 "E" St., Anchorage, AK 99501 (907) 272-2500
	Ch. 29, Laws of 1978	Creates Alternative Power Resource Revolving Loan Fund for development of energy production from sources other than fossil or nuclear fuel. Loans may not exceed $10,000.	Division of Business Loans, Department of Commerce and Economic Development, Pouch D, Juneau, AK 99811 (907) 465-2510 (907) 274-6693 (Anchorage) (907) 452-8182 (Fairbanks)
AZ	CH. 165, Laws of 1974; CH. 146, Laws of 1979	Provides exemption from property-tax increases through 1989 that may result from addition of solar system to new or existing housing.	State Department of Revenue, Box 29002, Phoenix, AZ 85038 (602) 271-3381
	CH. 42, Laws of 1977; CH. 146, Laws of 1979	Exempts solar devices from transaction privilege and use taxes through 1989.	State Department of Revenue, Box 29002, Phoenix, AZ 85038 (602) 271-3381
	CH. 93, Laws of 1975; CH. 129, Laws of 1976; CH. 112, Laws of 1979	Provides income-tax credit of 35% ($1000 maximum) of the cost of installing a solar device in taxpayer's residence. If credit exceeds tax liability, excess credit can be carried forward for up to 5 years. Credit percentage declines 5% per year starting in 1984 until law expires after 1989.	
AR	Act 535, Laws of 1977; Act 742, Laws of 1979; Act 59, Laws of 1980	Allows individual homeowner/taxpayer to deduct entire cost of solar heating/cooling equipment from gross income for year of installation through 1984. Deduction may be carried forward.	State Department of Revenue, Income Tax Section, 7th and Wolfe Sts., Little Rock, AR 77201 (501) 371-2193

Source: National Solar Heating and Cooling Information Center (1980, updated). *Solar Legislation*. Rockville, Md.: U.S. Department of Housing and Urban Development.

State	Bill Number	Description	Contact
CA	CH. 168, Laws of 1976; CH. 1082, Laws of 1977; CH. 1154, Laws of 1978; CH. 816, Laws of 1979; CH. 903, 904, and 906, Laws of 1980	Provides income-tax credit of 55% ($3000 maximum) of the cost of installing a solar system in a house. For any other building where the cost of the system exceeds $12,000, the credit is the greater of $3000 or 25% of the system cost. If the credit exceeds tax liability, excess credit can be carried forward until expended. The state credit is reduced so that combined state/federal credit equals 55%. Systems must meet criteria established by the Energy Resources Conservation and Development Commission.	Franchise Tax Board, Attn: Correspondence, Sacramento, CA 95807 (916) 335-0370
	CH. 1 and CH. 7, Laws of 1978	Provides an interest-free loan program ($2000 per dwelling) for installation of solar heating systems in homes damaged in areas where a state of emergency has been declared.	Department of Housing and Community Development, Division of Research and Policy Development, 921 10th St. 5th floor, Sacramento, CA 95814 (916) 445-4728
	CH. 1243, Laws of 1978; CH. 48, Laws of 1979	The maximum loan available to veterans for home mortgages is increased to $60,000 if the home is equipped with a solar energy system.	Department of Veterans' Affairs, 1227 "O" St. P. O. Box 1559, Sacramento, CA 95807 (916) 445-2347
	CH. 1366, Laws of 1978	Prohibits anyone from allowing a tree or shrub to cast a shadow on a solar collector between 9:30 A.M. and 2:30 P.M. Trees casting a shadow before the installation of a collector are excluded.	Local planning or zoning body
	CH. 1154, Laws of 1978	Declares that any restriction on real property purporting to prohibit the installation and use of a solar energy system is void and unenforceable except those provisions which ensure the public health. It recognizes solar easements and prescribes their contents.	Local planning or zoning body

SUMMARY OF STATE SOLAR LEGISLATION

CO	CH. 344, Laws of 1975; CH. 363, Laws of 1979	Provides that solar heating/cooling systems be assessed at 5% of their original value when property taxes are computed.	Local assessor or board of assessors
	CH. 512, Laws of 1977; CH. 374, Laws of 1979	Allows personal and corporate income-tax deduction equal to the cost of installing a solar energy device in a building.	Colorado Department of Revenue, Capitol Annex Bldg., 1375 Sherman St., Denver, CO 80203 (303) 839-2801
	CH. 326, Laws of 1975; CH. 358, Laws of 1979	Recognizes solar easements and prescribes their contents. Declares as void and unenforceable any unreasonable restrictions on real estate against solar energy devices.	Local zoning board or planning commission
	CH. 306, Laws of 1979	Authorizes local governments to assure access to direct sunlight for solar energy devices using planning and zoning regulations.	Local zoning board or planning commission
CT	PA 409 (1976), PA 490 (1977), PA 479 (1979)	Enables local taxation authorities to exempt property equipped with a passive or active solar system from increased assessment due to the system. Construction of the solar portion must occur before 10/1/91. Exemption extends for 15 years after construction and applies to new construction and retrofitting.	Commissioner of Revenue Services, 92 Farmington Ave., Hartford, CT 06115 (203) 566-7120
	PA 457 (1977)	Provides sales- and use-tax exemptions for solar collectors. Expires 10/1/82.	Commissioner of Revenue Services, 92 Farmington Ave., Hartford, CT 06115 (203) 566-7120
	PA 4 (1979), PA 509 (1979), PA 10 (1979)	Establishes a grant and loan program to assist local housing authorities in making energy-conservation improvements. Establishes an energy-conservation loan fund for energy conservation and alternate energy devices.	Department of Housing, 1179 Main St., P. O. Box 2910, Hartford, CT 06101
	PA 520 (1979)	Authorizes loans for industrial applications of energy-conservation techniques.	Department of Economic Development, 210 Washington St., Hartford, CT 06115 (203) 566-4555

(continued)

State	Bill Number	Description	Contact
	PA 578 (1979)	Authorizes insured loans for energy conservation and renewable energy devices.	Connecticut Housing Financing Authority, 190 Trumbull St., Hartford, CT 06103 (203) 525-9311
	PA 11 (1979)	Creates a grant program to assist towns in carrying out energy-conservation projects. To qualify for funds, towns must file a local winter-energy plan.	Office of Policy and Management, 20 Grand St., Hartford, CT 06115 (203) 566-5765
DE	CH. 512, Laws of 1978	Provides an income-tax credit of $200 for solar hot water systems. Systems must be warranted and must meet HUD Intermediate Minimum Property Standards Supplement for Solar Heating and Domestic Hot Water Systems.	Division of Revenue; State Office Bldg., 820 French St., Wilmington, DE 19801 (302) 571-3360
FL	CH. 699, Laws of 1979	Provides sales- and use-tax exemptions for solar energy systems.	Florida Department of Revenue, Sales Tax Division, Carlton Bldg., Tallahassee, FL 32304 (904) 488-6800
	CH. 309, Laws of 1978	Recognizes solar easements and prescribes their contents.	Local zoning or planning body
	CH. 361, Laws of 1974	Stipulates that all new single-family dwellings shall be designed to facilitate future installation of a solar hot water system.	Local building inspector
GA	Act 1030, Laws of 1976; Act 1309, Laws of 1978	Provides sales- and use-tax refunds to property owners for the purchase of equipment for solar systems. Expires 7/1/86.	State Department of Revenue, Sales Tax Division, 309 Trinity-Washington Bldg., Atlanta, GA 30334 (404) 656-4065
	GA Constitution, Article VII, Section I, Paragraph IV	Allows the exemption of solar heating and cooling equipment and machinery used to manufacture solar equipment from property taxes. Expires 7/1/86.	Local city council or county board of supervisors
	Act 1446, Laws of 1978	Recognizes solar easements and prescribes their contents.	Local zoning board or planning commission

SUMMARY OF STATE SOLAR LEGISLATION

HI	Act 189 (1976)	Provides income-tax credit for 10% of the cost of solar system for year of installation. System must be placed in service by 12/31/81. If credit exceeds tax liability, excess credit can be carried forward until exhausted. Also exempts solar systems from property taxation. Expires 12/31/81.	State Tax Department, P.O. Box 259, Honolulu, HI 96809 (808) 548-3270
ID	CH. 212, Laws of 1976	Allows an individual who installs a solar system in his or her residence to deduct the entire cost over a 4-year period. Deduction cannot exceed $5000 in any taxable year.	Idaho State Tax Commission, P.O. Box 36, Boise, ID 83722 (208) 384-3560
	CH. 294, Laws of 1978	Recognizes solar easements and prescribes their contents.	Local zoning board or planning commission
IL	PA 79-943 (1975), PA 80-430 (1977), PA 80-1218 (1978)	Provides that property with an installed solar energy system will be assessed twice; with the solar system and also as though it were equipped with a conventional system. The lesser of the two assessments will be used to compute property tax due.	Local assessor or board of assessors
IN	PL 15 (1974), PL 68 (1977)	Reduces property tax of solar building by the difference between the assessment of the property with the system and the assessment of the property without the system. Owner must apply to county auditor.	Local assessor or board of assessors.
	PL 20 (1980)	Creates a 25% income-tax credit for the installation of solar systems. The maximum credit is $3000 for single-family dwellings and $10,000 for other buildings. The credit may be carried forward. Expenditures must be made by 12/31/82.	Indiana Department of Revenue, 202 State Office Bldg., Indianapolis, IN 46204 (317) 232-2101
IA	Section 441.21, Code of 1979	Exempts solar energy systems from property-tax assessment. Expires 12/31/85.	Local assessor or board of assessors

(continued)

State	Bill Number	Description	Contact
	CH. 1086, Laws of 1978	Establishes a loan and grant fund for property improvements (including solar systems) for low-income families.	Iowa Housing Finance Authority, 218 Liberty Bldg., Des Moines, IA 50319 (515) 281-4058
KS	CH. 434, Laws of 1976; CH. 346, Laws of 1977; CH. 409, Laws of 1978	Allows a tax credit equal to 25% of the cost of a solar system. Maximum credit is $1000 for residences, and the least amount of that year's tax bill or $3000 for business or investment property. Installation cost can be amortized over 60 months for businesses.	State Department of Revenue, P.O. Box 692, Topeka, KS 66601 (913) 296-3909
	CH. 345, Laws of 1977; CH. 419, Laws of 1978	Provides for reimbursement of 35% of total property tax paid on entire building or building addition if solar system provides 70% of energy needed to heat/cool the building on an average annual basis. Reimbursement may be claimed for 5 successive years. System must be installed before 1/1/86.	State Department of Revenue, P.O. Box 692, Topeka, KS 66601 (913) 296-3909
	CH. 277, Laws of 1977	Recognizes solar easements and prescribes their contents.	Local zoning board or planning commission
LA	Act 591 (1978)	Exempts solar energy equipment installed in owner-occupied residential buildings or in swimming pools from property tax.	Local parish tax assessor
ME	CH. 542, Laws of 1977	Exempts solar heating from property taxation for 5 years from date of installation. Exemption must be applied for. Also provides refund of sales or use tax paid on solar equipment certified as such by the Office of Energy Resources. Both tax provisions expire 1/1/83.	Local assessor or board of assessors (for property tax). Office of Energy Resources, 55 Capitol St., Augusta, ME 04330 (207) 289-2196 (for sales tax)
	CH. 557, Laws of 1979	Creates an income-tax credit (the lesser of $100 or 20% of eligible costs) for solar, wind, and wood energy systems.	Bureau of Taxation, Department of Finance and Administration, State Office Bldg., Augusta, ME 04333 (207) 289-2076

	CH. 418, Laws of 1979; CH. 435, Laws of 1979	Enables local governments and planning boards to enact zoning, ordinances and various subdivision regulations to protect access to direct sunlight for solar energy use.	Local zoning board or planning commission
MD	CH. 509, Laws of 1975; CH. 509, Laws of 1978	Provides that buildings equipped with a solar heating/cooling system or a solar and a conventional heating/cooling unit shall be assessed for property-tax purposes at no more than the value of a conventional system needed to serve the structure.	Local assessor or board of assessors
	CH. 740, Laws of 1976	Allows Baltimore City, any city within a county, or any county to provide a credit against local real-property taxes for buildings using solar heating/cooling units. Amounts and definitions are at the discretion of local jurisdictions.	Local city or county department of revenue
	CH. 934, Laws of 1977	Recognizes solar easements as lawful restriction on land.	Local zoning board or planning commission
MA	CH. 734, Laws of 1975; CH. 388, Laws of 1978	Provides for real-estate tax exemption for solar system. Extends for 20 years after system installation.	Local assessor or board of assessors
	CH. 487, Laws of 1976	Provides that a corporation may deduct the cost of a solar heating system from taxable income for year of installation. Also provides that the system will be exempt from tangible property tax.	State Department of Corporations & Taxation, 100 Cambridge St., Boston, MA 02204 (617) 727-4201
	CH. 28, Laws of 1977; CH. 260, Laws of 1978	Authorizes banks and credit unions to make loans with extended maturation periods and increased maximum amounts for financing solar energy systems. Maximum amount is increased to $15,000 for banks and $12,000 for credit unions.	Local bank or credit union
	CH. 796, Laws of 1979	Exempts solar system used in an individual's principal residence from retail sales taxes.	State Department of Corporations & Taxation, 100 Cambridge St., Boston, MA 02204 (617) 727-4601

(continued)

State	Bill Number	Description	Contact
	CH. 796, Laws of 1979	Creates a personal income-tax credit of 35% (maximum $1000) of the cost of renewable energy equipment installed in the taxpayer's principal residence. Credit is reduced by federal-tax credit. Expires 12/31/83.	State Department of Corporations and Taxation, 100 Cambridge St., Boston MA 02204 (617) 727-4201
MI	PA 132 (1976)	Provides that receipts from sale of tangible property to be used in a solar system shall not be used to compute tax liability for business activities tax. Provision expires 1/1/85.	State Department of Treasury, State Tax Commission, State Capitol Bldg., Lansing, MI 48922 (517) 373-2910
	PA 133 (1976)	Provides that tangible property used for solar devices shall be exempt from excise (use) tax on personal property. Provision expires 1/1/85.	State Department of Treasury, State Tax Commission, State Capitol Bldg., Lansing, MI 48922 (517) 373-2910
	PA 135 (1976)	Exempts solar devices from real and personal property taxes. Applicant for exemption must obtain exemption certificate from State Tax Commission. Commission's authority to issue certificates ends 7/1/85.	Local Government Services, Treasury Bldg., Lansing, MI 48922 (517) 373-3232
	PA 605 (1978), PA 41 (1979)	Provides an income-tax credit for a residential solar, wind, or water energy device that is used for heating, cooling, or electricity. Energy-conservation measures are also eligible for the refundable credit. To be eligible, expenditures must be made by 12/31/83 and systems must meet Department of Commerce standards. The rate of the credit changes annually.	State Department of Treasury, State Tax Commission, State Capitol Bldg., Lansing, MI 48922 (517) 373-2910
MN	CH. 786, Laws of 1978	Excludes the market value of a solar system installed in a building prior to 1/1/84 from the market value of the building for purposes of computing property-tax liability.	Local assessor or board of assessors
	CH. 786, Laws of 1978	Recognizes solar easements and prescribes their contents. Allows depreciation resulting from easements for property-tax revaluation.	Local zoning board or planning commission

SUMMARY OF STATE SOLAR LEGISLATION

	CH. 303, Laws of 1979	Provides an income-tax credit of 20% of first $10,000 spent on renewable energy resource equipment installed on a building of 6 dwelling units or less. Excess credit can be carried forward through 1984. Expenditures must be made by 12/31/82.	Department of Revenue, Centennial Office Bldg., 658 Cedar St., St. Paul, MN 55145 (612) 296-3781
	CH. 2, Special Session, Laws of 1979	Prohibits local governments from preventing earth-sheltered construction as long as it otherwise complies with local zoning ordinances.	Minnesota Energy Agency, 980 American Center Bldg., 150 E. Kellogg Blvd., St. Paul, MN 55101 (612) 296-5120
MS	27-65-105 of the Mississippi Code	Exempts from sales tax labor, property, or services used in the construction of solar systems used by universities, colleges, or junior colleges. Expires 1/1/83.	Mississippi Tax Commission, Sales Tax Division, P.O. Box 960, Jackson, MS 39205 (601) 354-6274
MO	442.021 of the Missouri Code	Recognizes solar easements and mandates their contents. Declares that the right to use solar energy is a property right, but it cannot be acquired by eminent domain.	Local zoning board or planning commission
MT	CH. 548, Laws of 1975; CH. 574, Laws of 1977; CH. 652, Laws of 1979	Provides individual income-tax credit for installation of a solar system in taxpayer's residence prior to 12/31/82. Credit is for 5% of the first $1000 spent and 2½% of the next $3000. If federal government provides a similar tax credit, amount of credit is halved. If credit exceeds tax liability, excess credit can be carried forward for up to 4 years.	State Department of Revenue, Income Tax Section, Mitchell Bldg, Helena, MT 59601 (406) 449-2837
	CH. 576, Laws of 1977	Allows personal and corporate income-tax deduction for a capital investment in a building for a solar system. Maximum deduction is $1800 for a residence and $3600 for a non-residential building.	State Department of Revenue, Income Tax Section, Mitchell Bldg, Helena, MT 59601 (406) 449-2837

(continued)

State	Bill Number	Description	Contact
	Montana Code Annotated 15-32-107	Allows public utility to lend capital for installation of solar systems in dwellings at a rate not exceeding 7% per year. Financial institutions may lend money at no less than 2 percentage points below the Federal Reserve rate. Any interest foregone by not charging the prevailing rate of interest may be claimed as a credit against license tax by a utility, or against corporation license tax by a financial institution.	Local lending institution or utility company
	Montana Code Annotated 15-32-102; CH. 639, Laws of 1979	Provides a 10-year real estate tax exemption for capital investments in nonfossil forms of energy generation (maximum $20,000 for a single-family dwelling and $10,000 for other buildings).	State Department of Revenue, Property Assessment Division, Mitchell Bldg, Helena, MT 59601 (406) 449-2808
	CH. 524, Laws of 1979	Recognizes solar easements and prescribes their contents.	Local zoning board or planning commission
NE	Legislative Bill 353, 1979	Recognizes solar easements and prescribes their contents. Allows local governments to include solar access considerations in their zoning ordinances and development plans. Allows variances to facilitate solar access.	Local zoning board or planning commission
NV	CH. 345, Laws of 1977	Provides property-tax allowance for installation of a solar heating/cooling system in a residential building in amount equal to assessed value of property with system minus assessed value without system. Allowance may not exceed total value of property tax accrued or $2000, whichever is less.	Local county assessor
	CH. 314, Laws of 1979	Recognizes solar easements and prescribes their contents.	Local zoning board or planning commission

SUMMARY OF STATE SOLAR LEGISLATION   399

NH	CH. 391, Laws of 1975; CH. 5202, Laws of 1977	Allows each city and town to adopt (by local referendum) property-tax exemptions for solar heating/cooling systems.	Local assessor or board of assessors
NJ	CH. 256, Laws of 1977	Allows owner of real property with solar heating/cooling system annual exemption from property taxes equal to the remainder of assessed value of property with system minus assessed value without system. Exemption must be applied for and system must meet standards established by State Energy Office. Law expires 12/31/82.	Local assessor or board of assessors
	CH. 465, Laws of 1977	Exempts solar systems from sales and use taxes. To qualify for exemption, system must meet standards set by the State Department of Energy.	New Jersey Department of Energy, Office of Alternate Technology, 101 Commerce St., Newark, NJ 07102 (201) 648-6293
	CH. 152, Laws of 1978	Recognizes solar easements and prescribes their contents.	Local zoning board or planning commission
NM	CH. 12, Laws of 1975; CH. 170, Laws of 1978; CH. 353, Laws of 1979	Allows individual income-tax credit for 25% ($1000 maximum) of cost of solar heating/cooling system installed in taxpayer's residence or to heat a swimming pool. If credit exceeds tax liability, refund is paid. Excess credit will be refunded. System must meet performance criteria prescribed pursuant to the Federal Solar Demonstration Act of 1974.	State Department of Taxation and Revenue, Income Tax Division, P.O. Box 630, Santa Fe, NM 87503 (505) 827-3221
	CH. 114, Laws of 1977	Allows income-tax credit for solar system used for irrigation-pumping purposes if system reduces fossil fuel use by 75%. If credit claimed in any one year exceeds tax liability, refund is paid. Credit cannot be claimed if taxpayer claimed similar credit, deduction, or exemption on his or her federal-income-tax return, or claimed a credit provided by CH. 12, Laws of 1975. System must be certified by the Energy Board.	State Department of Taxation and Revenue, Income Tax Division, P.O. Box 630, Santa Fe, NM 87503 (505) 827-3221

(continued)

State	Bill Number	Description	Contact
	CH. 169, Laws of 1977	Recognizes the right to use solar energy as a property right. Disputes regarding access will be settled by rule of prior appropriation.	Local assessor or board of assessors
NY	CH. 322, Laws of 1977; CH. 220, Laws of 1979	Provides property-tax reduction for solar systems in amount of assessed value of property with system minus assessed value without system. All exempted systems must be approved by state energy office. Exemption extends for 15 years after approval. System must be installed prior to 7/1/88.	Local zoning board or New York State Energy Office, Agency Bldg. 2, Empire State Plaza, Albany, NY 12223 (518) 474-8181
	CH. 705, Laws of 1979; CH. 742, Laws of 1979	Recognizes solar easements and prescribes their contents. Permits the protection of solar access by zoning ordinances.	
NC	CH. 792, Laws of 1977; CH. 892, Laws of 1979	Allows personal and corporate income-tax credit equal to 25% ($1000 limit) of the cost of solar heating/cooling system installed in a building. If credit exceeds tax liability, excess credit can be carried forward for up to 3 years. System must meet certain performance criteria.	State Department of Revenue, Income Tax Division, P.O. Box 25000, Raleigh, NC 27640 (919) 733-3991
	CH. 965, Laws of 1977	Provides that buildings equipped with solar heating/cooling systems shall be assessed for property-tax purposes as if they were equipped with a conventional system only. Law expires 12/1/85.	Local assessor or board of assessors
ND	CH. 508, Laws of 1975	Exempts solar heating/cooling systems used in buildings from property taxation for 5 years following installation.	Local assessor or board of assessors
	CH. 537, Laws of 1977	Provides income-tax credit for installation of a solar system equal to 5% a year for 2 years.	State Tax Commission, Income Tax Division, Capitol Bldg., Bismarck, ND 58505 (701) 224-3450
	CH. 425, Laws of 1977	Recognizes solar easements and prescribes their contents.	Local zoning board or planning commission

SUMMARY OF STATE SOLAR LEGISLATION

OH	Amended Substitute House Bill 154, 1979	Exempts solar systems from sales and real-estate taxes through 12/31/85. Creates a corporate franchise tax credit of 10% of the cost of the solar system. Creates a personal income-tax credit of 10% of the cost of a solar heating system (maximum $1000). Credit may be carried forward for 2 years. Systems must meet certain guidelines.	For income and franchise tax information: Department of Taxation, Income Tax Division, 1030 Freeway Dr., Columbus, OH 43229 (614) 466-7910 For sales tax information: Department of Taxation, Sales Tax Division, 30 W. Broad St., Columbus, OH 43215 (614) 466-7350
	Amended Substitute House Bill 154, 1979	Recognizes solar easements and prescribes their contents.	Local zoning board or planning commission
OK	CH. 209, Laws of 1977	Provides income-tax credit of 25% ($2000 limit) of the cost of a solar system installed in a private residence. If credit exceeds tax liability, excess credit may be carried forward for up to 3 years. Law expires 1/1/88.	State Tax Commission, Income Tax Division, 2501 Lincoln Blvd., Oklahoma City, OK 73194 (405) 521-3125
OR	CH. 196, Laws of 1977; CH. 670, Laws of 1979	Exempts solar systems from real estate tax. Expires 1/1/98.	Local assessor or board of assessors
	CH. 196, Laws of 1977; CH. 670, Laws of 1979	Allows personal income-tax credit equal to 25% ($1000 limit) of cost of installing a solar system in a home. System must provide at least 10% of home's energy requirements and must meet performance criteria adopted by State Department of Energy. If credit exceeds tax liability, excess credit can be carried forward for up to five years.	State Department of Revenue, State Office Bldg., Salem, OR 97310 (503) 378-3366
	CH. 315, Laws of 1977	Permits veteran to obtain subsequent loan ($3000 limit) above the maximum amount allowed from Oregon War Veteran's Fund for installing a solar system in a home. System must provide at least 10% of home's energy requirements and must meet performance criteria adopted by State Department of Energy.	State Department of Veterans' Affairs, 3000 Market St. Plaza, Suite 552, Salem, OR 97310 (503) 378-6438

(continued)

State	Bill Number	Description	Contact
	CH. 483, Laws of 1979	Provides a corporation excise-tax credit for commercial lending institutions making loans (maximum $10,000) at 6½% interest or less for the installation of certified alternative energy devices. The tax credit is equal to the difference between 6½% and the average interest rate for home-improvement loans made in a previous calendar year.	State Department of Revenue, State Office Bldg., Salem, OR 97310 (503) 378-3366
	CH. 512, Laws of 1979	Creates a corporate income-tax credit of 35% of the cost of eligible energy-conservation facilities, including solar systems. The credit equals 10% for the first 2 years and 5% for each of the next 3 years. A carry-forward provision is included.	State Department of Revenue, State Office Bldg., Salem, OR 97310 (503) 378-3366
	CH. 732, Laws of 1977	Creates a loan fund for alternative energy projects and authorizes the director of the State Department of Energy to sell bonds to finance the loan fund.	Oregon Department of Energy, Room 111, Labor and Industries Bldg., Salem, OR 97310 (503) 378-4128
	CH. 671, Laws of 1979	Recognizes solar easements and prescribes their contents. Enables local governments to regulate solar access. Prohibits private restrictions preventing the use of solar energy.	Local zoning board or planning commission
RI	CH. 202, Laws of 1977	Provides that a solar heating/cooling system installed in a building be assessed at no more than the value of a conventional heating/cooling system necessary to serve the building. Law expires 4/1/97.	Local assessor or board of assessors

SD	CH. 74, Laws of 1978; CH. 84, Laws of 1980	Allows property-tax assessment credit for use of a solar system for either residential or commercial application. Residential credit is equal to the difference between assessed valuation of property with system and assessed value without system for 3 years. For the following 3 years the credit equals 75%, 50%, and 25% of the base credit. Commercial credit equals 50% of the difference in the assessments for the first 3 years, then 37.5%, 25%, and 12.5% for the following 3 years.	Local county auditor or Department of Revenue, Capitol Lake Plaza, Pierre, SD 57501 (605) 773-3311
TN	CH. 837, Laws of 1978	Exempts solar systems from property taxation. Law expires 1/1/88.	Local assessor or board of assessors
	CH. 884, Laws of 1978	Provides a loan program for installation of energy conserving improvements, including solar hot-water heaters in residential housing.	Tennessee Housing Development Authority, Hamilton Bank Bldg., Nashville, TN 37219 (615) 741-3023
	CH. 259, Laws of 1979	Recognizes solar easements and prescribes their contents. Empowers local government to protect solar access through zoning regulations.	Local zoning board or planning commission For sample easement: Tennessee Energy Authority, Suite 707, Capitol Boulevard Bldg., Nashville, TN 37219 (615) 741-2994
TX	CH. 719, Laws of 1975; CH. 584, Laws of 1977; CH. 107, Laws of 1979	Exempts solar devices from sales tax and real-estate tax assessments. Exempts corporations that exclusively manufacture, install, or sell solar devices from franchise tax. Corporation may deduct from taxable capital the amortized cost of a solar device over 60 months.	Comptroller of Public Accounts, Capitol Station, Drawer SS, Austin, TX 78775 (512) 475-2206

(continued)

State	Bill Number	Description	Contact
UT	CH. 66, Laws of 1980	Creates an individual income-tax credit (maximum $1000) of a residential solar system's cost. System must be installed before 6/20/85. The credit may be carried forward for 4 years. Businesses may claim a 10% credit (maximum $1000) against income or franchise tax and may pass the credit on to the purchaser of the property. Systems must be certified by the State Energy Office.	State Tax Commission, 200 State Office, Salt Lake City, UT 84114 (801) 533-5831
	CH. 82, Laws of 1979	Recognizes written solar easements as a property interest.	Local zoning board or planning commission
VT	Act 226, Laws of 1976	Allows towns to enact real and personal property-tax exemptions for solar systems.	Local assessor or board of assessors
	Act 210, Laws of 1978	Allows personal and business income-tax credits for installation of solar systems prior to 7/1/83. Credit is equal to the lesser of (1) 25% of cost of system installed on real property or (2) $1000 ($3000 for businesses). If credit exceeds tax liability, excess credit can be carried forward for up to 4 years.	State Tax Department, Income Tax Division, State St., Montpelier, VT 05602 (802) 828-2517
VA	CH. 561, Laws of 1977	Allows county, city, or local governing body to exempt certified solar equipment from property taxation for a minimum of five years.	Local tax-governing body
	CH. 351, Laws of 1979	Creates a separate class of tangible personal property for local taxation. The class includes energy-conservation equipment purchased by a manufacturer for the purpose of changing the energy source of a plant to alternative energy resources. This class of property may be taxed at a rate different from the rate on other tangible personal property.	Local taxing authority

	Citation	Description	Contact
	CH. 631, Laws of 1978	Directs the Virginia Housing Development Authority to establish a loan program to finance energy-conserving devices.	Virginia Housing Development Authority, 111 S. 6th St., Richmond, VA 23219 (804) 786-8241
	CH. 323, Laws of 1978	Subjects solar easements to the same legal requirements as other easements and mandates their contents.	Local zoning board or planning commission
WA	CH. 364, Laws of 1977	Exempts solar systems from property taxation. Exemption is valid for 7 years after installation and can only be applied to equipment meeting minimum standards promulgated by the Department of Housing and Urban Development. Opportunity to apply for exemption extends to 12/31/87.	Local assessor or board of assessors.
	CH. 239, Laws of 1979	Authorizes utility companies to establish programs to perform energy audits, recommend improvements, and to arrange the installation and financing of energy-conservation materials in residential buildings.	Local utility company
	CH. 170E-1, Laws of 1979	Permits local governments to regulate protection of solar access in comprehensive plans and zoning ordinances. It recognizes solar easements and mandates their contents.	Local zoning board or planning commission
WI	CH. 313, Laws of 1977	Allows businesses to deduct (in the year paid) depreciation or amortize (over 5 years) the cost of installing a solar system. Systems must meet Department of Industry, Labor, and Human Relations standards. Applies to expenses incurred before 12/31/84.	State Department of Revenue, Income Tax Division, P.O. Box 8910, Madison, WI 53708 (608) 226-1911

(continued)

State	Bill Number	Description	Contact
	CH. 34, Laws of 1979	Creates a program to subsidize the cost of an alternative energy system purchased by an individual. If the building on which the system is installed was on the local tax rolls before 4/20/77, the rate of refund is 24% in 1980, 18% in 1981 and 1982, and 12% in 1983 and 1984. For newer buildings the rate of refund equals 16% in 1980, 12% in 1981 and 1982, and 8% in 1983 and 1984. Eligible expenses must exceed $500 but will not be calculated on more than $10,000. All systems must meet standards of the Department of Industry, Labor, and Human Relations.	Department of Industry, Labor, and Human Relations, 201 E. Washington, Rm. 101, Safety and Buildings Division, Madison, WI 53702 (608) 266-1149

H How to Find More Information

WHERE TO LOOK **Libraries**

If your local library does not have a book you want, the librarian may be able to borrow it for you from another library. If you live near a college or university, check its libraries also. You may find information on solar energy scattered among the university's general, physical sciences, engineering, and architecture libraries.

Many of the papers and conference proceedings mentioned in the Reference lists and Bibliographies in this book are available in larger libraries on microfiche. Where possible the National Technical Information Service identification number or other identifying number is included in the publication information in this appendix. The best way to find a microfiche copy of a specfic article or find out what is available on microfiche is to tell the reference librarian what you are looking for and ask for help and instructions. Not every library receives every microfiche, so the article you need might be unavailable locally.

Where to Write for Publications and Information

Many of the published works listed in this book are available from the National Technical Information Service at the following address:

> U.S. Department of Commerce
> National Technical Information Service
> 5285 Port Royal Rd.
> Springfield, VA 22161

Many other publications can be purchased from local outlets of the U.S. Government Printing Office or from

> Superintendent of Documents
> U.S. Government Printing Office
> Washington, DC 20402

The National Solar Heating and Cooling Information Center provides free information on specific and general solar topics. It also has many free publications available. Write to

> National Solar Heating and Cooling Information Center
> P.O. Box 1607
> Rockville, MD 20850

or call toll free: (800) 523-2929 ([800] 462-4983 in Pennsylvania and [800] 523-4700 in Alaska and Hawaii).

The American Society of Heating, Refrigerating and Air-conditioning Engineers, Inc., (ASHRAE) has several technical papers and performance standards available. Write to

> ASHRAE
> Publications Sales Department
> 345 E. 47th St.
> New York, NY 10017

The American Section of the International Solar Energy Society (AS-ISES) is very active in all aspects of solar energy use. It sponsors many excellent conference workshops, and published proceedings are available to the general public. Membership in the American Section includes subscriptions to the official magazines, *Solar Age* and *Sunworld*, and to the journal *Solar Energy*. Members also receive substantial discounts on conference registrations and proceedings. Write to

> American Section, International Solar Energy Society
> American Technological University
> P.O. Box 1416
> Killeen, TX 76541

A Special Note on Conference Proceedings

Conference proceedings are often abundant sources of technical and general information about solar energy use and research. The difficulty is that usually proceedings have a limited circulation and are not widely advertised. Many proceedings are out of print but are available in libraries. Others are available through the National Technical Information Service or the American Section of the International Solar Energy Society.

WHAT TO LOOK FOR   General Information

Books

Anderson, B. (1977). *Solar Energy: Fundamentals in Building Design.* New York: McGraw-Hill.

Anderson, B., and M. Riordan (1976). *The Solar Home Book: Heating, Cooling and Designing with the Sun.* Harrisville, N.H.: Cheshire Books.

Daniels, F. (1974). *Direct Use of the Sun's Energy.* New York: Ballantine.

de Winter, F. (1975). *Solar Energy and Flat Plate Collectors: An Annotated Bibliography.* New York: ASHRAE, S-101.

Merrill, R., and T. Gage (1978). *Energy Primer—Solar, Water, Wind and Biofuels.* 2nd ed., rev. New York: Dell.

Shurcliff, W. A. (1976). *Solar Heated Buildings—A Brief Survey.* 13th ed. 19 Appleton St., Cambridge, Mass.: W. Shurcliff.

Shurcliff, W. A. (1978). *Solar Heated Buildings of North America: 120 Outstanding Examples.* Harrisville, N.H.: Brick House.

Sunset Books (1977). *Homeowner's Guide to Solar Heating.* Menlo Park, Calif.: Land.

Szokolay, S. V. (1975). *Solar Energy and Building.* New York: John Wiley.

Watson, D. (1977). *Designing and Building a Solar House—Your Place in the Sun.* Charlotte, Vt.: Garden Way.

Williams, J. R. (1977). *Solar Energy Technology and Applications.* 2nd ed., rev. Ann Arbor, Mich.: Ann Arbor Science Publishers.

Periodicals

Solar Age (published monthly). Church Hill, Harrisville, N.H.: SolarVision. (Official magazine of the American Section of the International Solar Energy Society.)

Solar Energy Digest (published monthly). San Diego, Calif.

Sunworld (published quarterly). Elmsford, N.Y.: Pergamon Press. (Publication of the International Solar Energy Society.)

Solar Architecture, Home Designs, and Conservation

A.I.A. Research Corp. (1976). *Solar Dwelling Design Concepts.* Washington, D.C.: U.S. Government Printing Office, 023-000-0334-1.

A.I.A. Research Corp. (1976). *Solar Heating and Cooling Demonstration: A Descriptive Summary of HUD Solar Residential Demonstrations, Cycle 1.* Washington, D.C.: U.S. Government Printing Office, 023-000-00338-4.

Department of Housing and Urban Development (1977). *In the Bank... or Up the Chimney? A Dollars and Cents Guide to Energy Saving Improvements.* 2nd ed. Washington, D.C.: U.S. Government Printing Office, 023-000-00411-9.

Eccli, E. (ed.) (1976). *Low-Cost Energy-Efficient Shelter—for the Owner and Builder.* Emmaus, Pa.: Rodale Press.

Gropp, L. (1978). *Solar Houses: 48 Energy-Saving Designs.* New York: Pantheon.

Hudson Home Guides (1978). *Practical Guide to Solar Homes.* New York: Bantam/Hudson.

Wade, A., and N. Ewenstein (1978). *30 Energy-Efficient Houses... You Can Build.* Emmaus, Pa.: Rodale Press.

Wells, M., and I. Spetgang (1978). *How to Buy Solar Heating... Without Getting Burnt!* Emmaus, Pa.: Rodale Press.

Passive Solar Energy Use

Mazria, E. (1978). *The Passive Solar Energy Book.* Emmaus, Pa.: Rodale Press.

McCullagh, J. C. (ed.) (1978). *The Solar Greenhouse Book.* Emmaus, Pa.: Rodale Press.

Yanda, B., and R. Fisher (1977). *The Food and Heat Producing Solar Greenhouse: Design, Construction, and Operation.* Santa Fe, N. Mex.: John Muir Publications.

The Science of Solar Energy

Books

ASHRAE (1974). *Handbook and Product Directory —1978 Applications.* New York: ASHRAE.

ASHRAE (1974). *Proceedings of the Workshop on Solar Heating and Cooling of Buildings.* New York: ASHRAE, NSF-RA-N-74-126.

ASHRAE (1977). *Applications of Solar Energy for Heating and Cooling of Buildings.* New York: ASHRAE, GRP-170.

ASHRAE (1978). *ASHRAE Journal*, reprint of solar energy issue. New York: ASHRAE.

Periodicals

Applied Solar Energy (Geliotekhnika) (published bimonthly). New York: Allerton Press. (Translated from the Russian.)

ASHRAE Journal (published monthly). New York: ASHRAE.

Solar Energy (published monthly). Elmsford, N.Y.: Pergamon Press. (Published for the International Solar Energy Society.)

Duffie, J. A., and W. A. Beckman (1974). *Solar Energy Thermal Processes.* New York: John Wiley.

Krieder, J. F., and F. Krieth (1976). *Solar Heating and Cooling: Engineering Practical Design and Economics*, Washington, D.C.: Hemisphere.

Meinel, A. B., and M. P. Meinel (1976). *Applied Solar Energy—An Introduction.* Reading, Mass.: Addison-Wesley.

Threlkeld, J. L. (1970). *Thermal Environmental Engineering.* 2nd ed. Englewood Cliffs, N.J.: Prentice-Hall.

Abstracts

CA Selects: Solar Energy (published biweekly). Columbus, Ohio: American Chemical Society.

Solar Energy Update (published monthly). Springfield, Va.: National Technical Information Service, NTISUB/E/145.

System Design and Installation

Books

ASHRAE (1976). *Handbook and Product Directory—1976 Systems.* New York: ASHRAE.

ASHRAE (1977). *Handbook and Product Directory—1977 Fundamentals.* New York: ASHRAE.

Beckman, W. A., S. A. Klein, and J. A. Duffie (1977). *Solar Heating Design by the f-Chart Method.* New York: John Wiley.

Los Alamos Scientific Laboratory (1976). *Pacific Regional Solar Heating Handbook.* Washington, D.C.: U.S. Government Printing Office, 060-000-00024-7.

SMACNA (1978). *Fundamentals of Solar Heating.* Washington, D.C.: U.S. Government Printing Office, 061-000-00043-7.

Solar Energy Applications Laboratory, Colorado State University (1977). *Solar Heating and Cooling of Residential Buildings—Design of Systems.* Washington, D.C.: U.S. Government Printing Office, 003-011-00084-4.

Solar Energy Applications Laboratory, Colorado State University (1977). *Solar Heating and Cooling of Residential Buildings—Sizing, Installation and Operation of Systems.* Washington, D.C.: U.S. Government Printing Office, 003-011-00085-2.

Solcost. Department-of-Energy-sponsored, computer-aided solar design and sizing (for a fee, usually $35–50). Information available from International Business Services, Inc., Solar Group, 1010 Vermont Ave. NW, Suite 1010, Washington, D.C. 20005.

For Do-It-Yourselfers

Campbell, S., with D. Taff (1978). *Build Your Own Solar Water Heater.* Charlotte, Vt.: Garden Way.

Lucas, T. (1975). *How to Build a Solar Heater.* Pasadena, Calif.: Ward Ritchie.

Popular Science Solar Energy Handbook 1979 (1979). New York: Times Mirror Magazines.

Periodicals

The Mother Earth News (published bimonthly). Hendersonville, N.C.: The Mother Earth News, Inc.

Popular Science (published monthly). New York: Times Mirror Magazines.

Business, Financial, and Legal Aspects of Solar Energy

Books

Buying Solar (1976). Washington, D.C.: U.S. Government Printing Office, 041-018-00120-4.

Barrett, D. et al. (1977). *Home Mortgage Lending and Solar Energy* (1977). Washington, D.C.: U.S. Government Printing Office, 023-000-0038M-2.

Eisenhard, R. M. (1977). *State Solar Energy Legislation of 1976: A Review of Statutes Relating to Buildings.* Washington, D.C.: U.S. Department of Commerce, NBSIR 77-1297.

Periodicals

Solar Energy Intelligence Report (published biweekly). P.O Box 1067, Silver Spring, Md.: Business Publishers.

Solar Law Reporter (published bimonthly). P.O. Box 5400, Denver, Colo.: Solar Energy Research Institute.

Solar Outlook (published weekly). Washington, D.C.: Observer.

Solar Engineering (published monthly). Dallas, Tex.: Solar Energy Industries Association.

Solar Heating and Cooling (published bimonthly). Morristown, N.J.: Gordon.

Solar Organizations and Information Centers

American Section of the International Solar Energy Society. American Technological University, P.O. Box 1416, Killeen, TX 76541.

Local Solar Energy Office.

National Solar Heating and Cooling Information Center, P.O. Box 1607, Rockville, MD 29850, toll free (800) 523-2929 (from Pennsylvania [800] 462-4983).

Solar Energy Industries Association, Inc., 1001 Connecticut Ave. NW, Suite 800, Washington, DC 20036.

State or Local Energy Office (see Appendix G, Summary of State Solar Legislation, for state office addresses).

Manufacturers and Products: Directories

Ann Arbor Science Publishers (1977). *The Solar Energy and Research Directory.* Ann Arbor, Mich.: Ann Arbor Science Publishers.

Pesko, C. (1975). *Solar Directory.* Ann Arbor, Mich.: Ann Arbor Science Publishers.

Shurcliff, W. A. (1976). *Informal Directory of the Organizations and People Involved in the Solar Heating of Buildings.* 3rd ed. Cambridge, Mass.: W. A. Shurcliff.

Solar Age Magazine (ed.) (1979). *Solar Products Specifications Guide.* Harrisville, N.H.: SolarVision.

Solar Energy Industries Association (1977). *Solar Industry Index.* Washington, D.C.: Solar Energy Industries Association.

Solar Energy Institute of America (1977). *Solar Energy Sourcebook.* Washington, D.C.: Solar Energy Institute of America.

SolarVision (1977). *The Solar Age Catalog.* Harrisville, N.H.: SolarVision.

Solar Usage Now (1978). *The Solar Usage Now Catalog.* Bascom, Ohio: Solar Usage Now.

I Who's Who in Solar Energy

Name	Specialty	Past and/or Present Affiliation
Abbot, Charles G.	Solar radiation measurement	Smithsonian Institution
Anderson, Bruce	Author of *Solar Home Book* (1976) and *Solar Energy: Fundamentals in Building Design* (1977)	Total Environmental Action, Inc.; *Solar Age Magazine*
Angstrom, K. J.	First truly precise device for measuring direct normal solar radiation (1893)	

Note: The selection of persons and accomplishments for inclusion in this chart was arbitrary and at the authors' discretion. Listing here is not to be taken as a recommendation of products or services.

Name	Specialty	Past and/or Present Affiliation
Baer, Steven	Passive solar design	Zomeworks, Albuquerque
Balcomb, Douglas J.	Passive-solar system sizing	Los Alamos Scientific Laboratory
Baum, Valentin A.	Solar energy-conversion methods	U.S.S.R.
Beckman, William A.	Coauthor of *Solar Energy Thermal Processes* (1974) and *Solar Heating Design by the f-Chart Method* (1977)	University of Wisconsin
Bliss, Raymond W., Jr.	Collector-performance prediction, night cooling	University of Arizona
Boer, Karl W.	Cadmium-sulfide photovoltaic cells	University of Delaware
Chahroudi, Day	Phase-change storage, opaque glazings	Massachusetts Institute of Technology
Close, Donald J.	Solar water heating, pebble-bed cooling	Commonwealth Scientific and Industrial Research Organization, Australia
Crowther, Richard L.	Solar architecture	Crowther/Solar Group, Denver
Daniels, Farrington	Author of *Direct Use of the Sun's Energy* (1964)	University of Wisconsin
de Winter, Francis	Heat exchanger performance, swimming pool heating	Copper Development Association
Duffie, John A.	Coauthor of *Solar Energy Thermal Processes* (1974) and *Solar Heating Design by the f-Chart Method* (1977)	University of Wisconsin
Farber, Erich A.	Solar water heating	University of Florida
Hay, Harold R.	Roof ponds	SkyTherm, Los Angeles
Hill, James E.	Testing methods and standards for solar heating systems	National Bureau of Standards

(continued)

WHO'S WHO IN SOLAR ENERGY

Name	Specialty	Past and/or Present Affiliation
Hottel, Hoyt C.	Collector-performance prediction	University of Minnesota
Klein, Sanford A.	System simulation, co-author of *Solar Heating Design by the f-Chart Method* (1977)	University of Wisconsin
Langley, Samuel P.	Solar radiation measurement (late 1800s), insolation measurement unit, langley (g cal/cm^2)	Smithsonian Institution
Liu, Benjamin Y. H.	Availability and use of solar energy	University of Minnesota
Lorsch, Harold G.	Collector-configuration testing, thermal storage	University of Pennsylvania; Franklin Institute
Löf, George O. G.	Air collectors, rock storage, solar economics	Colorado State University; Solaron, Denver
Mazria, Edward	Passive-system-performance prediction	University of Oregon; Mazria & Assoc., Albuquerque
Morse, Roger N.	Solar water heating	Commonwealth Scientific and Industrial Research Organization, Australia
Olgyay, Victor V.	Solar architecture and orientation of buildings	
Ruegg, Rosalie T.	Solar economics	National Bureau of Standards
Shurcliff, William A.	Author of *Solar Heated Buildings—a Brief Survey* (1976) and *Solar Heated Buildings of North America—120 Outstanding Examples* (1978) among others	
Tabor, Harry	Metal-oxide selective surfaces, solar radiation measurement	Israel
Telkes, Maria	Phase-change thermal storage	Massachusetts Institute of Technology; University of Delaware; New York University

Name	Specialty	Past and/or Present Affiliation
Thekaekara, Matthew P.	Solar radiation measurement	
Thomason, Harry E.	Trickle-down collectors	Thomason Solar Homes, Inc.
Trombe, Felix	Natural-convection air collectors, Trombe wall	Centre Nationale de la Recherche Scientifique, France
Tybout, Richard A.	Solar economics optimization	Ohio State University
Whillier, Austin	Collector-performance prediction	Massachusetts Institute of Technology; South Africa
Woertz, B. B.	Collector-performance prediction	Massachusetts Institute of Technology
Wright, David	Passive solar architecture	SEAgroup, Nevada City, California
Yanda, William F.	Solar greenhouses	New Mexico
Yellott, John I.	Solar radiation measurement, solar design	University of Arizona

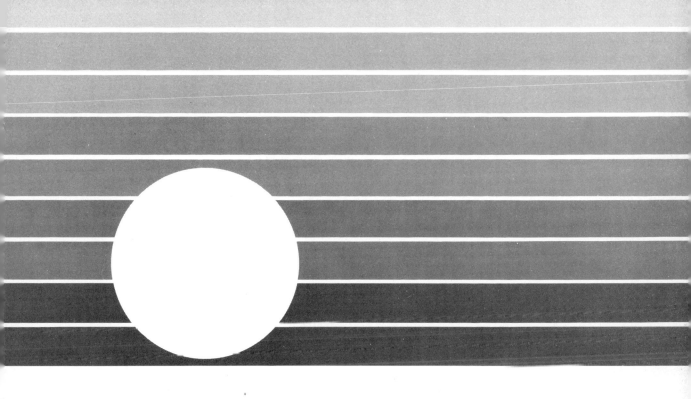

INDEX

INDEX

Abbot, Charles G., 413
Absorber plate, 40
  air collectors, 75–82
  liquid collectors, 83–90
  surface coatings, design factors, 90–91
  see also Selective surface
Absorptance, 25, 26, 47; *see also* Transmittance-absorptance product
Absorption cooling; *see* Air conditioning
Active system, overview, 40
Active vs. passive, 38, 45
Air cleaner, electronic, 170
Air collectors, 41, 43, 65
  absorber design factors, 81–82
  absorber plate, 75–82
  control sensors, 192–193
  simple system, 67
  sizing; *see* Sizing collector arrays
  system concepts, 151–155
  system layouts, 182–188
  see also Collector arrays, Flat-plate collectors

Air conditioning
  absorption, 173, 190–191, 329–331
  absorption layout, 190–191
  adsorption, 331
  definition, 172
  evaporative cooler, 124, 125, 172–273
  with phase-change storage, 143–144
  Rankine-cycle, 173, 329
  with rock storage, 124–125
Air-control unit; *see* Air handler
Air filter, definition, 170
Air handler, 163, 178, 186
Air vs. liquid systems, 61, 298, 299
Air mass, 11, 12
Air vent, definition, 166
Altitude angle, 5, 7, 8
Anaerobic digestion, 336
Analysis term, for sizing, 232
Anderson, Bruce, 413
Angstrom, K. J., 413
Annual-cycle energy storage (ACES), 146
Annual heat storage, 146

Antifreeze, 83, 126, 336; *see also* Closed-loop system
Antireflective glass, 72–73
Area-dependent costs, for sizing, 231
Area-independent costs, for sizing, 231
Attic pond, 266; *see also* Roof pond
Automatic-fill system, definition, 166
Azimuth, 5, 7

Baer, Steven, 414
Balcomb, Douglas J., 414
Baum, Valentin A., 414
Beadwall, 271–272
Beam radiation; *see* Direct radiation
Beckman, William A., 414
Biogas, 336
Biomass, 336
Blackbody, 24, 25
Bliss, Raymond W., Jr., 414
Boer, Karl W., 414
Boiler, definition, 159
Borax, 138
Buffering agent, 83
Building codes, 303; *see also* Solar standards
Building shape, 262, 286–288

Cadmium sulfide cell, 333
Carnot efficiency, 325–326
Chahroudi, Day, 414
Chemical heat storage, 146–147
Chrome black, 93; *see also* Selective surface
Close, Donald J., 414
Closed-loop system, 126, 155, 158–159
  penalty, 235–237
  system layout, 188–189
  see also Liquid collectors
Collector array
  definition, 158
  effectiveness at actual orientation, 230, 235–237
  effectiveness at actual tilt, 230, 235–237
  fluid flow, 95–102
  location, 292–300

  and reflected light, 295
  shading, 282–285
  sizing, *see* Sizing collector arrays
  see also Air collectors, Liquid collectors
Collector efficiency, 46–55
Collector panel, definition, 157
Combining active and passive features, 273, 289, 291
Comparing rock, water, and phase-change storage, 139–141
Concentrating collectors, 319–325
  circular, 316–317, 322
  and diffuse light, 324–325
  linear, 40, 44, 323
  misconceptions, 319–321
  reflectors, 324
  solar cooker, 316–317
  vs. other collector types, 324–325
Concentration ratio, 319–320
Condensation, 64, 74–75
Conditioned space, definition, 158
Conduction, 16–19
Conservation of energy, 32–33
Control circuit, definition, 169
Controls design, 191–200
Convection, 19–22
  forced convection, 20
  natural convection, 20
Convective loop, 20
Conventional heating system, basic concepts, 151
Conversion factors, 2
  table, 359–360
Cooling; *see* Air conditioning
Copper oxide, 93; *see also* Selective surface
Corrosion, 89–90, 128, 170, 171
Cover spacing, 75; *see also* Glazing
Crowther, Richard L., 414

Dampers
  backdraft, 164
  manual, 164
  motorized, 163
  placement, 181–182
Daniels, Farrington, 322, 414
de Winter, Francis, 415
Dead-air space, 38, 81
Declination, definition, 7

Degree day
  definition, 214
  modified, definition, 214
  modified, table, 367–375
Demand/energy rate, 352
Density of materials, 110, 136
Depreciation, for sizing, 230, 233
Design considerations
  commercial buildings, 277–278
  industrial buildings, 278
  institutional buildings, 278–279
  multifamily buildings, 277
  single-family buildings, 276–277
Design obsolescence, 200
Dielectric union, 171
Diffuse radiation, 9, 12, 13
  and concentrating collectors, 324–325
Direct-gain system
  definition, 247
  see also Passive system
Direct radiation, 9, 12, 13
Discount rate
  definition, 206
  for sizing, 232
Draindown system, 156, 160, 166
  control sensors, 193
  system layout, 188–189
  see also Liquid collectors
Drumwall, 255, 257–259
Drying, agricultural, 318–320
Ductwork, definition, 158
Duffie, John A., 414

Economic check, for sizing, 237–238, 366
Edge-defined-film grown silicon, 333, 334
Efficiency
  Carnot, 325–326
  collector, experimental, 52–54
  collector, factors affecting, 55–56
  collector, graph, 53
  collector, instantaneous, 46, 47, 49
  energy-conversion, 33
Electricity generation; *see* Ocean thermal, Power tower, Solar cell, Solar satellite, Thermionic generator, Thermoelectric generator, Wind power

Emittance, 25–26
Energy converters, performance, 34
Environmental radiation, 13, 14, 49
Eutectic
  mixture, 143
  salt, 135, 137
  see also Phase-change storage
Evacuated-tube collector, 43, 44, 102–104
Evaporative cooling; *see* Air conditioning
Expansion tank, definition, 165

F_R, definition, 50–52
f-Chart, 205, 208, 286
fan, 160
Farber, Erich A., 414
Federal tax benefits, 235, 237, 347–350, 363–365
Flat-plate collectors
  absorber; *see* Absorber plate
  actual problems, 64, 68
  design factors, 59–106
  frame, 62–64
  glazing; *see* Glazing
  insulation, 62–63
  sizing; *see* Sizing collector arrays
Fluid-control elements, 162–166
Fluid filter, 170–171
Fluid flow in collectors and arrays, 95–102
Fluid-mass flow rate, 52
Frame, collector, 62–64
Fresnel lens, 322–324, 334
Furnace, 159

G, definition, 52
Gallium arsenide cell, 333
Glass, 68–70
  antireflective, 72–73
  low-iron, 69
  scratch, 73
  tempered, 73
  transmittance, 69–70
  water-white, 69
  see also Glazing
Glauber's salt, 135–136, 138, 140, 141, 144; *see also* Phase-change storage

Glazing, 38, 68–75
  cover spacing, 75
  materials, properties of, 70–71
  number of covers, 69
  plastic, 71, 74
  see also Glass
Grashof number, 31
Greenhouse, 259, 268, 270–272
Greenhouse effect, 38, 39

Hay, Harold R., 414
Heat anticipator, 194
Heat capacity, 15, 109
  of materials, 110
Heat engines, 325–329
Heat-exchange ventilator, 288–289
Heat enchanger
  definition, 160–161
  correction factor for sizing, 235–237
Heat of fusion; *see* Latent heat of fusion
Heat load
  definition, 208
  calculation from old heating bills, 224–226
  calculation sheets, 209–212
  see also Heat loss
Heat loss
  basement floor, 217–218
  basement walls, 217–218
  ceiling, 216–217
  doors, 215
  ducts, 218–219
  floor, 216–217
  infiltration, 218
  slab, 217
  walls, 216–217
  windows, 215, 289
Heat loss, overall collector coefficient, 51, 52
Heat pipe, 339–340
Heat pump
  definition, 172
  system layout with liquid collector, 189–190
Heat-removal efficiency factor, 50–52
Heat sink, definition, 325
Heat storage, 107–149
  chamber, definition, 158

  how much?, 108, 117, 128, 140–142
  lid, 120
  location, 108
  tank, definition, 159
  why store heat?, 107
  see also Phase-change storage, Rock storage, Water storage
Heat and temperature, 14–15
Heat transfer, 15–31
  conduction, 16–19
  convection, 19–22
  convective coefficient, 21
  overall coefficient, 28–30, 214
  radiation, 22–27
  radiative coefficient, 27
Heat-transfer fluid, 19, 40, 83
Heat-transfer numbers, 30, 31
Hill, James E., 414
Horizontal collector performance, 295
Hot-water heater system
  active, 311–313
  passive, 309–311
  preheat system, definition, 161–162
  sizing, 310
  system layout with air system, 187–188
  system layout with draindown system, 188–189
Hot-water load, 219–220
Hottel, Hoyt C., 415
Hour angle, definition, 7
How solar heating works, overview, 38–46
Hybrid system, definition, 246
Hydrogen generator, 340

I, symbol for solar radiation, 10, 46, 47, 50–55
Income-tax rate, for sizing, 232
Indirect-gain system
  definition, 247
  see also Passive system
Infiltration, 218, 287–289
Inflating-discounting factor, 207, 227
  for sizing, 233
  table, 376–382
Inflation rates, for sizing, 232

Infrared light, 3, 13; *see also* Radiation heat transfer, Environmental radiation
Infrared-reflective coating, 73
Insolation, 8
   table, 367–375
Insulating connector, definition, 171
Insulation, 162
   flat-plate collectors, 62–63
   heat-storage unit, 118, 129
   recommended levels, 286–287
   and retrofitting, 279
   see also Beadwall
Insulator, 18, 19
Insurance costs, for sizing, 233
Internal energy, 14–15
Investment tax credit, 235
Ion-getter column, 90, 128, 171–172
Irradiance, 89

Kirchoff's law, 26
Klein, Sanford A., 415

Langley, Samuel P., 415
Latent heat of fusion, 15, 134, 136; *see also* Phase-change storage
Legal factors, 302–304
Legislation, summary by state, 388–406
Life-cycle costing, 205–206
Limit switch, definition, 171
Linear concentrator; *see* Concentrating collectors
Liquid collectors, 42, 43, 66
   absorber design factors, 85, 87–90
   absorber plate, 83–90
   sizing, *see* Sizing collector arrays
   system concepts, 155–156
   system layouts, 188–191
   see also Closed-loop system, Collector arrays, Draindown system, Flat-plate collectors
Liu, Benjamin Y. H., 415
Löf, George O. G., 77, 78, 317, 415
Lorsch, Harold G., 415

Maintenance costs, for sizing, 233
Make-up water, 129, 166
Mazria, Edward, 415

Mechanical components, location, 300–301
Melting point, of materials, 136
Modes, 45, 46
   consolidation, *see* System layout
   definition, 151, 152
   specification, 174–177
Morse, Roger N., 416
Mortgage factors, for sizing, 232

National Solar Heating and Cooling Information Center, 408
Net-benefits analysis, 205–206
Nickel black, 93; *see also* Selective surface
Nitinol engine, 341
Nuisance factor, 304
Nusselt number, 31

Ocean thermal energy conversion (OTEC), 338–339
Oil industry, government subsidies, 347–348
Olgyay, Victor V., 416
Orientation, collector, 230, 235–237, 278
Optimum area, 207, 226
   for sizing, 235
Ovshinsky cell, 333, 334

Paint, 91
Parabolic concentrator, *see* Concentrating collectors, circular
Paraffin, 135, 136, 141; *see also* Phase-change storage
Passive system, 246–274
   living with it, 272–273
   sizing, 262, 263
   temperature swing, 263, 273
   water heating, 309–311
Peak kilowatt, 334
Peak load, 350–353
Phase-change diagram, 133, 134, 137
Phase-change storage, 132–144
   air conditioning, 143–144
   compared to rock and water storage, 139–141
   containers, 142–143

   materials, properties of, 136
   seed crystal, 138
   sizing, 140–142
   stratification, 138–139
   supercooling, 137–138
   see also Heat storage
Photochemical collector, 340
Photovoltaic cell, *see* Solar cell
Pilaster, definition, 121
Pipes, liquid, 158
Power tower, 337–338
Prandtl number, 31
Present value, definition, 206
Pressure drop, definition, 113
Pressure-relief valve, definition, 171
Property tax, 232, 349, 388–406
Proportional control, 169–170, 196
Protective components, 170–172
pump, 160
pyranometer, 9
pyrheliometer, 9

R value
   definition, 213, 214
   of insulations, 213
   see also Insulation
Radiation, environmental, 13, 14, 49
Radiation heat transfer, 22–27
Rankine cycle, 173, 326–327, 329
Rayleigh number, 31
Reflectance, 47
Reflected radiation, 12, 13
   effect on collector performance, 295
Relay, 168, 198–200
Resistance, thermal
   conductive, 17, 18, 19
   convective, 21
   overall, 28–30
   radiative, 28
Retrofitting, 259, 279, 280
   hot-water heater, 312–313
Reverse return, 95–97
Reynolds number, 31
Ribbon-grown silicon, 333–334
Rock storage, 111–124
   air conditioning, 124, 125
   airflow direction, 123

bottom plenum, 121
compared to water and phase-change storage, 139–141
concrete container, 121–122
insulation, 118
lid, 120
metal tank, 122–123
rock selection, 123
sealing, 119
sizing, 117–118
temperature stratification, 113–117
wooden container, 119–121
see also Heat storage
Roof pond, 264–266
Ruegg, Rosalie T., 415

Sacrificial anode, see Ion-getter column
Salt pond, 144–145
Seed crystal, 138
Selective surface, 26, 77–79, 91–95
vs. nonselective surface, 91–93
Semiconductor, 331–333
Sensible-heat storage, 109–132
see also Rock storage, Water storage
Serpentine flow, 96–97
Shurcliff, William A., 415
Silicon cell; see Solar cell
Site planning, 281–286
Sizing collector arrays, 204–245
assumptions, 226, 228, 230
calculation sheets, 209–212, 227–231
derivation of method, 361–366
optimum area, definition, 205
summary chart, 242
why is sizing difficult?, 205–208
Sizing hot-water heaters, 310
Sizing passive systems, 262–263
Sizing phase-change storage, 140–142
Sizing rock storage, 117–118
Sizing water storage, 128
Sky radiation; see Environmental radiation
Sky temperature, 14, 36
Sodium sulfate decahydrate; see Glauber's salt

Softened water and corrosion, 89, 90
Solar angles; see Altitude angle, Azimuth angle
Solar architecture, 275–307
building design phases, 304–305
Solar cells, 331–334
cadmium sulfide, 333
cost, 334
efficiencies, 333
gallium arsenide, 333
Ovshinsky, 333, 334
silicon, 331–334
wet-type, 340
Solar constant, 12
Solar cooking, 316–318
Solar flux, 89
Solar fraction, 205, 237, 361–366
Solar noon, 7
Solar oven, 317–318
Solar pond; see Salt pond
Solar radiation, 8–14
at earth's surface, 9–14
table, 367–375
Solar satellite, 339
Solar skyspace easement, 303
Solar spectrum, 10, 11
Solar standards, 303, 386–387
organizations, 383–385
Solar still, 315–316
Solar time, 4, 6, 7
Solenoid valve, 163–164
Specific heat, 52; see also Heat capacity
Spectral curve of emitted radiation, 23
State income tax
credit, for sizing, 235–237, 363–365
see also Legislation, summary by state, Tax benefits
Stefan-Boltzman constant, definition, 24
Stirling-cycle engine, 326–328
Storage, heat; see Heat storage
Styrofoam bead insulation; see Beadwall
Sun, relationship to earth, 4, 5
Sun rights, 302–303
Supercooling, 137

Swimming-pool cover, 313
Swimming-pool heating, 313–315
System-control elements, 166–170
System cost, for sizing, 234–235
System elements, basic, 157–162
System-flow map, definition, 179
System layout, 177–191
examples, 185–191

$\gamma\alpha$ definition, 50–51
Tabor, Harry, 415
Tap water and corrosion, 89
Tax benefits
federal, 235, 237, 347–350, 363–365
property tax, 232, 349
state income, 235–237, 349–350, 363–365
state, summary, 388–406
Telkes, Maria, 135, 143, 317, 415
Temperature, 14–15
Temperature stratification, 113–117, 140, 152
Tempering valve, 171
Thekaekara, Matthew P., 417
Thermal conductivity, 16, 17, 111
of materials, 110
Thermal expansion of various materials, 63
Thermal mass, 111, 247
combination of materials, 260, 262
drumwall, 255, 257–259
greenhouse, 272
phase-change materials, 260
solid walls, 260, 261
walls and floors, 250–251
water walls, 254–255, 257–259
see also Trombe wall
Thermal properties of materials, 110
Thermal resistance; see Resistance
Thermic diode, 340
Thermionic generator, 341
Thermistor, 167–168
Thermochemical collector, 341
Thermocouple, 168, 340
Thermoelectric generator, 340
Thermopile, 168
Thermoresistor, 167–168

Thermosiphoning, 20
　air collector, 266–269
　hot-water heater, 310–311
　see also Trombe wall
Thermostat, 193–195
　definition, 167
　differential, definition, 167
　two-stage, definition, 168
Thermostatic mixing valve, 171
Thomason, Harry E., 416
Tilt, collector, 230, 235–237, 294, 295
Transformer, definition, 170
Transmittance, 47, 69–72; see also Trasmittance-absorptance product
Transmittance-absorptance product, 50–51
TRANSYS, 205
Trombe, Felix, 251, 268, 269, 417
Trombe wall, 251–254, 255, 256, 279, 280, 294
Turning vanes, 165
Tybout, Richard A., 416

U value, definition, 28–30, 214
U_L, definition, 51–52

Ultraviolet light, 3
　effect on plastic, 74
Unglazed collector, 313–315
Unions, labor, 304
Utility companies
　peak load problems, 350–353
　in the solar business, 353–354

Vacuum air vent, 166
Vacuum breaker, 166
Valve
　check, definition, 164
　manual, definition, 165
　solenoid, definition, 163–164
Vertical collector
　performance 294–295
　and reflected light, 295
Void fraction, 113

Ward, John C., 207, 208, 361
Water purification, 315–316
Water storage, 124–132
　compared to rock and phase-change storage, 139–141
　concrete tank, 129–130
　corrosion, 128
　fiberglass tank, 130–131
　insulation, 129
　location, 129
　make-up water, 129, 166
　metal tank, 131–132
　scaling, 127
　sizing, 128
　tank, pressurized vs. nonpressurized, 127
　temperature stratification, 117, 127
　see also Heat storage
Wavelengths, light, 4
Weatherstripping, 288
Whillier, Austin, 416
Windbreak, 281–282
Wind power, 335–336
Woertz, B. B., 416
Wright, David, 416

Yanda, William F., 416
Yellot, John I., 416